Lecture Notes in Computer Science 1267

Edited by G. Goos, J. Hartmanis and J. van Leeuwen

Advisory Board: W. Brauer   D. Gries   J. Stoer

Springer

Berlin
Heidelberg
New York
Barcelona
Budapest
Hong Kong
London
Milan
Paris
Santa Clara
Singapore
Tokyo

Eli Biham (Ed.)

Fast Software Encryption

4th International Workshop, FSE'97
Haifa, Israel, January 20-22, 1997
Proceedings

 Springer

Series Editors

Gerhard Goos, Karlsruhe University, Germany

Juris Hartmanis, Cornell University, NY, USA

Jan van Leeuwen, Utrecht University, The Netherlands

Volume Editor

Eli Biham
Technion — Israel Institute of Technology, Computer Science Department
Haifa 32000, Israel
E-mail: biham@cs.technion.ac.il

Cataloging-in-Publication data applied for

Die Deutsche Bibliothek - CIP-Einheitsaufnahme

Fast software encryption : 4th international workshop ;
proceedings / FSE '97, Haifa, Israel, January 20 - 22, 1997. Eli
Biham (ed.). - Berlin ; Heidelberg ; New York ; Barcelona ; Budapest
; Hong Kong ; London ; Milan ; Paris ; Santa Clara ; Singapore ;
Tokyo : Springer, 1997
   (Lecture notes in computer science ; Vol. 1267)
   ISBN 3-540-63247-6

CR Subject Classification (1991): F.3, F.2.1, E.4, G.2.1, G.4

ISSN 0302-9743
ISBN 3-540-63247-6 Springer-Verlag Berlin Heidelberg New York

© Springer-Verlag Berlin Heidelberg 1997
Printed in Germany

Typesetting: Camera-ready by author
SPIN 10548880   06/3142 – 5 4 3 2 1 0   Printed on acid-free paper

Preface

This fast software encryption workshop (FSE) follows the previous three workshops held in Cambridge in December 1993, in Leuven in December 1994, and in Cambridge in February 1996. The workshop was organized in cooperation with the International Association for Cryptologic Research (IACR), and with the kind support of Algorithmic Research and of Microsoft. It was held at the Technion (Haifa, Israel), January 20–22, 1997. The programme committee consisted of Eli Biham (Technion - chair), Ross Anderson (Cambridge University), Don Coppersmith (IBM Research), Cunsheng Ding (Turku), Dieter Gollmann (Royal Holloway), Jim Massey (ETH Zurich), Mitsuru Matsui (Mitsubishi), and Bart Preneel (Katholieke Universiteit Leuven). Next year's fast software encryption workshop will be organized by Serge Vaudenay and will be held in Paris.

This series of workshops concentrates on the theory and practice of fast cryptography, and in particular of blockciphers, stream ciphers, hash functions, and message authentication codes. The presentations deal with new suggestions of such cryptographic primitives, their design, and their analysis. Special preference is given to the applicability of these primitives in software, and their fast implementations. On the other hand, applications and analyses of other cryptographic primitives, and in particular of public key cryptosystems, are beyond the scope of the workshops, and their design and analysis is dealt with in other conferences and workshops on cryptography.

This year, 44 papers were submitted to the workshop. Each of these papers was refereed by at least three programme committee members. All the reports were later sent to the respective authors. Based on the reports, 23 papers were selected for presentation at the workshop, including seven papers on cryptanalysis, four papers suggesting new blockciphers, three dealing with stream ciphers, three with message authentication codes, three with modes of operation, and three papers with the core of fast software encryption, i.e., how to design fast encryption in software.

In addition, two discussion sessions were held: a discussion on the requirements and evaluation criteria for the Advanced Encryption Standard, whose development process was recently announced by the US National Institute of Standards and Technology (NIST), and a discussion on the security of cryptosystems and the relation between theory and practice. The minutes of the first discussion are included in these proceedings.

The workshop was organized almost entirely using email and WWW. A home page was created for the workshop, through which all the information on the workshop was distributed: the call for papers, registration and general information, acceptance of papers, and the workshop's program. All the papers were submitted using email (except for two papers submitted in paper form), and all the distribution of papers to the programme committee and the discussions of the programme committee were done using email. In addition, all the papers were processed directly from their LaTeX files, using the llncs style, and were automatically merged into these proceedings.

These proceedings follow the tradition of this series of workshops whose proceedings have been published in Springer-Verlag's Lecture Notes in Computer Science (LNCS) series: The proceedings of the first FSE workshop, held in Cambridge in 1993, were published as LNCS 809, the proceedings of the second FSE workshop, held in Leuven in 1994, were published as LNCS 1008, and the proceedings of the third FSE workshop, held in Cambridge in 1996, were published as LNCS 1039.

I would like to thank the authors for their submissions and the participants for attending the workshop. The programme committee deserves special thanks for their hard work. Simon Blackburn, Antoon Bosselaers, Karl Brincat, Mike Burmester, Lars Knudsen, Sean Murphy, Kenneth G. Paterson, Vincent Rijmen, Serge Vaudenay, and Peter Wild are acknowledged for their services as external referees. It is also a pleasure to thank the Department of External Studies of the Technion, and in particular Pnina Sasson, who made all the local arrangements, and to thank Yvonne Sagi for her help in preparing some of the material for the workshop. Finally, special thanks go to the sponsors for their generous support.

May 1997   Eli Biham

Contents

χ² Cryptanalysis of the SEAL Encryption Algorithm

Helena Handschuh *
Gemplus PSI
1, Place de la Méditerranée
95200 Sarcelles
France

Henri Gilbert
France Télécom
CNET PAA-TSA-SRC
38-40 Rue du Général Leclerc
92131 Issy-les-Moulineaux
France

Abstract. SEAL was first introduced in [1] by Rogaway and Coppersmith as a fast software-oriented encryption algorithm. It is a pseudorandom function which stretches a short index into a much longer pseudorandom string under control of a secret key pre-processed into internal tables. In this paper we first describe an attack of a simplified version of SEAL, which provides large parts of the secret tables from approximately 2^{24} algorithm computations. As far as the original algorithm is concerned, we construct a test capable of distinguishing SEAL from a random function using approximately 2^{30} computations. Moreover, we describe how to derive some bits of information about the secret tables. These results were confirmed by computer experiments.

1 Description of the SEAL Algorithm

SEAL is a length-increasing "pseudorandom" function which maps a 32-bit string n to an L-bit string $SEAL(n)$ under a secret 160-bit key a. The output length L is meant to be variable, but is generally limited to 64 Kbytes. In this paper, we assume it is worth exactly 64 Kbytes (2^{14} 32-bit words), but all our results could be obtained with a smaller output length.

The key a is only used to define three secret tables R, S, and T. These tables respectively contain 256, 256 and 512 32-bit values which are derived from the Secure Hash Algorithm (SHA) [2] using a as the secret key and re-indexing the 160-bit output into 32-bit output words.

SEAL is the result of the two cascaded generators shown below.

---

* The study reported in this paper was performed while Helena Handschuh was working at France Télécom-CNET.

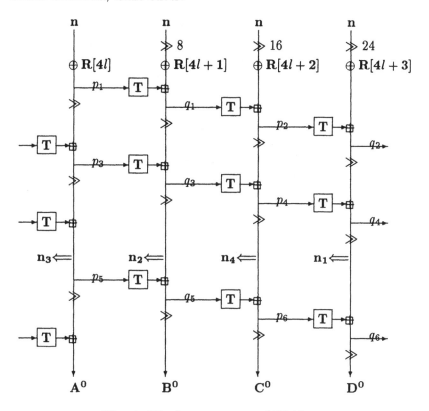

Fig. 1. The first generator of SEAL

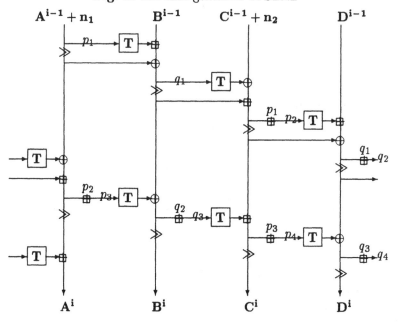

Fig. 2. The second generator of SEAL (i^{th} iteration)

The first generator uses a routine depending on the a-derived tables R and T depicted in figure 1. It maps the 32-bit string n and the 6-bit counter l to four 32-bit words A^0, B^0, C^0, D^0 and another four 32-bit words n_1, n_2, n_3, n_4. These eight words are to be used by the second generator.

The second generator uses a routine depending on the a-derived tables depicted in figure 2. There are 64 iterations of this routine, indexed by $i = 1$ to 64. $A^0 B^0 C^0 D^0$ serves as an input to the first iteration, producing an $A^1 B^1 C^1 D^1$ block. For the next iterations, the input block is alternately $(A^{i-1} + n_1, B^{i-1}, C^{i-1} + n_2, D^{i-1})$ for even i values and $(A^{i-1} + n_3, B^{i-1}, C^{i-1} + n_4, D^{i-1})$ for odd i values. At iteration i, the output block $(Y_1^i, Y_2^i, Y_3^i, Y_4^i)$ is deduced from the intermediate block (A^i, B^i, C^i, D^i) using the a-derived table S as shown below in figure 3.

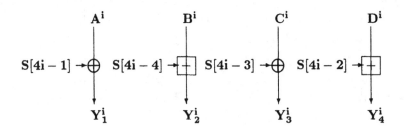

Fig. 3. Deriving the generator output

In the above figures :

- $\oplus$ stands for the XOR function;
- $+$ stands for the sum   (mod 2^{32});
- $\gg$ stands for a right rotation of 9 bits ($\gg$ has precedence over $+$ and $\oplus$);
- $\gg N$ stands for a right rotation of N bits;
- p_1 through p_4 and q_1 through q_4 stand for the inputs of table T obtained from the 9 bits 2 to 11 of the 32-bit intermediate values A, B, C and D; for instance in figure 1, $p1 = A\&0x7fc$.

Concerning the definition of SEAL, more details can be found in [1] and in [2].

The algorithm is divided into three steps.

- First we compute the internal tables under the secret key a. The security of this step relies on SHA which is assumed to be highly secure. Therefore, R, S and T are pseudorandom tables.
- Second we compute A^0, B^0, C^0, D^0, n_1, n_2, n_3 and n_4 from n, l and table R. This is what we already called the first generator. Let us assume the output is pseudorandom as well.
- Finally, the second generator computes iteratively the $A^i B^i C^i D^i$ blocks, from which the Y_1^i, Y_2^i, Y_3^i, Y_4^i values are derived. We change the original notation as follows :

- $Y_1^i = A^i \oplus S_1^i$
- $Y_2^i = B^i + S_2^i$
- $Y_3^i = C^i \oplus S_3^i$
- $Y_4^i = D^i + S_4^i$.

In this part we found certain weaknesses which are investigated in Sections 3 and 4.

2   Preliminaries

2.1   Role of mod 2^{32} additions

Although the combined use of the $+$ and $\oplus$ operations probably strengthens SEAL as compared with a situation where only one of these operations would be used, we do not believe that this represents the main ingredient of the security of SEAL, which is essentially a table-driven algorithm.

As a matter of fact, any $x + y$ sum can be written :

$$x + y = x \oplus y \oplus c(x, y)$$

where the carry word $c(x, y)$ is far from being uniformly distributed thus $+$ just introduces an additional, unbalanced term, as compared with $\oplus$.

This remark led us to assume that replacing in SEAL (more precisely in the second generator of figure 2) all $+$ operators by xors would not fundamentally modify the nature of the algorithm, and that cryptanalytic results obtained with such a simplified version could at least partially be transposed to the real cipher. The results of our analysis of this simplified version of SEAL are summarised in Section 3 hereafter.

2.2   The three words D^{i-1}, C^i and D^i are correlated

Let us consider the function depicted in figure 2. Given a fixed value of the iteration index i (say $i = 3$), the input and output to this function are known from the generator outputs $(Y_1^{i-1}, Y_2^{i-1}, Y_3^{i-1}, Y_4^{i-1})$ and $(Y_1^i, Y_2^i, Y_3^i, Y_4^i)$ up to the following unknown words :

- the 8 words $(S_1^{i-1}, S_2^{i-1}, S_3^{i-1}, S_4^{i-1})$ and $(S_1^i, S_2^i, S_3^i, S_4^i)$, the value of which does not depend on the considered initial value (n, l).
- the 2 words n_1 and n_2, the value of which depends on (n, l).

The involvement of the IV-dependent words n_1 and n_2 in the function considerably complicates the analysis of the ith iteration because of the randomisation effect on the input to output dependency.

In order to find statistics applicable to any IV value, we investigate how to "get rid" of any dependency on n_1 and n_2 in some relations induced by the equation of iteration i.

Let us consider the D^{i-1} input word and the C^i and D^i input words. Denote the output tables involved in the right part of figure 2 by : $T_1 = T[p_2]$, $T_2 = T[q_3]$ and $T_3 = T[p_4]$. It is easy to establish the relation :

(1)

$$(D^{i-1} + T_1) \oplus (C^i \ll 9 + T_2) = (D^i \ll 18) \oplus (T_3 \ll 9)$$

This relation does not involve n_1 and n_2. The T_1, T_2, T_3 terms in this relation can be seen as three random values selected from the T table. Since there are only 2^9 values in the T table, given any two words out of the (D^{i-1}, C^i, D^i) triplets, there are at most 2^{27} possible values for the third word of the triplet instead of 2^{32} if D^{i-1}, C^i and D^i were statistically independent. This gives some evidence that the D^{i-1} input and the C^i and D^i output are statistically correlated, in a way which does not depend upon n_1 and n_2. In other words, the SEAL generator derives from an (n, l) initial value three slightly correlated output words Y_4^{i-1}, Y_3^i and Y_4^i.

Relation (1) above represents the starting point for the various attacks reported in Sections 3 to 5 hereafter.

3   An Attack of a simplified version of SEAL

In this Section we present an attack of the simplified version of SEAL obtained by replacing in figure 2 all mod 2^{32} additions by xors. The attack is divided into four steps.

3.1   Step 1

We derive the unordered set of values of the T table, up to an unknown 32-bit constant Δ^i. Relation (1) above represents the starting point for this derivation. After replacing $+$ by $\oplus$ in (1) and introducing the notation $X_4 = Y_4^{i-1}$, $Y_3 = Y_3^i$ and $Y_4 = Y_4^i$, we obtain the relation :

(2)

$$Y_4 \oplus Y_3 \gg 9 \oplus X_4 \gg 18 = T_3 \gg 9 \oplus (T_1 \oplus T_2) \gg 18 \oplus \Delta^i \gg 9$$

where the Δ^i constant depends on the S table. T_1 and T_2 are 2 among 512 values of the table T. Statistically speaking, once in 2^9, $T_1 = T_2$, thus $T_1 \oplus T_2 = 0$. If we compute 2^{18} samples, each of the 512 values of the table $T \oplus \Delta^i$ should appear once on average.

We collect the combination of the generator output words given by the left term of (2) for about 2^{21} (n, l) samples. Whenever one value appears more than 4 times, we assume this is a value of the table $(T \oplus \Delta^i) \gg 9$. All the other values have a probability of about $\frac{2^{21}}{2^{32}}$ to appear. This way, we find about 490 out of 512 values of the table T up to a constant value.

3.2   Step 2

The purpose of the second step is to compute a constant α^i which is needed in the third step in order to find out statistics involving B^{i-1}, D^i, A^i and B^i (see Fig. 2.). Consider the following equation (3) established in a similar way to (2) from the relation between B^{i-1} and the output words :

(3) $Y_4 \gg 9 \oplus Y_2 \oplus Y_1 \gg 9 \oplus X_2 \gg 18 \oplus T_3 \gg 18 = (T_1' \oplus T_2') \gg 18 \oplus (T_3' \oplus T_4') \gg 9 \oplus (S_4^i \gg 9 \oplus S_2^{i-1} \gg 18 \oplus S_2^i \oplus S_1^i \gg 9)$

where

$$\alpha^i = S_4^i \gg 9 \oplus S_2^{i-1} \gg 18 \oplus S_2^i \oplus S_1^i \gg 9 \oplus \Delta^i \gg 18$$

and

$$T_1' = T[p_1], T_2' = T[q_2], T_3' = T[p_3] and T_4' = T[q_4].$$

For each sample, we can find out $T_3 \oplus \Delta^i$ by searching exhaustively the right combination (T_1, T_2, T_3) in equation (2). In order to save time, we compute once and for all a table with the 2^{18} values of $(T_1 \oplus T_2) \gg 18$ and search for the right third value. We perform this search as well as the computation of the left term of (3) for 2^{21} samples. Once in 2^{18}, $T_1' = T_2'$ and $T_3' = T_4'$. This way the constant value α^i we are looking for appears at least 4 times.

3.3   Step 3

The purpose of this step is to find out various values of n_1. Once we have these values, we can find the relation between the inputs and outputs of table T up to a constant value. Let us consider equation (4) established from the relation between A^{i-1} and the output words :

(4)

$$X_1 \gg 18 \oplus Y_1 \oplus Y_4 \oplus T_2' \gg 9 \oplus T_4' \oplus T_3 \gg 9 = n_1 \gg 18 \oplus S_1^{i-1} \gg 18 \oplus S_4^i \oplus S_1^i$$

We can find out $T_2' \oplus \Delta^i$ and $T_4' \oplus \Delta^i$ by searching the right combination of (T_1', T_2', T_3', T_4') in equation (2) using the value α^i we found in step 2. For each sample we compute, we get about 16 possibilities, as (T_1', T_2', T_3', T_4') gives 2^{36} possible values for a 32-bit word.

In order to find the right combination, let us consider two distinct iteration indexes i and j : we know that for a given l value, if i is even (or odd), we always xor the same n_1 (or n_3) to the input A. Let us therefore take two rounds i and j that are both odd (or even). We need to know table $T \oplus \Delta^i$, table $T \oplus \Delta^j$, α^i and α^j.

We collect samples of the combination of the generator output words given by the left term of (4) in order to find out possible values of :

- $n_1 \gg 18 \oplus \beta^i$
  where $\beta^i = S_1^{i-1} \gg 18 \oplus S_4^i \oplus S_1^i \oplus \Delta^i$;
- $n_1 \gg 18 \oplus \beta^j \oplus \Delta^j \gg 9 \oplus \Delta^i \gg 9$
  as value T_3 is found through table $T \oplus \Delta^i$ and values T_2' and T_4' through table $T \oplus \Delta^j$.

We xor all the samples of round i with all the samples of round j. One of these values is the right combination of $\beta^i \oplus \beta^j \oplus (\Delta^i \oplus \Delta^j) \gg 9$

Then we find all the samples for rounds i and j of another value n_1 (i.e. of round l). We compare these two sets of samples and find the right value of $n_1 \gg 18 \oplus \beta^i$.

This step can be repeated various times to collect values of n_1 while computing only once the tables $T \oplus \Delta$ and the constants α.

3.4   Step 4

In this step we finally derive the inputs and outputs of table T from equation (5):

(5)
$$p_1 = ((X_1 \oplus n_1 \oplus S_1^{i-1}) \& 0x7fc)/4$$

In this equation p_1 is the input of table T. We have seen in the first three steps that we can derive the value of T_1 from input and output samples of SEAL.

So we finally derive several values of :

$T[p \oplus \delta^i] \oplus \Delta^i$

where $\delta^i = ((S_1^{i-1} \oplus \beta^i \ll 18) \& 0x7fc)/4$.

3.5   Summary

Summing up the four steps we have just described, we can break the T table up to a constant value using about 2×2^{21} samples of (n, l) for step 1, 2×2^{21} samples of (n, l) for step 2 and about 2^9 values of (n, l) for steps 3 and 4. This means, the T table can be broken using about 2^{24} samples of (n, l).

We could probably go on breaking the simplified version of SEAL by finding out sets of values (n_1, n_2, n_3, n_4), then trying to break the first generator and find table R, but this is not our purpose here.

4   A Test of the real version of SEAL

In this Section we use some of the ideas of Vaudenay's Statistical Cryptanalysis of Block Ciphers [3] to distinguish SEAL from a truly random function.

4.1 χ^2 Cryptanalysis

The purpose of Vaudenay's paper is to prove that statistical analysis on ciphers such as DES may provide as efficient attacks as linear or differential cryptanalysis. Statistical analysis enables to recover very low biases and a simple χ^2 test can get very good results even without knowing exactly what happens inside the inner loops of the algorithm or the S-boxes.

We intend to use this property to detect low biases of a certain combination of the output words of SEAL suggested by the analysis made in Section 3 in order to prove SEAL is far from being undistinguishable from a pseudo-random function. This provides a first test of the SEAL algorithm.

4.2   Number of samples needed for the χ^2 test to distinguish a biased distribution from an unbiased one

We denote by N the number of samples computed. We assume the samples are drawn from a set of r values. We call n_i the number of occurences of the ith of the r values among the N samples and S_{χ^2} the associated indicator :

$$S_{\chi^2} = \frac{\sum_{i=1}^{r} \left(n_i - \frac{N}{r} \right)^2}{\frac{N}{r}}$$

The χ^2 test compares the value of this indicator to the one an unbiased distribution would be likely to provide. If the n_i were drawn according to an unbiased multinomial distribution of parameters $(\frac{1}{r}, \frac{1}{r}, ..., \frac{1}{r})$, the expectation and the standard deviation of the S_{χ^2} estimator would be given by :

- $E(S_{\chi^2}) \to \mu = r - 1$
- $\sigma(S_{\chi^2}) \to \sigma = \sqrt{2(r-1)}$

If the distribution of the n_i is still multinomial but biased, say with probabilities $p_1, ..., p_r$, then we can compute the new expected value μ' of the S_{χ^2} :

$$\mu' = E\left(\frac{\sum_{i=1}^{r}(n_i - \frac{N}{r})^2}{\frac{N}{r}}\right) = \frac{N}{r}\sum_{i=1}^{r} E\left((n_i - p_iN + p_iN - \frac{N}{r})^2\right)$$

It can be easily shown that :

$$\mu' \to \mu + r(N-1)\sum_{i=1}^{r}(p_i - \frac{1}{r})^2$$

An order of magnitude of the number N of samples needed by the χ^2 test to distinguish a biased distribution from an unbiased one with substantial probability is given by the condition :

$$\mu' - \mu \gg \sigma$$

which gives us the following order of magnitude for N :

$$N \gg \frac{\sqrt{2(r-1)}}{r\sum_{i=1}^{r}(p_i - \frac{1}{r})^2}$$

4.3   Model of the test

Let us consider equation (2) with the real scheme (including the sums). We can rewrite it :

(6) $Y_4 \oplus Y_3 \gg 9 \oplus X_4 \gg 18 = T_3 \gg 9 \oplus (T_1 \oplus T_2) \gg 18 \oplus (r_1 \oplus r_2 \oplus r_3) \gg 18 \oplus \Delta^i \gg 9 \oplus r_4$

where r_1 and r_2 are the carry bits created by the addition of T_1 and T_2, and r_3 and r_4 the ones of the addition of S_4^{i-1} and S_4^i.

We apply the χ^2 test to the four leftmost bits of $Y_4 \oplus Y_3 \gg 9 \oplus X_4 \gg 18$ suspecting a slight bias in this expression. Without having carefully analysed the exact distribution of the sum of the four carries, we intend to prove that its convolution with the biased distribution of $T_3 \gg 9 \oplus (T_1 \oplus T_2) \gg 18 \oplus \Delta^i \gg 9$ does result in a still slightly unbalanced distribution.

As we take 4-bit samples, we apply the χ^2 test with $r - 1 = 15$ degrees of freedom. Detailed information about this test can be found in [4].

4.4   Results

Whenever we analyse at least 2^{33} samples, the test proves with probability $\frac{1}{1000}$ to be wrong, that SEAL has a biased distribution. We have made several tests, and each time the value of the S_{χ^2} estimator we obtained for this order of magnitude of N was greater than 40 for the 4 least significant bits and greater than 320 for the 8 least significant bits. In other words, the test proves that the distribution is a biased one.

Figure 4 shows the value of the S_{χ^2} estimator for tests made with $a = 0x67452301$.

S_{χ^2}	2^{23}	2^{24}	2^{25}	2^{26}	2^{27}	2^{28}	2^{29}	2^{30}	2^{31}	2^{32}	2^{33}	2^{34}	2^{35}
4 bits	14.27	25	13.97	9.41	26.96	16.5	16.78	29.65	21.05	30.15	45.74	44.69	55.96
8 bits	261	293	274	238	229	227	246	225	278	313	331	378	453

Fig. 4. Results of the tests with up to 2^{35} samples of (n, l).

The former test test can be slightly improved as follows : let us denote the four least significant bits of S_4^i by s_4^i. For each of the 16 possible values of s_4^i, apply the S_{χ^2} test to the 4 or the 8 least significant bits of $(Y_4 - s_4^i) \oplus Y_3 \gg 9 \oplus X_4 \gg 18$. The test with the correct s_4^i value detects a bias with 2^{30} (n, l) values only (see Figure 5). Note that a significant bias is also detected whenever the two least significant bits of s_4^i are correct.

S_{χ^2}	2^{25}	2^{26}	2^{27}	2^{28}	2^{29}	2^{30}	2^{31}	2^{32}	2^{33}	2^{34}	2^{35}
4 bits	9.45	13.56	20.62	25.59	32.77	71.90	83.63	130	250	438	928
8 bits	253	271	292	321	276	357	379	438	569	838	1520

Fig. 5. Results of the test with the correct value of the four s_4^i bits.

4.5   Deriving first information on SEAL

The interpretation of the above improved test is straightforward : when the two least significant bits of s_4^i are correct, r_4 is partly known and the χ^2 test gives much better results than with wrong bits of s_4^i. Therefore we can derive at least two bits of information on table S. If the test is applied to more than the four leftmost bits of the samples, more than 2 bits can be derived from secret table T. Whenever these bits are right, the χ^2 rises much faster than for wrong values.

As the evolution of the χ^2 indicator is quite close to a straight line when the divergence starts, the results can be checked applying the test to 2^{20} through 2^{32} samples. Divergence becomes obvious when about 2^{30} samples have been computed.

5   Deriving information on the T table

In this Section we give some evidence that the initial step of the attack on the simplified version of SEAL introduced in Section 3 can be adapted to provide large parts of the T table for the real algorithm.

Let us consider relation (6).

As seen in Section 3.1, the distribution of the $T_3 \gg 9 \oplus (T_1 \oplus T_2) \gg 18 \oplus \Delta^i \gg 9$ value at the right of (6) is unbalanced. The most frequent values are provided by the 512 $T \oplus \Delta^i$ words.

On the other hand the distribution of the carry words $r_1 \oplus r_2 \oplus r_3$ and r_4 is also unbalanced. More precisely, due to the fact that in any carry word r each bit $r[j]$ has a $\frac{3}{4}$ probability of being equal to the next bit $r[j+1]$, the number of 'inversions', i.e. j values s.t. $r[j] \neq r[j+1]$ is likely to be small when r is a carry word or an exclusive or of carry words.

Thus we can expect the 512 $T \oplus \Delta^i$ values to give rise to 'spread' probability maxima in the distribution probability of the left term of (6).

Based on 2^{32} (n, l) values, we did the following experiment :

We analysed the probability distribution of the 23 lowest weight bits of the left combination of (6) in order to reduce the memory requirements. So in the rest of this Section, though we do not introduce any new notation, we implicitly refer to 23-bit words instead of 32-bit words.

For several $T \oplus \Delta^i$ values (about 25), we computed the sum of the probabilities of the neighbours of $T \oplus \Delta^i$, i.e. the values of the form $T \oplus \Delta^i + r$, where r is one of the approximately 2^{22} 23-bit words with at most 11 inversions.

We computed the same sum of approximately 2^{22} probabilities around arbitrarily chosen values other than the 512 $T \oplus \Delta^i$ values.

For more than half of the $T \oplus \Delta^i$ values, the obtained sum was larger than all the sums associated to the arbitrarily chosen values.

The complexity of the search of the $T \oplus \Delta^i$ values is quite high if an independent computation of a sum of 2^{22} probabilities is made for each of the 2^{23} candidate values. This approach leads to a 2^{45} complexity, far over the computing capabilities of the computer we used for the experiments ; however, substantial gains might be achieved by reusing appropriately selected partial sums of probabilities.

Thus in summary, we believe that with slightly more than 2^{32} (n, l) values, it should be possible to recover a substantial part of the information on the T table, up to an unknown constant.

6   Conclusion

We have shown in some detail in Section 3 that the simpler scheme with xors instead of sums can be attacked with the generator output corresponding to about 2^{24} samples of (n, l), e.g. 2^{18} n values and 2^6 l values for each n value.

The test of Section 4 can be applied with about 2^{30} samples of (n, l), e.g. 2^{24} n values and 2^6 l values for each n value. Moreover, information about table S can be derived from this test with 2^{30} samples of (n, l) as well, and large amounts of information contained in table T can be derived from approximately 2^{32} samples of (n, l) or slightly more.

Despite of their relatively low time and space complexity, which enabled us to perform the computer simulations mentioned in Sections 4 and 5, the attacks reported in this paper do not seriously endanger the practical security of SEAL, because a too large amount of keystream samples (corresponding to more than 2^{30} (n, l) initialisation vectors) is required. These attacks suggest however that simple modifications of some design features of SEAL, e.g. the detail of the involvement of the IV-dependent values n_1 to n_4 in the second generator, would probably strengthen the algorithm without significant impact upon its performance.

7   Acknowledgements

We thank Thierry Baritaud and Pascal Chauvaud for the elaboration of the LaTeX version of the figures and the paper. We would like to address special thanks to François Allègre who gave us access to a quite powerful computer that made the tests reasonably quick and to Alain Scheiwe who helped us write the C source code of the tests.

References

1. P. Rogaway and D. Coppersmith, "A Software-Optimized Encryption Algorithm", Proceedings of the 1993 Cambridge Security Workshop, Springer-Verlag, 1994.
2. B. Schneier, Applied Cryptography, Second Edition, John Wiley & Sons, 1996.
3. S. Vaudenay, "Statistical Cryptanalysis of Block Ciphers - χ^2 Cryptanalysis", 1995.
4. J. Bass, Eléments de Calcul des Probabilités, 3e édition, Masson Et Cie, 1974.

Partitioning Cryptanalysis

Carlo Harpes, James L. Massey

ETH Zürich, Signal and Info. Proc. Lab., CH-8092 Zürich
CETREL S.C., L-2956 Luxemburg
email: harpes@cetrel.lu, massey@isi.ee.ethz.ch

Abstract. Matsui's linear cryptanalysis for iterated block ciphers is generalized to an attack called partitioning cryptanalysis. This attack exploits a weakness that can be described by an effective partition-pair, i.e., a partition of the plaintext set and a partition of the next-to-last-round output set such that, for every key, the next-to-last-round outputs are non-uniformly distributed over the blocks of the second partition when the plaintexts are chosen uniformly at random from a particular block of the first partition. The last-round attack by partitioning cryptanalysis is formalized and requirements for it to be successful are stated. The success probability is approximated and a procedure for finding effective partition-pairs is formulated. The usefulness of partitioning cryptanalysis is demonstrated by applying it successfully to six rounds of the DES.

Keywords. Iterated block ciphers, linear cryptanalysis, partitioning cryptanalysis, DES.

1 Introduction

In cryptography, frequent use is made of iterated block ciphers in which a keyed function, called the round function, is iterated r times. Linear cryptanalysis, introduced by Matsui in [Mat94b, Mat94a] is a known-plaintext attack that requires the existence of "unbalanced linear expressions". In [HKM95], linear cryptanalysis was generalized by replacing linear expressions with "input/output (I/O) sums". In [Har96, Har94], an even more general attack called *partitioning cryptanalysis* was introduced. This attack is based on the same principle as the statistical attacks independently developed in [MPWW94] and [Vau96]. Similarly to linear and differential cryptanalysis [BS93], partitioning cryptanalysis can be used to evaluate the strength of iterated block ciphers or to detect the existence of backdoors in such ciphers. This paper is intended to provide a thorough treatment of partitioning cryptanalysis and, as a side benefit, to give additional insight into linear cryptanalysis.

In a *last-round attack*, many plaintext/ciphertext-pairs, hereafter called p/c-pairs, are considered. For every considered ciphertext, one guesses the next-to-last-round output by decrypting the last-round with a guessed key. Then, one computes an *empirical decision metric* for this guessed key, which is an estimate of the expectation of some function of some random variable that depends on the plaintext and the guessed next-to-last round output. One repeats this computation for all last-round key guesses and chooses the key with largest empirical decision metric as the "cryptanalyst's guess" for the actual last-round key.

In the last-round attack by the generalization of linear cryptanalysis, the empirical decision metric is the sample-mean estimate of the "imbalance" of an "I/O sum". The *imbalance* of a binary random variable V is defined to be $I(V) := |2P[V = 0] - 1|$, where $P[V = 0]$ denotes the probability that V is 0. An *I/O sum* is a modulo-two sum of a balanced binary-valued function of the plaintext random variable and a balanced binary-valued function of the guessed next-to-last-round output random variable.

In partitioning cryptanalysis, only p/c-pairs whose plaintexts lie in some fixed block of a partition of the plaintext set, called the *input partition*, are considered. Let $J(\tilde{k})$ be the random variable specifying the block of some chosen *output partition* containing the guessed next-to-last round output, where $\tilde{k}$ is the guessed key used to decrypt the last round. The decision metric is an estimate of the "imbalance" of $J(\tilde{k})$, where an *imbalance* of an m-ary random variable V is a measure for how non-uniformly distributed V is. The weakness exploited in partitioning cryptanalysis is thus described by a *partition-pair*, i.e., a pair consisting of an input partition and an output partition, and we will introduce an *imbalance* to measure the effectiveness of a partition-pair in partitioning cryptanalysis. The success of partitioning cryptanalysis relies on the fact that $J(\tilde{k})$ is less balanced when $\tilde{k}$ is the true last-round key than when $\tilde{k}$ is a wrong guess.

In Section 2, we introduce some preliminaries. The last-round attack by partitioning cryptanalysis is developed in Section 3. In Section 4, we give conditions for a successful attack. In Section 5, we consider ciphers in which a part of the round key is inserted by means of a group operation at the inputs to the rounds and we define coset-partitions as partitions whose elements are the cosets of some subgroup with respect to this group operation. To apply partitioning cryptanalysis, there must exist a sufficiently effective partition-pair and the cryptanalyst must have a practical method to find it; in Section 6, we discuss such a method. In Section 7, we apply partitioning cryptanalysis successfully to 6-round DES. We close in Section 8 with a summary of the main results.

In Appendix A, we formulate *combined partitioning cryptanalysis* as an attack that combines several attacks by partitioning cryptanalysis, which attacks exploit the same partition-pair but use p/c-pairs with plaintexts from different blocks of the plaintext partition. In Appendix B, we approximate the success probability of partitioning cryptanalysis. In Appendix C, we provide some details of the partitioning cryptanalysis of 6-round DES.

2   Preliminaries

In this section, Y denotes the output of some keyed function ϕ whose input is X and whose key is Z, i.e., $Y = \phi_Z(X)$. It may be that ϕ is the round function of a cipher or the composition of several round functions. It may also be that ϕ

is unkeyed and not invertible, as when ϕ is the function realized by an S-box of DES.

A *partition* of a set S is a finite set whose elements are pairwise-disjoint non-empty subsets of S whose union is S. These subsets are called the *blocks* of the partition.

Definition 1. Let $\mathcal{F} = \{\mathcal{F}_0, \mathcal{F}_1, \ldots, \mathcal{F}_{l-1}\}$ and $\mathcal{G} = \{\mathcal{G}_0, \mathcal{G}_1, \ldots, \mathcal{G}_{m-1}\}$ be partitions of the input set and the output set, respectively, of a keyed function ϕ_Z. The pair $(\mathcal{F}, \mathcal{G})$ is a *partition-pair* for ϕ if all blocks of $\mathcal{F}$ contain the same number (at least two) of elements, as also do all blocks of $\mathcal{G}$, and if both l and m are at least two.

The blocks of the *input partition* $\mathcal{F}$ will be called *input blocks* and the blocks of the *output partition* $\mathcal{G}$ will be called *output blocks*. The function from the input set of ϕ onto $\{0, 1, \ldots, l-1\}$ that maps an element x to the index i of the block $\mathcal{F}_i$ containing x will be called the *partitioning function* of $\mathcal{F}$ and denoted by f. Similarly, g will denote the partitioning function of $\mathcal{G}$. Note that f and g are always balanced functions, i.e., functions that take on each of their possible values for the same number of arguments.

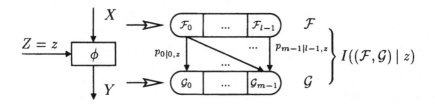

Fig. 1. Representation of a partition-pair $(\mathcal{F}, \mathcal{G})$ for ϕ given the key z.

We use capital letters X, Y, Z, etc., to denote random variables and the corresponding lowercase letters x, y, z, etc., to denote specific values of these random variables. The experiment on which partitioning cryptanalysis relies is the random experiment in which the plaintext and all round keys are chosen independently and uniformly at random over the appropriate sets. Note that, because all blocks of $\mathcal{F}$ have the same size, $f(X)$ is uniformly distributed.

A function of an m-ary random variable, taking on real values between 0 and 1, inclusive, and measuring how non-uniformly the random variable is distributed, will be called an *imbalance*. The imbalance of the random variable V will be denoted $I(V)$. We will consider two imbalance measures, namely, the *peak imbalance*

$$I_{\mathrm{p}}(V) \quad := \quad \frac{m}{m-1}\left(\max_{0 \le i < m} P[V = i] - \frac{1}{m}\right)$$

and the *squared Euclidean imbalance*

$$I_2^2(V) \; := \; \frac{m}{m-1} \sum_{i=0}^{m-1} \left(P[V=i] - \frac{1}{m} \right)^2 \; = \; \frac{m}{m-1} \sum_{i=0}^{m-1} (P[V=i])^2 - \frac{1}{m-1}.$$

The usefulness of a partition-pair for partitioning cryptanalysis will be characterized by a partition-pair imbalance, which also lies between 0 and 1, inclusive.

Definition 2. Let $(\mathcal{F}, \mathcal{G})$ be a partition-pair for the function with input X and output Y and let $I(.)$ be an imbalance measure for m-ary random variables where $m = |\mathcal{G}|$. Let $I(g(Y) \mid f(X) = i)$ denote the imbalance of the random variable $g(Y)$ when conditioned on the event that $f(X) = i$. The *imbalance* $I((\mathcal{F}, \mathcal{G}))$ of the partition-pair $(\mathcal{F}, \mathcal{G})$ is the quantity

$$I((\mathcal{F}, \mathcal{G})) \; := \; \frac{1}{l} \sum_{i=0}^{l-1} I(g(Y) \mid f(X) = i) \; ,$$

where $l = |\mathcal{F}|$ and where f and g are the partitioning functions of $\mathcal{F}$ and $\mathcal{G}$, respectively.

The notion of key-dependent imbalance is of special importance. Consider the keyed function ϕ_Z with key Z taken from a set $\mathcal{Z}$. Let $I(g(Y) \mid f(X) = i, Z = z)$ denote the imbalance of the m-ary random variable $g(Y)$ when conditioned on the joint event that $f(X) = i$ and $Z = z$. We will call this quantity the *key-dependent input-block-dependent imbalance* of the partition-pair $(\mathcal{F}, \mathcal{G})$ for ϕ. The *key-dependent imbalance* of the partition-pair $(\mathcal{F}, \mathcal{G})$ given the key z is the quantity

$$I((\mathcal{F}, \mathcal{G}) | z) \; := \; \frac{1}{l} \sum_{i=0}^{l-1} I(g(Y) \mid f(X) = i, Z = z) \; .$$

The *average-key imbalance* of $(\mathcal{F}, \mathcal{G})$ is the quantity

$$\overline{I}((\mathcal{F}, \mathcal{G})) \; := \; \frac{1}{|\mathcal{Z}|} \sum_{z \in \mathcal{Z}} I((\mathcal{F}, \mathcal{G}) \mid z) \; = \; \frac{1}{l} \sum_{i=0}^{l-1} \overline{I}(g(Y) \mid f(X) = i) \; ,$$

where $\overline{I}(g(Y) \mid f(X) = i)$ denotes the *average-key input-block-dependent imbalance* of $(\mathcal{F}, \mathcal{G})$ for the input block $\mathcal{F}_i$.

The imbalance $\overline{I}((\mathcal{F}, \mathcal{G}))$ of the partition-pair $(\mathcal{F}, \mathcal{G})$ can be calculated from the *key-dependent transition probabilities* $p_{j|i,z}$ for $0 \leq i < l$, $0 \leq j < m$, and z in $\mathcal{Z}$, where $p_{j|i,z}$ is the conditional probability that the output $Y = \phi_Z(X)$ lies in the output block $\mathcal{G}_j$ given that the input X is chosen uniformly at random in the input block $\mathcal{F}_i$ and that $Z = z$ (cf. Fig. 1).

The partition-pair $(\mathcal{F}, \mathcal{G})$ will be said to have *guaranteed transitions* if its average-key imbalance is 1. Guaranteed transitions mean that, for each key, the block of $\mathcal{G}$ in which the output lies is uniquely determined by the key and the block of $\mathcal{F}$ in which the input lies, i.e., *the key and the input block uniquely determine the output block.* The partition-pair is *effective* if its average-key imbalance is substantially greater than zero. This means that the output block is determined with substantially large probability by the key and by the input block.

In the following, we apply partitioning cryptanalysis to *iterated block ciphers* as defined in Fig. 2.

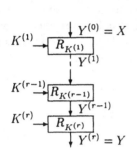

X is the *plaintext*, Y is the *ciphertext*, $K^{(1)}, \ldots, K^{(r)}$ are the round keys, $K^{(1..r)} := (K^{(1)}, \ldots, K^{(r)})$ is the *full key*, and R is the *round function*. The round keys take values in a set $\mathcal{K}$, the *round key set*. X and Y take values in $\mathbb{B}^n$, the set of binary n-tuples. For each value k of the round key, the keyed round function R_k is an invertible function on $\mathbb{B}^n$. Let $Y^{(i)}$ denote the output n-tuple of the i-th round so that $Y = Y^{(r)}$ and let $Y^{(0)} = X$.

Fig. 2. Structure and notation for an r-round iterated block cipher.

3   Last-Round Attack by Partitioning Cryptanalysis

In the last-round attack by partitioning cryptanalysis, we consider a partition-pair $(\mathcal{F}, \mathcal{G})$ for the keyed function consisting of the first $r - 1$ rounds of a cipher. Such a partition-pair will be called an $(r - 1)$-*round partition-pair*. $\mathcal{F}$ and $\mathcal{G}$ are partitions of $\mathbb{B}^n$ so that the numbers m and l of blocks in these partitions must both be powers of 2. Typically, m will be 2, 4, 8, or 16, and $l \geq m$. For the last-round attack, $Z = K^{(1..r-1)}$ and this key lies in the set $\mathcal{K}^{r-1}$. Let $\mathcal{F}_i$ be the input block used in the attack and suppose that N p/c-pairs with plaintexts in $\mathcal{F}_i$ are known. The attack proceeds as follows.

0. For each $\tilde{k}$ in the set $\tilde{\mathcal{K}}$ of possible last-round keys, set up m counters with one counter $c[\tilde{k}, j]$ for each j, $0 \leq j < m$, and initialize all counters to 0.
1. Consider a known p/c-pair (x, y) with plaintext x in $\mathcal{F}_i$.
2. For each $\tilde{k}$ in $\tilde{\mathcal{K}}$, evaluate $\tilde{y}^{(r-1)} := R_{\tilde{k}}^{-1}(y)$ and increment the counter $c[\tilde{k}, g(\tilde{y}^{(r-1)})]$ of the output block in which $\tilde{y}^{(r-1)}$ lies by 1.
3. Repeat Steps 1 and 2 for all N known p/c-pairs (x, y) with x in $\mathcal{F}_i$.
4. Output the key(s) $\tilde{k}$ maximizing $\hat{\mu}(\tilde{k}) = I\left(\frac{c[\tilde{k},0]}{N}, \frac{c[\tilde{k},1]}{N}, \ldots, \frac{c[\tilde{k},m-1]}{N}\right)$ (which, with slight abuse of notation, denotes the imbalance of the random variable whose probability distribution is shown as the argument) as the candidate(s) for the actual key in the last round.

Note that $I(.)$ can be either the peak imbalance or the Euclidean imbalance. The quantity $\hat{\mu}(\tilde{k})$ is an empirical estimate of the decision metric

$$\mu(\tilde{k}) \quad = \quad I(\, g(R_{\tilde{K}}^{-1}(Y)) \mid f(X) = i, \; K^{(1..r)}\tilde{K} = k^{(1..r)}\tilde{k}\,) \;,$$

which is the key-dependent input-block dependent imbalance of the partition-pair $(\mathcal{F}, \mathcal{G})$ for the keyed function whose input is X, whose key is the concatenation of $K^{(1..r)}$ and $\tilde{K}$, and whose output is $R_{\tilde{K}}^{-1}(Y)$.

The last-round attack must in practice be speeded up by exploiting "key equivalence". Two last-round keys k and k' are *equivalent* if there is a bijection ψ of $\{0, 1, \ldots, m-1\}$ such that $g(R_k^{-1}(y)) = \psi(g(R_{k'}^{-1}(y)))$ for all y in $\mathbb{B}^n$. Keys belonging to the same *key equivalence class* produce counter lists $(c[k, 0], c[k, 1], \ldots, c[k, m-1])$ that differ only by a permutation and hence yield the same empirical decision metric so that they are indistinguishable by the attack. Therefore, we need to consider in Step 2 only one representative of each key (equivalence) class. We will write $\tilde{\mathcal{K}}$ to denote a set containing exactly one representative $\tilde{k}$ of each key class. Partitioning cryptanalysis determines only the class in which the true last-round key $k^{(r)}$ lies. This class is called the *right class* and its representative is the *right key* $\tilde{k}_r$. The other key classes are *wrong classes* and their representatives are *wrong keys* $\tilde{k}_w$ lying in $\tilde{\mathcal{K}} \setminus \{\tilde{k}_r\}$.

The *success probability* p is the probability that the last-round attack outputs only the right key when the round keys are chosen independently and uniformly at random. The *key-dependent success probability* $p_{k^{(1..r)}}$ is the conditional probability of this event conditioned on the event that $K^{(1..r)} = k^{(1..r)}$. Note that $p_{k^{(1..r)}}$ depends on the input block used in the attack, but the success probability p is generally independent of this input block. In most cases of practical interest, $p_{k^{(1..r)}}$ is also independent of the actual value of $k^{(r)}$.

The generalization of linear cryptanalysis in [HKM95] exploits an $(r-1)$-round I/O sum $S = f(X) \oplus g(Y^{(r-1)})$ where f and g are balanced binary-valued functions on $\mathbb{B}^n$. The key-dependent imbalance of S is defined as $I(S|k^{(1..r-1)}) := |2 \cdot P[S = 0 \mid K^{(1..r-1)} = k^{(1..r-1)}] - 1|$ [HKM95]. Let $\mathcal{F}$ and $\mathcal{G}$ be the partitions whose partitioning functions are f and g, respectively. Then, $I((\mathcal{F}, \mathcal{G}) \mid k^{(1..r-1)}) = I(S \mid k^{(1..r-1)})$. Thus, the last-round attack by partitioning cryptanalysis performs a last-round generalized linear cryptanalysis attack with one difference: partitioning cryptanalysis can use only half of the plaintexts, either those in $\mathcal{F}_0$ or those in $\mathcal{F}_1$, whereas the generalization of linear cryptanalysis can use all plaintexts. Since $|\mathcal{F}| = |\mathcal{G}| = 2$, this difference can be removed by modifying Steps 1 and 2 in the procedure of partitioning cryptanalysis as follows:

1'. Consider a known p/c-pair (x, y) with arbitrary plaintext.

2'. For each $\tilde{k}$ in $\tilde{\mathcal{K}}$, calculate $\tilde{y}^{(r-1)} := R_{\tilde{k}}^{-1}(y)$; increment $c[k, g(\tilde{y}^{(r-1)})]$ by 1 if x is in $\mathcal{F}_0$, and increment $c[k, g(\tilde{y}^{(r-1)}) \oplus 1]$ by 1 if x is in $\mathcal{F}_1$.

4   Success Probability of Partitioning Cryptanalysis using Peak Imbalance

Partitioning cryptanalysis using peak imbalance can be applied successfully if one can find an $(r - 1)$-round partition-pair $(\mathcal{F}, \mathcal{G})$ and an input block $\mathcal{F}_i$ that satisfy the following conditions:

1) Effectiveness: $(\mathcal{F}, \mathcal{G})$ is effective.

2) Smallness of the number of key classes: The partition $\mathcal{G}$ is such that the number $\kappa := |\tilde{\mathcal{K}}|$ of key classes is reasonably small. (The computational complexity of the attack will be proportional to this number.)

3) Hypothesis of wrong-key randomization: The key-dependent input-block-dependent peak imbalance $I_p(g(Y^{(r-1)}) \mid f(X) = i,\ K^{(1..r-1)} = k^{(1..r-1)})$ of the partition-pair $(\mathcal{F}, \mathcal{G})$ for X and $Y^{(r-1)}$ (i.e., for the first $r - 1$ rounds) is substantially larger than the maximum over wrong keys $\tilde{k}_w$ of this same imbalance for X and the guess $R_{\tilde{k}_w}^{-1}(Y)$ for $Y^{(r-1)}$ computed from the ciphertext Y by using the wrong key $\tilde{k}_w$ in the last round. More precisely, let the *minimum wrong-key peak imbalance decrease* be defined by

$$\Delta I_p(k^{(1..r)}) \quad := \quad I_p(g(Y^{(r-1)}) \mid f(X) = i,\ K^{(1..r-1)} = k^{(1..r-1)})$$
$$- \max_{\tilde{k}_w \in \tilde{\mathcal{K}} \setminus \{k_r\}} I_p(g(R_{\tilde{K}}^{-1}(Y)) \mid f(X) = i,\ K^{(1..r)}\tilde{K} = k^{(1..r)}\tilde{k}_w).$$

Then, the hypothesis is that there exists a positive real number $\Delta_{\min}$, substantially larger than 0, such that, for virtually all $k^{(1..r)}$ that can result from the cipher's key scheduling algorithm, $\Delta I_p(k^{(1..r)}) > \Delta_{\min}$.

It is insightful to consider $I_p(g(R_{\tilde{K}}^{-1}(Y)) \mid f(X) = i,\ K^{(1..r)}\tilde{K} = k^{(1..r)}\tilde{k}_w)$ as the key-dependent input-block-dependent imbalance of the partition-pair $(\mathcal{F}, \mathcal{G})$ for an $(r+1)$-round "iterated" block cipher obtained by appending to the original cipher an $(r + 1)$-th round with round function R^{-1} and round key $\tilde{K}$. If $\tilde{K}$ is the right key $\tilde{k}_r$, then the $r + 1$ rounds collapse to $r - 1$ rounds, and the above imbalance equals $I_p(g(Y^{(r-1)}) \mid f(X) = i,\ K^{(1..r-1)} = k^{(1..r-1)})$. Otherwise, for good ciphers, we would naturally expect the $(r + 1)$-round partition-pair to have substantially lower key-dependent imbalance than an $(r - 1)$-round partition-pair for virtually all keys $K^{(1..r)} = k^{(1..r)}$.

By assuming hypotheses similar to those that Matsui used for approximating the success probability of linear cryptanalysis, one finds that the success probability of partitioning cryptanalysis can be approximated as

$$(1) \quad p \quad \approx \quad \int_{-\sqrt{N(m-1)\overline{I}_r^{\,2}}}^{\infty} \frac{1}{\sqrt{\pi}} e^{-\frac{t^2}{2}}\, Q\Big(-(1 - I)(t + \sqrt{N(m-1)\overline{I}_r^{\,2}})\Big)^{\kappa-1} dt$$

where $m = |\mathcal{G}|$, where $Q(\alpha) := \frac{1}{\sqrt{2\pi}} \int_\alpha^\infty e^{-\frac{t^2}{2}} dt$, and where $\overline{I}_r := \overline{I}_p(g(Y^{(r-1)}) \mid f(X) = i)$ (cf. Appendix B). This approximation gives p as an increasing function of $\overline{I}_r$, which suggests that the average-key peak imbalance $\overline{I}_p(.)$ is a good measure for the usefulness of the partition-pair $(\mathcal{F}, \mathcal{G})$.

5   Coset-Partition-Pairs

Since there is generally an infeasibly large number of partitions with blocks of equal sizes, we must concentrate on partitions with properties that suggest their usefulness for partitioning cryptanalysis. Many ciphers use a group operation to insert the round keys at the input of each round. For such ciphers, it is natural to consider "coset-partitions" of $\mathbb{B}^n$ with respect to this group operation.

Definition 3. Let "$\otimes$" be a group operation in $\mathbb{B}^n$ and e the neutral element for "$\otimes$". A *coset-partition for* "$\otimes$" is a partition $\mathcal{F}$ for which the block containing e, $\mathcal{F}(e)$, is a subgroup of $(\mathbb{B}^n, \otimes)$ and whose other blocks are the cosets of this subgroup; i.e., a coset-partition is a partition that can be written as

$$\mathcal{F} \;=\; \{x \otimes \mathcal{F}(e) \,:\, x \in \mathbb{B}^n\} \;.$$

A *coset-partition-pair* is a partition-pair both of whose components are coset-partitions. A coset-partition for the component-wise XOR operation on n-tuples will be called a *linear partition* and a partition-pair whose input and output partitions are both linear will be called a *linear partition-pair*. The following lemma gives a fundamental property of coset-partitions.

Lemma 4. *Let ϕ_z be the automorphism $\phi_z : \mathbb{B}^n \to \mathbb{B}^n$, $x \mapsto z \otimes x$. A partition is a coset-partition for the group operation "$\otimes$" in $\mathbb{B}^n$ if and only if, for every z in $\mathbb{B}^n$, the automorphism ϕ_z maps all elements of each block of $\mathcal{F}$ onto one block of $\mathcal{F}$, i.e., if and only if the block containing $z \otimes x$ is uniquely determined by z and by the block containing x, or, again equivalently, if and only if $(\mathcal{F}, \mathcal{F})$ has guaranteed transitions for "$\otimes$", i.e., for every ϕ_z with z in $\mathbb{B}^n$.*

6   Finding Effective Partition-Pairs

We now suppose that the round function R of an iterated block cipher is defined by

$$Y^{(i)} \;=\; R_{K^{(i)}}(Y^{(i-1)}) \;=\; \phi(Y^{(i-1)} \otimes K_{\mathrm{L}}^{(i)}, K_{\mathrm{R}}^{(i)}) \;,$$

where "$\otimes$" is a group operation in $\mathbb{B}^n$, where $K_{\mathrm{L}}^{(i)}$ and $K_{\mathrm{R}}^{(i)}$ denote the left and the right part of $K^{(i)}$, and where $\phi(., k_{\mathrm{R}}^{(i)})$ is invertible for all $k_{\mathrm{R}}^{(i)}$.

We propose a method for finding effective coset-partition-pairs for such ciphers. For linear cryptanalysis and the attack in [HKM95], there exists a procedure for finding effective linear expressions and effective I/O sums. With the help of Matsui's piling-up lemma, the imbalance of multi-round homomorphic I/O sums can be lower bounded in terms of the imbalances of one-round homomorphic threefold sums [HKM95]. Fortunately, the most effective I/O sums are often those for which this lower bound is the largest. For partitioning cryptanalysis, there is no general way to compute a good lower bound on the imbalance of a multi-round partition-pair given imbalances of the one-round partition-pairs that it comprises. We must settle for *approximating* the average-key imbalance of a multi-round partition-pair.

Piling-up hypothesis for partition-pairs. *Consider a cascade of ρ rounds with round function R as defined above. Consider a list $\mathcal{F}^{(0)}, \mathcal{F}^{(1)}, \ldots, \mathcal{F}^{(\rho)}$ of coset-partitions for "$\otimes$" and let $\overline{I}^{(i)}(\mathcal{F}^{(i-1)}, \mathcal{F}^{(i)})$ be the average-key imbalance of $(\mathcal{F}^{(i-1)}, \mathcal{F}^{(i)})$ for ϕ. Then, the average-key imbalance $\overline{I}^{(1..\rho)}((\mathcal{F}^{(0)}, \mathcal{F}^{(\rho)}))$ for $X^{(1)}$ and $Y^{(\rho)}$ can be well approximated by*

(2)
$$\tilde{I}^{(1..\rho)}((\mathcal{F}^{(0)}, \mathcal{F}^{(\rho)})) := \max \ \overline{I}^{(\phi)}((\mathcal{F}^{(0)}, \mathcal{F}^{(1)})) \cdot \ \ldots \ \cdot \overline{I}^{(\phi)}((\mathcal{F}^{(\rho-1)}, \mathcal{F}^{(\rho)}))$$

where the maximum is taken over all coset-partitions $\mathcal{F}^{(1)}, \mathcal{F}^{(2)}, \ldots \mathcal{F}^{(\rho-1)}$ for $\otimes$ for which $|\mathcal{F}^{(0)}| \geq |\mathcal{F}^{(1)}| \geq \cdots \geq |\mathcal{F}^{(\rho)}|$.

We have not been able to *prove* any result of this kind, but we have found experimentally that the "piling-up approximation" (2) is often so descriptive of the actual average-key imbalance that it can safely be used for finding effective multi-round partition-pairs.

Procedure for finding effective ρ-round coset-partition-pairs

1. Find the set $\mathcal{S}_\otimes$ of all coset-partitions for "$\otimes$".
2. For all $(\mathcal{F}, \mathcal{G})$ in $\mathcal{S}_\otimes^2$ where $|\mathcal{F}| \geq |\mathcal{G}|$, find the imbalance of the partition-pair $(\mathcal{F}, \mathcal{G})$ for the input and the output of a round. Discard the partition-pairs with small imbalance from further consideration.
3. For each (l, m), where l and m are powers of two with $l \geq m$, consider all retained partition-pairs $(\mathcal{F}, \mathcal{G})$ in $\mathcal{S}_\otimes^2$ for the input of the first and the output of the ρ-th round with $|\mathcal{F}| = l$ and $|\mathcal{G}| = m$; use the piling-up approximation (2) to estimate their average-key imbalance; find the most effective such partition-pairs and their approximate average-key imbalances.
4. Use (1) to approximate the success probability of partitioning cryptanalysis. Decide for which l and for which m the most successful attack can be obtained.

The complexity of this procedure depends crucially on the number of coset-partitions. For cyclic group operations of $\mathbb{B}^n$, there are only $n - 2$ non-trivial coset-partitions, whereas there exist many more linear partitions. There seems to be little chance of finding an effective partition-pair for ciphers using cyclic group operations on $\mathbb{B}^n$ and partitioning cryptanalysis seems not to be powerful against such ciphers. For ciphers using bitwise XOR to insert the keys, there are so many coset-partitions that it is generally infeasible to compute the maximum specified in the piling-up hypothesis.

7   Partitioning Cryptanalysis of DES

We applied the last-round attack by partitioning cryptanalysis to six rounds of the Data Encryption Standard (DES). Details of this attack are given in Appendix C.

For a certain partition-pair, we analyzed in our attack all 256 p/c-pairs with plaintexts in one input block. The success probability was about 95% for partitioning cryptanalysis using the Euclidean imbalance but was only 74% for partitioning cryptanalysis using the peak imbalance [Har96]. Our better success probability is about the same as that obtained by Biham and Shamir for differential cryptanalysis of 6-round DES (i.e., a success probability of 95% using 240 p/c-pairs with chosen plaintexts [BS93, page 31]), and it is slightly worse than the attack of [Knu95], but a new weakness of 6-round DES is exploited in partitioning cryptanalysis.

8   Conclusions

Partitioning cryptanalysis of iterated block ciphers was introduced. This is a generalization of linear cryptanalysis and exploits a weakness that can be described by an effective partition-pair, i.e., a pair of partitions such that, for every key, the next-to-last-round outputs are substantially non-uniformly distributed over the blocks of the second partition when the plaintexts are chosen uniformly from a particular block of the first partition. The success probability of partitioning cryptanalysis was approximated by assuming hypotheses similar to those that Matsui assumed to estimate the success probability of linear cryptanalysis. The crucial problem of partitioning cryptanalysis, namely the problem of finding effective partition-pairs, was addressed. Our procedure for finding effective coset-partition-pairs requires extensive computation for ciphers in which the round keys are inserted with the XOR operation. When the keys are inserted with group operations for large cyclic groups, our procedure is considerably faster, but the chance to find an effective partition-pair is generally small. To illustrate the potential usefulness of partitioning cryptanalysis, we applied it to DES. For 6-round DES, attacks by partitioning cryptanalysis are chosen-plaintext attacks that are about as successful as differential cryptanalysis, although they are based on a different weakness. We close by remarking that ciphers that are very weak against partitioning cryptanalysis, but quite strong against both linear and differential cryptanalysis have been designed in [Har96].

References

[BS93]   Eli Biham and Adi Shamir. *Differential Cryptanalysis of the Data Encryption Standard*. Springer, New York, 1993. ISBN 3-540-97930-1.

[Har96]   Carlo Harpes. *Cryptanalysis of iterated block ciphers*, volume 7 of *ETH Series in Information Processing*. Hartung-Gorre Verlag Konstanz, J. L. Massey editor, 1996. ISBN 3-89649-079-6.

[Har94]   Carlo Harpes. *Success probability of partitioning cryptanalysis*. In H. C. A. van Tilborg and F. M. J. Willems, editors, *Proceedings of the EI-DMA Winter Meeting on Coding Theory, Information Theory and Cryptology*. December 1994.

[HKM95]   Carlo Harpes, Gerhard G. Kramer, and James L. Massey. A generalization of linear cryptanalysis and the applicability of Matsui's piling-up lemma. In L. C. Guillou and J.-J. Quisquater, editors, *Advances in Cryptology - Eurocrypt '95*, Lecture Notes in Computer Science No. 921, pages 24-38. Springer, 1995.

[KR95]   Burton S. Kaliski and Matthew J. B. Robshaw. Linear cryptanalysis using multiple approximations. In A. De Santis, editor, *Advances in Cryptology - Eurocrypt '94*, Lecture Notes in Computer Science No. 950, pages 26-39. Springer, 1995.

[Knu95]   Lars Knudsen. Truncated and high order differentials. In B. Preneel, editor, *Fast Software Encryption*, Lecture Notes in Computer Science No. 1008, pages 196-211. Springer, 1995.

[Mat94a]   Mitsuru Matsui. The first experimental cryptanalysis of the Data Encryption Standard. In Y. G. Desmedt, editor, *Advances in Cryptology - Crypto '94*, Lecture Notes in Computer Science No. 839, pages 1-11. Springer, 1994.

[Mat94b]   Mitsuru Matsui. Linear cryptanalysis method for DES cipher. In T. Helleseth, editor, *Advances in Cryptology - Eurocrypt '93*, Lecture Notes in Computer Science No. 765, pages 386-397. Springer, 1994.

[MPWW94] Sean Murphy, Fred Piper, Michael Walker, and Peter Wild. Likelihood estimation for block cipher keys. Submitted for publication, May 1994.

[Vau96]   Serge Vaudenay. An experiment on DES: Statistical Cryptanalysis. In *Proceedings of the 3rd ACM Conferences on Computer Security*, pages 139-147. ACM Press, 1996.

A   Combined Partitioning Cryptanalysis

Partitioning cryptanalysis is a chosen-plaintext attack because only p/c-pairs with plaintexts from one input block can be considered, but any input block in $\mathcal{F}$ can generally be used. We may extend partitioning cryptanalysis to form a known-plaintext attack by combining partitioning cryptanalysis attacks that use different input blocks in the following way.

0. Set up a counter $c[i, \tilde{k}, j]$ for each i such that $0 \leq i < l$, for each j such that $0 \leq j < m$, and for each $\tilde{k}$ in $\tilde{\mathcal{K}}$; and initialize all counters to 0.

1. Choose a known p/c-pair (x, y) with arbitrary plaintext.

2. For each $\tilde{k}$ in $\tilde{\mathcal{K}}$, calculate $\tilde{y}^{(r-1)} := R_{\tilde{k}}^{-1}(y)$ and
   increment $c[f(x), \tilde{k}, g(\tilde{y}^{(r-1)})]$ by 1.

3. Repeat Steps 1 and 2 for all known p/c-pairs.

4. For each i such that $0 \leq i < l$ and each $\tilde{k}$ in $\tilde{\mathcal{K}}$, reorder the values of the counters $c[i, \tilde{k}, 0], c[i, \tilde{k}, 1], \ldots, c[i, \tilde{k}, m-1]$ so that $c[i, \tilde{k}, 0] \geq \ldots \geq c[i, \tilde{k}, m-1]$.

5. For each $\tilde{k}$ in $\tilde{\mathcal{K}}$ and each j such that $0 \leq j < m$,
   set $c'[\tilde{k}, j] = \sum_{i=0}^{l-1} c[i, \tilde{k}, j]$.

6. Output the key(s) $\tilde{k}$ maximizing $\hat{\mu}(\tilde{k}) = I(\frac{c'[\tilde{k},0]}{N}, \frac{c'[\tilde{k},1]}{N}, \ldots, \frac{c'[\tilde{k},m-1]}{N})$ as candidate(s) for the key actually used in the last round.

We will call this attack *combined partitioning cryptanalysis* (CPC). The combined partitioning cryptanalysis attack is similar to the attack by linear cryptanalysis using multiple approximations [KR95]. Note that combined partitioning cryptanalysis is a known-plaintext attack and that it will not be successful if l is large as there will be generally too few p/c-pairs with plaintexts in the same input block unless the plaintexts have been chosen such that they lie in a small number of input blocks. The complexity of the attack is considerably reduced if peak imbalance is used.

B   Approximation of the Success Probability

We assume that the imbalance used in partitioning cryptanalysis is the peak imbalance $I_p(.)$. We first propose a model of how the various counters in the last-round attack increment. Let $\tilde{J}(\tilde{k})$ denote the random variable indicating which counter is incremented for the key guess $\tilde{k}$ when a certain p/c-pair is analyzed, i.e., $\tilde{J}(\tilde{k}) = g(\tilde{Y}^{(r-1)}(\tilde{k}))$, and let $J = g(Y^{(r-1)})$ indicate which counter is incremented for the right key $\tilde{k}_r$.

Model for counter incrementing.   *For each wrong key $\tilde{k}_w$ in $\tilde{\mathcal{K}} \setminus \{\tilde{k}_r\}$, independently for all $\tilde{k}_w$, $\tilde{J}(\tilde{k}_w)$ is the output of an m-ary symmetric channel with dominant probability q, where $q \geq \frac{1}{m}$, and input J, where $J = \tilde{J}(\tilde{k}_r)$, i.e.,*

$$P[\,\tilde{J}(\tilde{k}_w) = \tilde{j} \mid J = j\,] \quad = \quad \begin{cases} q & if \ \tilde{j} = \pi_{\tilde{k}_w}(j) \\ \frac{1-q}{m-1} & if \ \tilde{j} \neq \pi_{\tilde{k}_w}(j) \ , \end{cases}$$

where $\pi_{\tilde{k}_w}$ is some permutation of $\{0, 1, \ldots, m-1\}$. Moreover, $q = \frac{m-1}{m}I + \frac{1}{m}$ and I, which will be called the imbalance *of the m-ary symmetric channel with dominant probability q, is well approximated as*

(3) $$I \quad \approx \quad \frac{E[\,I_p(J(\tilde{K}_w))\,]}{I_p(J)} \ .$$

The parameter I or, equivalently, the parameter q, characterizes the randomization caused by the last round.

To find I in partitioning cryptanalysis using peak imbalance, we may approximate $I_p(J)$ and $I_p(\tilde{J}(\tilde{k}_w))$ by obvious estimates, namely by the empirical decision metrics $\hat{\mu}(\tilde{k}_r)$ and $\hat{\mu}(\tilde{k}_w)$, respectively. Moreover, if we assume that $I_p(\tilde{J}(\tilde{k}_w)) \approx E[\,I_p(J(\tilde{K}_w))\,]$ for all $\tilde{k}_w$, then (3) follows from the following lemma.

Lemma 5. *Let the random variable J be the input to an m-ary symmetric channel with dominant probability q or, equivalently, with imbalance $I = \frac{m}{m-1}(q - \frac{1}{m})$, and let $\tilde{J}(\tilde{k}_w)$ be the output. Then, for any probability distribution for J,*

$$I_p(\tilde{J}(\tilde{k}_w)) \quad = \quad I_p(J) \cdot I \ .$$

Note that if the model for counter incrementing is satisfied for an effective partition-pair and if I is substantially smaller than 1, then the hypothesis of wrong-key randomization is also fulfilled. The following theorem is based on the given model for counter incrementing and provides a good approximation of the success probability of partitioning cryptanalysis.

Theorem 6. (Success probability of partitioning cryptanalysis)   *Consider the last-round attack by partitioning cryptanalysis for the partition-pair $(\mathcal{F}, \mathcal{G})$, using peak imbalance, analyzing N p/c-pairs with plaintexts in an input block $\mathcal{F}_i$, and distinguishing κ key classes. Suppose first that N is sufficiently large. Suppose second that the above model for counter incrementing holds. Suppose third that, for each of the κ key guesses, the count that is the most likely to be incremented dominates the other counts. Then, the success probability of this attack, when the key in use is $k^{(1..r)}$, is well approximated by*

$$(4) \quad p_{k^{(1..r)}} \approx \int_{-\sqrt{N(m-1)I_r^2}}^{\infty} \frac{1}{\sqrt{\pi}} e^{-\frac{t^2}{2}} Q\left(-(1-I)(t + \sqrt{N(m-1)I_r^2})\right)^{\kappa-1} dt \;,$$

where $m = |\mathcal{G}|$, where $Q(\alpha) := \frac{1}{\sqrt{2\pi}} \int_{\alpha}^{\infty} e^{-\frac{t^2}{2}} dt = \frac{1}{2}(1 - \mathrm{erf}(\frac{\alpha}{\sqrt{2}}))$, and where I_r is the key-dependent input-block-dependent imbalance of $(\mathcal{F}, \mathcal{G})$ for the first $r-1$ rounds given that $\mathcal{F}_i$ is the input block used in the attack and that $K^{(1..r-1)} = k^{(1..r-1)}$.

In the following figure, we show the approximate success probability as a function of $N(m-1)I_r^2$ for different values of κ and I, where I is well estimated by the quotient of the wrong-key and the right-key empirical decision metrics.

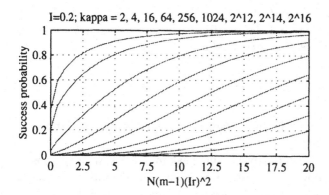

I=0.2; kappa = 2, 4, 16, 64, 256, 1024, 2^12, 2^14, 2^16

N(m−1)(Ir)^2

Now, if for virtually all keys $k^{(1..r-1)}$ in $\mathcal{K}^{r-1}$ and for all i, $0 \leq i < l$,

$$I_p(g(Y^{(r-1)}) \mid f(X)=i, \; K^{(1..r-1)} = k^{(1..r-1)}) \quad \approx \quad \overline{I}_p(g(Y^{(r-1)}) \mid f(X)=i)$$

— this assumption may be called the *hypothesis of fixed-key equivalence in partitioning cryptanalysis* — then (1) follows. In practice, however, the hypothesis of fixed-key equivalence for partitioning cryptanalysis is generally not well satisfied. We observe that the approximation on the right of (1) is often a convex-$\cup$

function of the key-dependent imbalance I_r for small imbalances and a convex-∩ (or concave) function of I_r for large imbalances. By Jensen's inequality, we then expect this approximation to be smaller than the true success probability when the key-dependent imbalances are low (or, equivalently, when the success probability is low) and vice versa. We now confirm these expectations for DES.

C   Partitioning Cryptanalysis of DES

A linear partition $\mathcal{F}$ can be described by a *parity-check matrix* $H_{\mathcal{F}}$ whose rows are the n-tuples $\alpha_1, \alpha_2, \ldots, \alpha_d$. Given $H_{\mathcal{F}}$, the partition $\mathcal{F}$ is the set of cosets of the linear code $\mathcal{F}(0)$ for which $H_{\mathcal{F}}$ is a parity-check matrix, i.e., $\mathcal{F}$ is the set of cosets of the subgroup $\mathcal{F}(0) = \{x : x \in \mathbb{B}^n, xH_{\mathcal{F}}^T = 0\}$. The partitioning function of $\mathcal{F}$ can be chosen as $f : x \mapsto xH_{\mathcal{F}}^T$. The parity-check matrix $H_{\mathcal{F}}$ is called *reduced* if its rows are linearly independent. In this case, the blocks of $\mathcal{F}$ can be labeled with $d = \log_2(|\mathcal{F}|)$ binary digits.

A linear partition with parity-check matrix $H_{\mathcal{F}}$ such that the Hamming weight of each row is 1 will be called *bit-selecting*. Its subgroup can be written as $\mathcal{F}(0) = \{x : x \in \mathbb{B}^n, \alpha \& x = 0\}$ for some n-tuple α with Hamming weight d, where "&" denotes bitwise AND and 0 represents the all-zero n-tuple. The n-tuple α will be called the *bit-selecting n-tuple* of $\mathcal{F}$. The i-th bit of a bit-selecting n-tuple α will be called *constrained by* $\mathcal{F}$ if the i-th bit of α is 1, and called *free* otherwise. A block of a linear partition is uniquely determined by specifying the constrained bits for an element of this block.

Our procedure for finding effective partition-pairs for DES is similar to the procedure given in Section 6, but we use a simpler approximation of the partition-pair average-key imbalances. Since it is infeasible for 5-round DES to compute the product in (2) for all lists $(\mathcal{F}^{(0)}, \mathcal{F}^{(1)}, \ldots, \mathcal{F}^{(5)})$ such that $|\mathcal{F}^{(0)}| \geq |\mathcal{F}^{(1)}| \geq \cdots \geq |\mathcal{F}^{(5)}|$, we reduce the number of lists $(\mathcal{F}^{(0)}, \mathcal{F}^{(1)}, \ldots, \mathcal{F}^{(5)})$ for which we actually compute the product as follows. The S-boxes from which constrained bits emerge will be called *active*. We restricted ourselves to lists of bit-selecting linear partitions in which all bits that connect active S-boxes are constrained. Moreover, we used an appropriate approximation for the one-round partition-pair peak imbalances. Our procedure provided us with many effective partition-pairs. We empirically computed the peak imbalances of the most effective partition-pairs found and we verified that the empirical peak imbalance depends on the imbalance estimated in our procedure (cf. Fig. 3 left), which shows that the imbalance estimate can be used for finding effective partition-pairs. We then approximated the success probability of attacks exploiting these partition-pairs, performed the attacks many times, and empirically estimated the success probability.

In Fig. 3 (right), we compared the approximate success probability and the empirically estimated success probability. Note that to evaluate (1), I is given by (3) in which the peak imbalances are replaced by empirical estimates of average-key peak imbalances. The difference between the quantities shown in Fig. 3 can be explained mainly by the fact that the hypothesis of fixed-key equivalence is not well satisfied (cf. Appendix B). Thus, our results suggest that Theorem 6 gives a

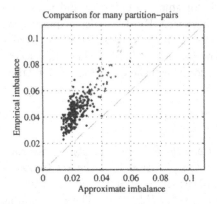 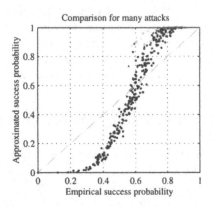

Fig. 3. Left: comparison of the approximate imbalances to the empirical imbalances of partition-pairs for 5-round DES; right: comparison of the empirical success probability and the approximate success probability of partitioning cryptanalysis using the peak imbalance for 6-round DES. For many attacks all using different partition-pairs whose input partition has 2^{64-l^*} blocks where $l^* = 8$ (light-grey points), 10, 12, or 14 (dark points), the empirical partition-pair imbalance and the empirical success probability are estimated after attacking 100 000 random keys. The number of known p/c-pairs in the attacks was $N = 256$ for $l^* = 8$ and $N = 1024$ otherwise.

good approximation to the true success probability, although the approximation (1) may be crude.

We also performed attacks by partitioning cryptanalysis using different imbalance measures. Partitioning cryptanalysis attacks using Euclidean imbalance were generally slightly stronger than attacks using peak imbalance.

The most successful partitioning cryptanalysis attack on 6-round DES using 256 chosen p/c-pairs was obtained when the attack used Euclidean imbalance, when the bit-selecting 64-tuple of the input partition was `81fff9ff ffffffff` and when the bit-selecting 64-tuple of the round four output partition was `00000000 0000f000` (i.e., when the output block was defined only by the bits that are linked to the S-box S5 of the fourth round). The approximate imbalance computed in the procedure for finding effective bit-selecting linear partition-pairs is then 0.0638 and the empirical imbalance is 0.0970. The empirical success probability is 95% for partitioning cryptanalysis using Euclidean imbalance but is only 74% for partitioning cryptanalysis using peak imbalance when all 256 p/c-pairs with plaintexts in one block are used. According to our approximation with q empirically determined as 0.39, the latter success probability should be 95% instead of 74%, which indicates that the approximation is good for most purposes.

The Interpolation Attack on Block Ciphers*

Thomas Jakobsen[1], Lars R. Knudsen[2]

[1] Department of Mathematics, Building 303, Technical University of Denmark, DK-2800 Lyngby, Denmark, email:jakobsen@mat.dtu.dk.
[2] Katholieke Universiteit Leuven, Dept. Electrical Engineering-ESAT, Kardinaal Mercierlaan 94, B–3001 Heverlee, Belgium, email:knudsen@esat.kuleuven.ac.be.

Abstract. In this paper we introduce a new method of attacks on block ciphers, the interpolation attack. This new method is useful for attacking ciphers using simple algebraic functions (in particular quadratic functions) as S-boxes. Also, ciphers of low non-linear order are vulnerable to attacks based on higher order differentials. Recently, Knudsen and Nyberg presented a 6-round prototype cipher which is provably secure against ordinary differential cryptanalysis. We show how to attack the cipher by using higher order differentials and a variant of the cipher by the interpolation attack. It is possible to successfully cryptanalyse up to 32 rounds of the variant using about 2^{32} chosen plaintexts with a running time less than 2^{64}. Using higher order differentials, a new design concept for block ciphers by Kiefer is also shown to be insecure. Rijmen et al presented a design strategy for block ciphers and the cipher SHARK. We show that there exist ciphers constructed according to this design strategy which can be broken faster than claimed. In particular, we cryptanalyse 5 rounds of a variant of SHARK, which deviates only slightly from the proposed SHARK.

1 Introduction

In an r-round iterated cipher the ciphertext is computed by iteratively applying in r rounds a *round function* G to the plaintext, s.t.

$$C_i = G(K_i, C_{i-1}),$$

where C_0 is the plaintext, K_i is the ith round key, and C_r is the ciphertext. A special kind of iterated ciphers are the **Feistel** ciphers. A Feistel cipher with block size $2n$ and r rounds is defined as follows. Let C_0^L and C_0^R be the left and right hand halves of the plaintext, respectively, each of n bits. The round function G operates as follows

$$C_i^L = C_{i-1}^R$$
$$C_i^R = F(K_i, C_{i-1}^R) + C_{i-1}^L,$$

---

* The work in this paper was initiated while the authors were visiting the Isaac Newton Institute, Cambridge, U.K., February 1996.

and the ciphertext is the concatenation of C_r^R and C_r^L. Note that F can be any function taking as arguments an n-bit text and a round key K_i and producing n bits. '+' is a commutative group operation on the set of n bit blocks. For the remainder of this paper we will assume that '+' is the exclusive-or operation ($\oplus$).

Based on the use of a quadratic function over a Galois field, Knudsen and Nyberg demonstrated in [10] how to construct a cipher which is provably secure against differential cryptanalysis [1]. The cipher is a Feistel cipher with the function F given by $F : \mathrm{GF}(2^{32}) \to \mathrm{GF}(2^{32})$ with

$$F(k, x) = d(f(e(x) \oplus k)),$$

where $f : \mathrm{GF}(2^{33}) \to \mathrm{GF}(2^{33})$, $f(x) = x^3$, $k \in \mathrm{GF}(2^{33})$, $e : \mathrm{GF}(2^{32}) \to \mathrm{GF}(2^{33})$ is a function which extends its argument by concatenation with an affine combination of the input bits, and $d : \mathrm{GF}(2^{33}) \to \mathrm{GF}(2^{32})$ discards one bit from its argument. We call this cipher $\mathcal{KN}$.

Also, we will consider the cipher with round function given by $F_k(x) = f(x \oplus k)$ where $f : \mathrm{GF}(2^{32}) \to \mathrm{GF}(2^{32})$, $f(x) = x^3$, i.e., the cubing function's input is not extended and the output not truncated as in the previous case. We call this cipher $\mathcal{PURE}$.

Both ciphers are secure against differential attacks [10]. Also, both ciphers are secure against the linear attack [7], which follows from [9].

In [10] the cipher $\mathcal{KN}$ is defined to be used with 6 rounds and since $f(x)$ is differentially 2-uniform, it is possible to prove that this yields a provably secure cipher (secure against conventional differential cryptanalysis). The same holds for $\mathcal{PURE}$. However, in both cases the non-linear order of the output is low with respect to the input and this can be exploited to mount an attack.

In the following, $x = (x_L, x_R)$ denotes the plaintext where x_L and x_R denote the left and right hand side of x, respectively. Similarly, $y = (y_L, y_R)$ denotes the ciphertext. By the *reduced cipher*, we denote the cipher that one gets by removing the final round of the original cipher. The output from this cipher is denoted $\tilde{y} = (\tilde{y}_L, \tilde{y}_R)$.

The attacks presented in this paper are classified according to the taxonomy of [4]. That is, by a *key-recovery attack* we mean that an attacker finds the secret key. By a *global deduction* we mean that an attacker finds an algorithm, which encrypts any plaintext into a valid ciphertext without knowing the secret key. By an *instance deduction* we mean that an attacker finds an algorithm, which encrypts a subset of all plaintexts into valid ciphertexts without knowing the secret key. In the key-recovery attacks we try to guess the last-round key. The guess is then used to decrypt the ciphertext by one round and in this way one (hopefully) obtains the output from the reduced cipher. If there exists a method to distinguish whether this is the actual output from the reduced cipher or not, then we can find the last-round key. Once this key has been found, attacks similar to the ones we present can be mounted on a cipher one round shorter than the original. As the measurement of the time needed by an attack, we use the total number of encryptions of the attacked block cipher.

This paper is organised as follows. In § 2 we give new attacks based on higher order differentials. We apply the attacks to the cipher $\mathcal{KN}$ by Knudsen and

Nyberg [10] and to a cipher by Kiefer [3]. In § 3 we present our new attack on block ciphers, the interpolation attack. We apply the attack to a cipher, provably secure against differential and linear attacks. Also, we apply our methods to a slightly modified version of the cipher SHARK [11]. We conclude in § 4.

2   Attacks Using Higher Order Differentials

In [6] Lai gave a definition of higher order derivatives of discrete functions. Later Knudsen used higher order differentials to cryptanalyse ciphers presumably secure against conventional differential attacks, i.e. attacks based on first order differentials [5]. In this section we give an extension of Knudsen's attacks and apply it in an attack on the cipher $\mathcal{KN}$. We refer to [6, 5] for the definitions of higher order differentials.

Consider a Feistel cipher with block size $2n$. Suppose that x_R is kept constant and consider the right hand side $\tilde{y}_R$ of the output from the reduced cipher. Since x_R is a constant, the bits in $\tilde{y}_R$ are all expressible as polynomials $GF(2)[x_1, x_2, \ldots, x_n]$ in the bits of $x_L = (x_1, x_2, \ldots, x_n)$. Assume that these polynomials have degree not higher than d. Then according to [6, Proposition 2] (see also [5]), we have

$$\sum_{x_L \in \mathcal{L}_d} p(x_L) = c, \tag{1}$$

where $\mathcal{L}_d$ denotes a d-dimensional subspace of $GF(2)^n$, c is the same for any space parallel to $\mathcal{L}_d$, and p is a function which computes the output from the reduced cipher. It follows that

$$\sigma(w) = \sum_{x_L \in \mathcal{L}_{d+1}} p(x_L + w) = 0 \text{ for all } w \in GF(2)^n \tag{2}$$

if and only if $p(x)$ is a polynomial of degree d or lower. In the following algorithm, the variables $x = (x_L, x_R)$ and $y = (y_L, y_R)$ hold the plaintext and the ciphertext, respectively. L is a full rank $(d+1) \times n$ matrix over $GF(2)$ and F the round function.

1. Let x_R and w be n-bit constants.
2. For all $a \in GF(2)^{d+1}$:
   (a) Let $x_L = aL + w$.
   (b) Obtain the ciphertext $y(a)$ of plaintext (x_L, x_R).
3. For all values, k, of the last-round key:
   (a) Let $\sigma = 0$.
   (b) For all $a \in GF(2)^{d+1}$:
      i. Let $y = y(a)$.
      ii. Let $\tilde{y}_R = y_L \oplus F(k, y_R)$.
      iii. Let $\sigma = \sigma \oplus \tilde{y}_R$.

The key for which σ ends up being zero is the correct last-round key with a high probability. Consequently, for every possible value k of the last-round key, we check whether the corresponding value of σ is zero, and if it is, then we have found the correct key with high probability. If one wants a higher level of certainty, the algorithm is simply repeated with another choice of w. This method is easily generalised to an iterated cipher, and we get the following result, extending that of [5, Th. 11].

Theorem 1. *Given an iterated block cipher, let d denote the polynomial degree of the ciphertext bits of the round next to the last as a function of the plaintext bits. Furthermore, let b denote the number of last-round key bits. Assume that the polynomial degree of the ciphertext bits increases with the number of rounds. Then there exists a d-th order differential attack of average time complexity 2^{b+d} requiring 2^{d+1} chosen plaintexts which will successfully recover the last-round key.*

Proof. We give the proof in the case of a Feistel cipher, from which the general case follows. Consider the iteration (3b). Let k denote the correct value of the last-round key, and let k' denote any wrong value. Then

$$\tilde{y}_R = y_L \oplus F(k, y_R)$$
$$\tilde{y}'_R = y_L \oplus F(k', y_R)$$
$$= \tilde{y}_R \oplus F(k, y_R) \oplus F(k', y_R).$$

The difference between $\tilde{y}_R$, obtained using the correct key, and $\tilde{y}'_R$, obtained with a wrong key, is two applications of the function F. Since by assumption the polynomial degree increases with the number of rounds, one can expect that σ will be zero only for the correct value of the last-round key with a high probability. Running an algorithm similar to the one above takes 2^{d+1} steps for each value of the last-round key. On the average, we have to test half of the keys before finding the correct one, from which the time complexity follows. $\square$

The attack can be improved by a factor of two, if the constant of Equation (1) can be predicted. In that case the iterations (2) and (3b) of the above algorithm are performed only for all $a \in \mathrm{GF}(2^d)$. The key for which $\sigma = c$ will be the correct key with a high probability. For most ciphers, depending on the F-function, there are possible extensions to the above attack. It may be possible to perform the attack for only a subset of the last-round key, and also it may be possible to search for (a subset of) the first-round key.

In the following we apply the attack to the cipher $\mathcal{KN}$. We choose plaintexts where the right halves are fixed. Since the output bits from the round function are only quadratic in the input bits, the polynomials in the attack described above on the 6 round version have degree not higher than 8. Therefore the attack requires only $2^{8+1} = 512$ chosen plaintexts and an average running time of order 2^{41}. A variant of the attack guessing for the keys in the last two rounds requires about 32 chosen plaintexts and an average running time of order 2^{70}. Similarly, there are attacks on the 7 and 8 rounds versions of $\mathcal{KN}$, the complexities are

# Rounds	# Chosen plaintexts	Running time
6	2^9	2^{41}
6	2^5	2^{70}
7	2^{17}	2^{49}
7	2^9	2^{74}
8	2^{17}	2^{82}

Table 1. Higher order differential attacks on the Knudsen-Nyberg cipher.

given in Table 1. The attack on $\mathcal{KN}$ using higher order differentials has been implemented, and it recovers the last round key as predicted. Note that these attacks are applicable to ciphers with any block size $2n$, as long as the number of chosen plaintexts is less than 2^n. The bigger the block size the more rounds can be attacked.

We now attack the scheme by Kiefer [3] by the use of higher order differentials[3]. The cipher is probabilistic and uses the following encryption rule:

$$m_i \mapsto (F(k) \oplus r_i, f_k(r_i) \oplus m_i), \tag{3}$$

where $F : \mathrm{GF}(2^n) \to \mathrm{GF}(2^n)$ is a one-way function, $f_k : \mathrm{GF}(2^n) \to \mathrm{GF}(2^n)$ is a function depending on the key $k \in \mathrm{GF}(2^n)$ in some complex way, $r_i \in \mathrm{GF}(2^n)$ is a random value, and $m_i \in \mathrm{GF}(2^n)$ is a message block. The function f_k has the form $f_k = \pi_k \circ g$ where $\pi_k : \mathrm{GF}(2^n) \to \mathrm{GF}(2^n)$ is a bitwise linear transform depending on k and $g : \mathrm{GF}(2^n) \to \mathrm{GF}(2^n)$ is a public, almost perfectly non-linear function of the form $g(x) = x^{2^s+1}$ for some s.

Assume that we know enough plaintext to have four pairs on the form

$$(a_i, b_i) = (F(k) \oplus r_i, f_k(r_i)), \; i = 1, \ldots, 4 \tag{4}$$

such that $a_1 \oplus a_2 = a_3 \oplus a_4$. Define $\beta = \bigoplus_{i=1}^4 b_i$ and $\gamma = \bigoplus_{i=1}^4 g(r_i)$. Then

$$\beta = \bigoplus_{i=1}^4 b_i = \pi_k \left(\bigoplus_{i=1}^4 g(r_i) \right) = \pi_k(\gamma). \tag{5}$$

Since $\{a_1, \ldots, a_4\}$ is a two-dimensional subspace of $\mathrm{GF}(2^n)$, the elements in $\{r_1, \ldots, r_4\}$ also constitute a two-dimensional subspace. Note also that the Hamming weight of the exponent in the definition of g expressed as a binary number is only two, implying that the output bits are only quadratic in the input bits. By Equation (1), this implies that we can compute the value of γ.

If repeated n times, we will have n corresponding pairs of β and γ. This makes it possible to solve Equation (5) with respect to the unknown function π_k (it is a linear transform). After having found π_k, we can invert f_k and thus obtain a value of r_i. Using this, we compute $F(k)$ and the system is broken.

It remains to compute the minimum number t of known plaintexts needed to obtain n times four pairs (a_i, b_i) with the required property; recall that the

---

[3] This attack was presented at the rump session of Pragocrypt'96.

cipher is probabilistic and thus we have no control over the values of r_i. By using a birthday paradox type argument it can be shown that $t \approx (n \cdot 2^{n+2})^{\frac{1}{4}}$. For a typical block size of $n = 64$ this gives $t \approx 2^{18}$.

3   The Interpolation Attack

In this section, we introduce a new attack on block ciphers. The attack is based on the following well-known formula.

Let R be a field. Given $2n$ elements $x_1, \ldots, x_n, y_1, \ldots, y_n \in R$, where the x_is are distinct. Define

$$f(x) = \sum_{i=1}^{n} y_i \prod_{1 \leq j \leq n, j \neq i} \frac{x - x_j}{x_i - x_j}. \tag{6}$$

Then $f(x)$ is the only polynomial over R of degree at most $n - 1$ such that $f(x_i) = y_i$ for $i = 1, \ldots, n$. Equation (6) is known as the *Lagrange interpolation formula* (see e.g. [2, page 185]).

In the *interpolation attacks* presented in this paper we construct polynomials using pairs of plaintexts and ciphertexts. We will assume that the time needed to construct these polynomials is small compared to the time needed to do the encryptions of the plaintexts needed in the attack.

3.1   Global and instance deduction

Consider the cipher $\mathcal{PURE}$ with r rounds. We exploit the fact that the exclusive-or operation used in the cipher corresponds to addition over a finite field with characteristic 2. Consequently, the cipher consists of simple algebraic operations only, and hence each of the two halves of the ciphertext y, e.g., the left hand part, can be described as a polynomial $p(x_L, x_R) \in \mathrm{GF}(2^{32})[x_L, x_R]$ of the plaintext with at most $3^{2r-1} + 3^r + 3^{r-1} + 1$ coefficients. Note, that degrees of x_R and x_L are at most 3^r and 3^{r-1}, respectively. Thus, we can reconstruct this polynomial by considering at most $3^{2r-1} + 3^r + 3^{r-1} + 1$ plaintext/ciphertext pairs (p/c-pairs) using, e.g., Lagrange interpolation. With $r = 6$ the attack needs at most 2^{18} known p/c-pairs, which yields an algorithm for a global deduction. Note that the number of coefficients will be lower than specified, since not all elements $x_L^i x_R^j$ for $0 \leq i \leq 3^r$ and $0 \leq j \leq 3^{r-1}$ will appear in the polynomial.

We have the following more general theorem.

Theorem 2. *Consider an iterated block cipher with block size m. Express the ciphertext as a polynomial of the plaintext and let n denote the number of coefficients in the polynomial. If $n \leq 2^m$, then there exists an interpolation attack of time complexity n requiring n known plaintexts encrypted with a secret key K, which finds an algorithm equivalent to encryption (or decryption) with K.*

In a chosen plaintext variant of this attack it is possible for an attacker to establish polynomials with a reduced number of coefficients by fixing some of the bits in the chosen plaintexts. In that case, the result is an instance deduction, since the obtained algorithm can only encrypt plaintexts for which a number of bits are fixed to a certain value. As as example, $\mathcal{PURE}$ can be attacked in such a way using only 730 chosen p/c-pairs. Subsequently, the attacker has an algorithm, which encrypts 2^{32} plaintexts without knowing the secret key.

3.2   Key-recovery

In this section we extend the method of the previous section to a key-recovery attack.

Consider first a known plaintext attack. Instead of specifying the ciphertext as a function of the plaintext, we express the output from the reduced cipher $\tilde{y}$ as a polynomial $p(x) \in \mathrm{GF}(2^m)[x]$ of the plaintext. Assume that this polynomial has degree d and that $(d+1)$ known p/c-pairs are available. Then for all values of the last-round key one decrypts the ciphertexts one round and tries to construct the polynomial. With one extra p/c-pair one checks whether the polynomial is correct. If this is the case, then the correct value of the last-round key has been found with a high probability, by reasoning similarly as in the proof of Theorem 1.

The chosen plaintext variant of this attack is quite similar. Let us illustrate the method with an example. Once again, consider the cipher $\mathcal{PURE}$ with 6 rounds. Assume that the right hand half x_R of the plaintext is fixed (that is, we consider a chosen plaintext attack), and consider the right hand side of the output $\tilde{y}_R = p(x_L)$ from the reduced cipher expressed as a polynomial $p(x_L) \in \mathrm{GF}(2^{32})[x_L]$. This polynomial has degree at most $3^3 = 27$ since the degree does not increase in the first round and since ty_R equals the left half of the output of the fourth round. Consequently, 28 pairs of corresponding values of x_L and $\tilde{y}$ are enough to determine it uniquely (using Lagrange interpolation).

We then test whether $\tilde{y}$ is actually output from the reduced cipher or not. This is done by verifying whether a 29-th p/c-pair agrees with the obtained polynomial. If it does, then we assume that we have found the correct key. The average time complexity is $29 \times 2^{32-1} \approx 2^{36}$.

More generally, we have the following theorem.

Theorem 3. *Consider an iterated block cipher of size m. Express the output from the round next to the last as a polynomial of the plaintext and let n denote the number of coefficients in the polynomial. Furthermore, let b denote the number of last-round key bits. Then there exists an interpolation attack of average time complexity $2^{b-1}(n+1)$ requiring $n+1$ known (or chosen) plaintexts which will successfully recover the last-round key.*

Similar to the attack of Theorem 1 it may be possible to perform the attack for only a subset of the last-round key, and also it may be possible to search for (a subset of) the first-round key, depending on the structure of the round function.

3.3   Meet-in-the-middle approach

The attacks described in this section are extensions of the attacks in the previous sections using a meet-in-the-middle technique. We describe only the extension of the key-recovery attack; the extension of the global and instance deductions follow easily.

Once more, we try guessing the correct last-round key and use this to (hopefully) obtain $\tilde{y}$, the output from the reduced cipher. In the following, only the verification of $\tilde{y}$ is described. Given an iterated cipher of r rounds, let z denote the output of round s, where $s \leq (r - 1)$. The value of z is expressible via the plaintext x as a polynomial $g(x) \in \mathrm{GF}(2^m)[x]$ where m is the block size. Similarly, z can be expressed as a polynomial $h(\tilde{y}) \in \mathrm{GF}(2^m)[\tilde{y}]$ of the output $\tilde{y}$ of the reduced cipher. Let the degree of $g(x)$ be d_g, the degree of $h(\tilde{y})$ be d_h and let $d_{gh} = d_g + d_h$. Thus, the following equation

$$g(x) = h(\tilde{y}) \tag{7}$$

has at most $d_{gh} + 2$ unknowns. The equation is solvable up to a multiplication and an addition of both g and h with a constant. Therefore, to ensure that we obtain a non-trivial and unique solution, we set the coefficient corresponding to the highest exponent equal to 1 and the constant term equal to 0. After this, we solve the equation by using d_{gh} known or chosen plaintexts. We then check whether yet another p/c-pair $(x, \tilde{y})$ obeys $g(x) = h(\tilde{y})$. If it does, then we assume that we have guessed the correct value of the last-round key.

Again, let us illustrate the attack on the cipher $\mathcal{PURE}$ with 6 rounds. Assume that the right hand half x_R of the plaintext is fixed (that is we consider a chosen plaintext attack.) Let z_L denote the left half of the output from round four. The value of z_L is expressible via the plaintext as a polynomial $g(x_L) \in \mathrm{GF}(2^{32})[x_L]$. This polynomial has degree at most 3^2, i.e. there are at most 10 non-zero coefficients in $g(x_L)$. Similarly, z_L can be expressed as a polynomial $h(\tilde{y}_L, \tilde{y}_R) \in \mathrm{GF}(2^{32})[\tilde{y}_L, \tilde{y}_R]$ of the output from the reduced cipher. It follows that $h(\tilde{y}_L, \tilde{y}_R) = \tilde{y}_L^3 \oplus a\tilde{y}_L^2 \oplus b\tilde{y}_L \oplus c \oplus \tilde{y}_R$, where $a, b,$ and c are some key-dependent constants. Thus, there are at most $10 + 3 = 13$ unknown coefficients of the equation

$$g(x_L) = h(\tilde{y}_L, \tilde{y}_R) \tag{8}$$

Setting the constant term of g to equal 0 (the coefficient corresponding to the highest exponent in h has already been found to equal 1), we proceed to solve the resulting system of equations by using 12 p/c-pairs from the reduced cipher. This gives us the polynomials g and h. We then check whether yet another p/c-pair $(x, \tilde{y})$ obeys $g(x_L) = h(\tilde{y}_L, \tilde{y}_R)$. If it does, then we assume that we have guessed the correct key.

Similar attacks can be applied to versions of $\mathcal{PURE}$ with up to 32 rounds. Consider the version with 32 rounds. Let $g(x_L) \in \mathrm{GF}(2^{32})[x_L]$ be an expression of the left half z_L of the output from round 22. The degree of this polynomial is at most 3^{20}. Let $h(\tilde{y}_L, \tilde{y}_R) \in \mathrm{GF}(2^{32})[\tilde{y}_L, \tilde{y}_R]$ be an expression of z_L from the output of the reduced cipher. In the algebraic normal form of $h(\tilde{y}_L, \tilde{y}_R)$, the number of exponents in $\tilde{y}_L$ and $\tilde{y}_R$ is at most $(3^9 + 1)$ and $(3^{10} + 1)$, respectively. Thus the number of coefficients in $h(\tilde{y}_L, \tilde{y}_R)$ is at most $(3^9 + 1)(3^{10} + 1) \approx 3^{19}$. This means that the number of coefficients in Equation (8) is at most $3^{20} + 3^{19} \approx 2^{32}$. I.e., the average time complexity for this attack is about 2^{63} and it requires about 2^{32} chosen plaintexts.

We obtain the following general result.

Theorem 4. *Consider an iterated block cipher of block size m with r rounds. Express the output from round s, $s \leq r-1$, as a polynomial of the plaintext and let n_1 denote the number of coefficients in the polynomial. Also, express the output from round s as a polynomial of the output from round $(r-1)$, and let n_2 denote the number of coefficients in the polynomial. Furthermore, set $n = n_1 + n_2$ and let b denote the number of last-round key bits. Then there exists an interpolation attack of average time complexity $2^{b-1}(n-1)$ requiring $(n-1)$ known (chosen) plaintexts which will successfully recover the last-round key.*

In the following section we describe a variant of the interpolation attack.

3.4   Attacks on modified SHARK

The iterated cipher SHARK was described by Rijmen, Daemen, et al in [11]. The cipher has block size nm bits and each round has a non-linear layer and a diffusion layer. The non-linear layer consists of n parallel m-bit S-boxes. The diffusion layer consists of an nm-bit linear mapping constructed from the Reed-Solomon code. There are two suggested ways to introduce the keys into the cipher. The first is by a simple exclusive-or with the inputs to the S-boxes, the other uses a key-dependent affine mapping. Also, an output transformation is applied after the last round of SHARK. The transformation consists of a key addition and an inverse diffusion layer.

The design strategy of SHARK is to consider each component of the cipher separately. It is argued "The non-linear layer has uniform non-linear properties, such that when measuring the resistance of the cipher against cryptanalysis we don't have to take the details of the interaction between the non-linear and the diffusion layer into account."[11]. Furthermore, "If, for example, the S-boxes are replaced by other S-boxes, with equivalent non-linearity properties, the resistance of the cipher remains constant"[11].

We will denote by SHARK(n, m, r) the version with block size nm bits using n parallel m-bit S-boxes in r rounds. In [11] an implementation SHARK$(8, 8, r)$ (64 bit blocks) is given. The 8 S-boxes are identical and constructed from the permutation $f : \mathrm{GF}(2^m) \rightarrow \mathrm{GF}(2^m)$ given by $f(x) = x^{-1}$. The cipher is analysed with respect to linear and differential attacks, and it is argued that 8 rounds of SHARK$(8, 8, r)$ give a security level comparable to that of triple-DES, and from [11, Table 1] it follows that 4 rounds of this version give a security level comparable to that of DES.

In the following we will show that there are many instances of SHARK that can be broken significantly faster than expected.

First of all, the number of rounds of SHARK must be determined with respect to the non-linear order of the S-boxes. Assume that the outputs of the S-box have non-linear order d in the input bits. Since the S-boxes represent the only non-linear component in SHARK, the non-linear order of the ciphertexts after r rounds of encryption will be at most d^r. To avoid attacks based on higher order differentials it must be ensured that d^r is high, preferably that $d^r \geq nm$. Thus, for a 64 bit block cipher, if $d = 2$, e.g. using the cubing function in a Galois field, the number of rounds must be at least 6.

We consider in the following versions of SHARK where the keys are mixed with the texts by the exclusive-or operation. Once again, we make use of the fact, that exclusive-or is equivalent to addition over a finite field of characteristic 2. We will show that there are instances of SHARK(n, m, r), for which the interpolation attacks are applicable. We consider 64-bit versions using as S-box $f(x) = x^{-1}$ in GF(2^m). This is the S-box suggested in [11], but, as it is also said "To remove the fixed points $0 \to 0$ and $1 \to 1$ an invertible transformation is applied to the output bits of the S-box." In what we are about to show, these fixed points play no role, so according to the design strategy of SHARK, variants with $f(x)$ as S-box without the invertible transformation should give equivalent security. We stress that the attacks we are about to present are not applicable to the specific instance of SHARK presented in [11].

The interpolation attack described so far in this paper work well for ciphers of low algebraic degree. The inverse permutation in a Galois field has a high algebraic degree, note that $f(x) = x^{-1} = x^{2^m - 2}$ in GF(2^m). However, as we will show, there are variants of the interpolation attack, which work for these functions. These attacks depend only on the number of S-boxes and of the number of rounds in the cipher.

Consider first a version with $n = 1$. It follows by easy calculations that the ciphertext y after any number of rounds can be expressed as a fraction of polynomials of the plaintext x (or similarly, x can be expressed as a polynomial of y) as follows

$$y = \frac{x \oplus a}{bx \oplus c} \tag{9}$$

where a, b, c are key-dependent constants. These three constants can be found using the interpolation attack with only 4 known p/c-pairs[4] by considering and solving $y \cdot (bx \oplus c) = (x \oplus a)$. The result is a global deduction, i.e. an algorithm that encrypts (decrypts) any plaintext (ciphertext).

Next consider a version with $n = 2$. Let x_L and x_R denote the left and right halves of the plaintext, respectively, and let $y_{i,L}$ and $y_{i,R}$ denote the left and right halves of the ciphertext after i rounds of encryption. In general we get

$$y_{i,L} = \frac{p_{i,1}(x_L, x_R)}{p_{i,2}(x_L, x_R)} \tag{10}$$

and similarly for $y_{i,R}$, where $p_{i,1}, p_{i,2} \in$ GF$(2^{32})[x_L, x_R]$. It remains to show how many coefficients there are in the two polynomials. First note that the number of coefficients in $p_{i,1}$ is at most the number of coefficients in $p_{i,2}$. Consider the

---

algebraic normal form of $p_{i,2}$ and assume that the largest exponents of x_L and x_R are $e^i_{x_L}$ respectively $e^i_{x_R}$. Then the number of coefficients in $p_{i,2}$ is at most $(e^i_{x_L} + 1) \cdot (e^i_{x_R} + 1)$. From the description of SHARK [11] it follows that

$$y_{i,L} = \frac{a_1}{y_{(i-1),L} \oplus k_{i,1}} \oplus \frac{a_2}{y_{(i-1),R} \oplus k_{i,2}} \tag{11}$$

$$= \frac{p_{i,1}}{(y_{(i-1),L} \oplus k_{i,1}) \cdot (y_{(i-1),R} \oplus k_{i,2})}, \tag{12}$$

where $k_{i,j}$ are the round keys and a_1, a_2 some constants. Now it is easy to see that for $i > 1$ $e^i_{x_L} \leq 2 \cdot e^{i-1}_{x_L}$ and since $e^1_{x_L} = e^1_{x_R} = 1$, one gets $e^i_{x_L} \leq 2^{i-1}$ and $e^i_{x_R} \leq 2^{i-1}$. Therefore, the number of coefficients in $p_{i,2}$ is at most $(2^{i-1} + 1)^2$, which also upper bounds the number of coefficients in $p_{i,1}$. In order to be able to solve Equation (10) one would need at most $2 \cdot (2^{i-1} + 1)^2$ plaintexts and their corresponding ciphertexts. Note that the same pairs can be used to solve a similar equation for $y_{i,R}$. Consider versions of the cipher with n S-boxes. One finds by calculations similar as above that the number of known plaintexts needed to solve Equation (10) is $2 \cdot (n^{i-1} + 1)^n$. The number of coefficients in the polynomials used in our attacks increases with the number of diffusion layers in the cipher. Note that because of the inverse diffusion layer in the output transformation there are only $r - 1$ diffusion layers in an r-round version of SHARK. To sum up, the number of known plaintexts for the interpolation attack on an r-round version yielding a global deduction is

$$2 \cdot (n^{r-2} + 1)^n.$$

It follows that the attack is independent of the sizes of the S-boxes, and it depends only the number of S-boxes and the number of rounds.

The interpolation attack with the meet-in-the-middle technique can be applied also for these ciphers. We consider the interpolation attack with known plaintexts. One first establishes

$$\frac{q_{j,1}(y_1, \ldots, y_n)}{q_{j,2}(y_1, \ldots, y_n)} = \frac{p_{i,1}(x_1, \ldots, x_n)}{p_{i,2}(x_1, \ldots, x_n)}, \tag{13}$$

i.e., expressions of the ciphertexts in one middle round, where $i + j = r - 1$, using polynomials of both the plaintext and the ciphertext. Subsequently, one can solve the following systems of equations

$$q_{j,1}(y_1, \ldots, y_n) \cdot p_{i,2}(x_1, \ldots (, x_n) = p_{i,1}(x_1, \ldots x_n) \cdot q_{j,2}(y_1, \ldots, y_n). \tag{14}$$

The number of known plaintexts required to solve (14) is

$$2 \cdot (n^{r_1-1} + 1)^n \cdot (n^{r_2-1} + 1)^n,$$

where $r_1 + r_2 = r - 1$ and $r_1, r_2 \geq 1$.

The round keys for SHARK are typically quite big, so the general key-recovery attack described earlier in this paper may be impractical. However, it is possible to perform the attack for only a subset of the first-round and/or last-round keys.

# Rounds	# S-boxes	Known plaintexts	
any	1	3	
6	2	2^9	
6	4	2^{27}	$(+)$
3	8	2^{17}	$(+)$
4	8	2^{35}	$(+)$
5	8	2^{52}	$(+)$
6	8	2^{75}	$(+)$
7	8	2^{98}	$(+)$
8	8	2^{121}	$(+)$

Table 2. Complexities of the interpolation attack on variants of SHARK using as S-box $f(x) = x^{-1}$. $(+)$ Meet-in-the-middle approach.

As an example, one can repeat the attack for all values of the first s words of the first-round key and express the ciphertext (of a middle round) as a polynomial $p_{i,1}(S(x_1 \oplus k_1), \ldots, S(x_s \oplus k_s), x_{s+1}, \ldots x_n)$, where $S(\cdot)$ are the S-boxes and x_i are the plaintext words. The values of the key words for which the interpolation succeeds are candidates for the secret key, and the attack is repeated sufficiently many times until one value of the secret key is found.

In Table 2 we give the complexities of the interpolation attack on variants of SHARK using as S-box $f(x) = x^{-1}$ in $GF(2^m)$. It follows that using 8 S-boxes, the 64-bit variant with up to 5 rounds and the 128-bit variant with up to 8 rounds are (theoretically) vulnerable to our attacks. The number of required plaintexts of the key-recovery attack is a little less than indicated variants and the workload of the attack is a little higher. We will not go into further details here.

In a chosen plaintext attack the number of coefficients in the polynomials used in the attack can be reduced by fixing some plaintext bits. As examples, there exist interpolation attacks on the variant with 8 S-boxes and 4 rounds using about 2^{21} chosen plaintexts and on the variant with 8 S-boxes and 7 rounds using about 2^{61} chosen plaintexts. In this attack we fix four of the eight plaintext words, so for a 64-bit block cipher the interpolation will work only if the needed number of plaintexts is less than 2^{32} and for a 128-bit block cipher less than 2^{64} plaintexts.

We have demonstrated that certain instantiations of SHARK are insecure. Our results also demonstrate a case where the use of bigger and fewer S-boxes does not result in more secure ciphers. Finally, we note that the designers of SHARK expressed their concern with the use of the inverse in a Galois field as S-boxes: "This may create uneasy feelings, but we are not aware of any vulnerability caused by this property. For the time being we challenge cryptanalysts to demonstrate any vulnerability caused by this property." [11]. Challenge taken!

4   Concluding Remarks

We introduced a new attack on block ciphers, the interpolation attack. We demonstrated the attack on slightly modified versions of a cipher proposed by Knudsen and Nyberg and of a cipher proposed by Rijmen, Daemen et al. These modifications do not violate the design principles of the original ciphers and are as secure with respect to the security measures proposed by the authors. Also, we presented an improved variant of differential attacks based on higher order differentials, which was used to cryptanalyse the (unmodified) cipher by Knudsen and Nyberg and a cipher by Kiefer.

One might try to find a probabilistic version of the interpolation attack that would also work when the output of the cipher is expressible as a polynomial of low degree in only a fraction of the cases. However, it looks like this attack would require an effective maximum likelihood decoding algorithm for higher order Reed-Muller codes and such an algorithm is not known to exist.

Finally, it should be mentioned that with the use of Newton interpolation instead of Lagrange interpolation one can speed up the attacks slightly.

References

1. E. Biham and A. Shamir. *Differential Cryptanalysis of the Data Encryption Standard*. Springer Verlag, 1993.
2. P.M. Cohn. *Algebra, Volume 1*. John Wiley & Sons, 1982.
3. K. Kiefer. A New Design Concept for Building Secure Block Ciphers. In J. Pribyl, editor, *Proceedings of the 1st International Conference on the Theory and Applications of Cryptology, PRAGOCRYPT'96, Prague, Czech Republic*, pages 30–41. CTU Publishing House, 1996.
4. L.R. Knudsen. *Block Ciphers – Analysis, Design and Applications*. PhD thesis, Aarhus University, Denmark, 1994.
5. L.R. Knudsen. Truncated and higher order differentials. In B. Preneel, editor, *Fast Software Encryption - Second International Workshop, Leuven, Belgium, LNCS 1008*, pages 196–211. Springer Verlag, 1995.
6. X. Lai. Higher order derivatives and differential cryptanalysis. In *Proc. "Symposium on Communication, Coding and Cryptography", in honor of James L. Massey on the occasion of his 60'th birthday, Feb. 10-13, 1994, Monte-Verita, Ascona, Switzerland*, 1994.
7. M. Matsui. Linear cryptanalysis method for DES cipher. In T. Helleseth, editor, *Advances in Cryptology - Proc. Eurocrypt'93, LNCS 765*, pages 386–397. Springer Verlag, 1993.
8. K. Nyberg. Differentially uniform mappings for cryptography. In T. Helleseth, editor, *Advances in Cryptology - Proc. Eurocrypt'93, LNCS 765*, pages 55–64. Springer Verlag, 1993.
9. K. Nyberg. Linear approximations of block ciphers. In A. De Santis, editor, *Advances in Cryptology - Proc. Eurocrypt'94, LNCS 950*, pages 439–444. Springer Verlag, 1994.
10. K. Nyberg and L.R. Knudsen. Provable security against a differential attack. *The Journal of Cryptology*, 8(1):27–38, 1995.
11. V. Rijmen, J. Daemen, B. Preneel, A. Bosselaers, and E. De Win. The cipher SHARK. In Gollmann D., editor, *Fast Software Encryption, Third International Workshop, Cambridge, U.K., February 1996, LNCS 1039*, pages 99–112. Springer Verlag, 1996.

Best Differential Characteristic Search of FEAL

Kazumaro Aoki[1], Kunio Kobayashi[2]*, and Shiho Moriai[3]**

[1] NTT Laboratories
[2] School of Science and Engineering, Waseda University
[3] Information & Communication Security Project,
Telecommunications Advancement Organization of Japan

Abstract. This paper presents the results of the best differential characteristic search of FEAL.

The search algorithm for the best differential characteristic (best linear expression) was already presented by Matsui, and improvements on this algorithm were presented by Moriai et al. We further improve the speed of the search algorithm. For example, the search time for the 7-round best differential characteristic of FEAL is reduced to about 10 minutes (Pentium/166 MHz), which is about $2^{12.6}$ times faster than Matsui's algorithm. Moreover, we determine all the best differential characteristics of FEAL for up to 32 rounds assuming all S-boxes are independent.

As a result, we confirm that the N-round ($7 \leq N \leq 32$) best differential characteristic probability of FEAL is 2^{-2N}, which was found by Biham. For N = 6, we find 6-round differential characteristics with a greater probability, 2^{-11}, than that previously discovered, 2^{-12}.

1 Introduction

Since the introduction of differential cryptanalysis [BS91], evaluating the security of symmetric block ciphers against differential cryptanalysis has become an important research topic. Roughly speaking, we can evaluate the security of differential cryptanalysis using the best differential characteristic probability. The search algorithm for the best differential characteristic (best linear expression) of DES-like cryptosystems was already presented by Matsui [M95], and improvements on this algorithm were presented by Moriai et al. [MAO96]. However, if we apply these algorithms to the search for the best differential characteristic of FEAL [SM88, MKOM90], we cannot complete a search in a practical amount of time due to the enormous number of search candidates.

This paper proposes improvements on the algorithms introduced in [M95, MAO96] and presents the results of applying this improved algorithm to the search for the best differential characteristic of FEAL[4]. For example, the search for the 7-round best differential characteristic of FEAL requires about 10 minutes (Pentium/166 MHz), which is about $2^{12.6}$ times faster than Matsui's algorithm.

---

* Part of this research was done during his 1995 summer intern at NTT Laboratories.
** This research was done at NTT Laboratories.

[4] The best linear expression of FEAL is obtained in [MAO96].

As a result, we confirm that the N-round ($7 \leq N \leq 32$) best differential characteristic probability of FEAL is 2^{-2N}, which was already known.

Strictly speaking, the security of Markov ciphers[5] against differential cryptanalysis would be better evaluated by the maximum average of the differential probability[6] than the best differential characteristic probability [LMM91].
However, it is very difficult to determine the maximum average of the differential
probability of a block cipher and to evaluate the upper bound of the maximum
average of the differential probability in a practical range. Therefore, at present,

- the security of block ciphers against differential cryptanalysis is evaluated by
  the best differential characteristic probability, or
- some block ciphers are designed in a manner that facilitates evaluating the
  maximum average of differential probability.

Requiring a large amount of time is not particularly problematic when applying the search algorithm for the best differential characteristic to an existing
cipher. However, *designers* must repeatedly apply it to draft ciphers in the design
process, for example, in order to select better S-boxes. Thus, it is indispensable
that the time complexity of the search algorithm be reduced. It is important for
the fast search algorithm to determine the *best* differential characteristic (linear
expression). For example, [LSK95] proposed a fast search algorithm that finds
effective linear expressions, but they are not particular about the optimality of
the linear probability. Our search algorithm can determine the best differential
characteristic probability faster than ever.

2   Notation

$$\text{BEST}_N\text{: N-round best differential characteristic probability}$$
$$\text{CAND}_N\text{: temporal value of BEST}_N$$
$$\Delta X_r\text{: } r\text{-th round input difference}$$
$$\Delta Y_r\text{: } r\text{-th round output difference}$$
$$p_r = p_r(\Delta X_r, \Delta Y_r)\text{: } r\text{-th round differential characteristic probability}$$
$$\oplus\text{: bit-wise } exclusive \; or \text{ operation}$$
$$S_d(x, y)\text{: S-box of FEAL; } x + y + d \bmod 2^8$$

We define operator "$\oplus$" as having a higher priority than operator "+," and
operator "+" as having a higher priority than operator "mod."

3   Previous Results

3.1   Biham's Result

Biham found several differential characteristics of FEAL. The best differential
characteristic probabilities of FEAL he showed are listed in Table 1 [BS93].

---

[5] FEAL is a Markov cipher.
[6] Note that the maximum average of the differential probability is greater than the best
differential characteristic probability.

Round (N)	Differential characteristic probability
$N \leq 3$	1
$N = 4$	2^{-3}
$N = 5$	2^{-4}
$N \geq 6$	2^{-2N}

Table 1. Biham's results

3.2 Matsui's Algorithm

Matsui proposed a search algorithm for DES-like cryptosystems that determines BEST_N ($N \geq 3$) and the corresponding differential characteristics using the knowledge of BEST_r ($r < N$) [M95].

Algorithm 1 (Matsui's Algorithm).

Preparation: *Determine the initial value of* CAND_N, *so that* CAND_N *will be as large as possible but smaller than the best differential characteristic probability.*

1-st round search: *For each candidate for* ΔX_1, *do the following:*
 - *Choose* ΔY_1 *which maximizes* p_1.
 - *Go to the 2-nd round search if* $p_1 \geq \dfrac{\text{CAND}_N}{\text{BEST}_{N-1}}$ *holds.*

2-nd round search: *For each candidate for* ΔX_2 *and* ΔY_2, *do the following:*
 - *Go to the 3-rd round search if* $p_1 p_2 \geq \dfrac{\text{CAND}_N}{\text{BEST}_{N-2}}$ *holds.*

r-th round search $(3 \leq r \leq N-1)$: *Let* $\Delta X_r = \Delta Y_{r-1} \oplus \Delta X_{r-2}$ *and for each candidate for* ΔY_r, *do the following:*
 - *Go to the* $(r+1)$-*th round search if* $\displaystyle\prod_{i=1}^{r} p_i \geq \dfrac{\text{CAND}_N}{\text{BEST}_{N-r}}$ *holds.*

N-th round search:
 - *Let* $\Delta X_N = \Delta Y_{N-1} \oplus \Delta X_{N-2}$.
 - *Choose* ΔY_N *which maximizes* p_N.
 - *Let* $\text{CAND}_N = \displaystyle\prod_{i=1}^{N} p_i$ *if* $\displaystyle\prod_{i=1}^{N} p_i \geq \text{CAND}_N$ *holds.*

As a result, we have $\text{BEST}_N = \text{CAND}_N$.

3.3 Improved Algorithm (Moriai et al.)

Moriai et al. introduced the concept of *search patterns*. These patterns reduce the search complexity by detecting the unnecessary search candidates before the search. This method works because detection process requires only the probability of differential characteristics for each round and no differential characteristics.

A search pattern used in the search for the N-round best differential characteristic is a set of N probabilities, and each probability is a differential characteristic probability for each round. Their algorithm examines whether differential
characteristics with probability $CAND_N$ ($N \geq 3$) exist and finds the differential
characteristics, if any, using the knowledge of $BEST_r$ ($r < N$) [MAO96]. Their
algorithm is shown below.

Algorithm 2 (Moriai et al. Algorithm).

1. *Generate all the search patterns* $(q_1, q_2, \ldots, q_N)$ *according to the conditions
   for Algorithm 1 which concern only the differential characteristics probabilities.*
2. *Discard the search patterns which satisfy the following conditions.*

$$\exists i, r \ (1 \leq i \leq N, i + r - 1 < N); \ \prod_{j=i}^{i+r-1} q_j > BEST_r$$

3. *Discard either search pattern* $(q_1, q_2, \ldots, q_N)$ *or* $(q_N, q_{N-1}, \ldots, q_1)$ *whichever
   has more search candidates.*
4. *Search the differential characteristics corresponding to search pattern* $(q_1, q_2,
   \ldots, q_N)$ *using Algorithm 1 with all the inequalities replaced by equalities.*

4   Proposed Algorithms

Algorithm 2 discards only the unnecessary search patterns that satisfy the two
conditions described above. However, we can discard more search patterns without
loss of search exhaustivity. In sections 4.1 and 4.2 we present our improved algorithms and apply them to FEAL[7]. They are useful especially when the differential characteristic probability is of the powers of 2, for example, those for addition
and (almost) bent functions, but they are applicable to other DES-like cryptosystems. Section 4.3 describes the generalization of our algorithm, particularly how
to treat search patterns.

4.1   Using Presearch Based on Algorithm 2

Algorithm 2 introduces the search patterns to consider the probability of each
round, but does not use the pattern itself. Taking this into consideration, we
improve the search algorithm for the best differential characteristics.

Consider the search for 5-round differential characteristics of FEAL for example. In this case, we know all the 4-round search patterns corresponding to
differential characteristics with probability 2^{-3} shown below when we complete
a search for the 4-round differential characteristics.

$$(2^{-1}, 1, 2^{-1}, 2^{-1}), \ (2^{-1}, 2^{-1}, 1, 2^{-1}) \tag{1}$$

---

[7] Here we assume that all S-boxes are independent.

On the other hand, all search patterns of the 5-round with probability 2^{-3} which should be searched for by Algorithm 2 are as follows:

$$(1,1,1,2^{-3},1), \ (1,1,2^{-1},2^{-2},1), \ (1,1,2^{-2},2^{-1},1) \ ,$$
$$(1,1,2^{-3},1,1), \ (1,2^{-1},1,2^{-2},1), \ \text{and} \ (1,2^{-1},2^{-1},2^{-1},1) \ .$$

All these search patterns have interior 4-round search patterns with probability 2^{-3}. However, none of the 4-round search patterns (1) matches them. Thus the differential characteristics corresponding to these search patterns do not exist, and we do not have to search these search patterns. We also know that there is no 5-round differential characteristic with probability 2^{-3} without conducting a search.

Next, consider 5-round differential characteristics with probability 2^{-4}. In this case, search patterns for which Algorithm 2 searches are as follows.

1:$(1,1,1,2^{-3},2^{-1})$	**2**:$(1,1,1,2^{-4},1)$	**3**:$(1,1,2^{-1},2^{-2},2^{-1})$
4:$(1,1,2^{-1},2^{-3},1)$	**5**:$(1,1,2^{-2},2^{-1},2^{-1})$	**6**:$(1,1,2^{-2},2^{-2},1)$
7:$(1,1,2^{-3},1,2^{-1})$	**8**:$(1,1,2^{-3},2^{-1},1)$	**9**:$(1,1,2^{-4},1,1)$
10:$(1,2^{-1},1,2^{-2},2^{-1})$	**11**:$(1,2^{-1},1,2^{-3},1)$	**12**:$(1,2^{-1},2^{-1},2^{-1},2^{-1})$
13:$(1,2^{-1},2^{-1},2^{-2},1)$	**14**:$(1,2^{-1},2^{-2},1,2^{-1})$	**15**:$(1,2^{-1},2^{-2},2^{-1},1)$
16:$(1,2^{-2},1,2^{-1},2^{-1})$	**17**:$(1,2^{-2},1,2^{-2},1)$	**18**:$(2^{-1},1,1,2^{-2},2^{-1})$
19:$(2^{-1},1,1,2^{-3},1)$	**20**:$(2^{-1},1,2^{-1},2^{-1},2^{-1})$	**21**:$(2^{-1},1,2^{-1},2^{-2},1)$
22:$(2^{-1},1,2^{-2},1,2^{-1})$	**23**:$(2^{-1},2^{-1},1,2^{-1},2^{-1})$	

Note that the search patterns are numbered **1** to **23**. Search patterns numbered **1, 3, 5, 7, 10, 12, 14, 16, 18, 20, 22**, and **23** have probability 2^{-3} from the 1-st round to 4-th round. Considering the results of the search for the 4-round differential characteristic (1), we have only to search search patterns **20** and **23**. Moreover, search patterns **19, 20, 21**, and **23** have probability 2^{-3} from the 2-nd round to 5-th round. Similarly, as a result of the 4-round differential characteristic search (1), only search pattern **23** survives. From the discussion above, the number of search patterns which should be searched decreases from 23 to 10. The remaining search patterns are **2, 4, 6, 8, 9, 11, 13, 15, 17**, and **23**.

Furthermore, the information which is used for discarding search patterns need not be that of the best differential characteristic. Not only the search patterns of the best differential characteristic, but also those of the differential characteristics with lower probabilities are useful. We know that no 3-round differential characteristic with probability 2^{-1} exists as the result of the search, for example. If we use this condition, search patterns **4** and **11** need not be searched.

We summarize this algorithm as follows.

Algorithm 3.

1. Search r-round $(r < N)$ differential characteristics with various probabilities, and compile information to the extent possible whether or not the search patterns exist for each round and probability. (presearch phase)

2. In Algorithm 2, discard the search patterns which do not exist using the information from the presearch phase.

The information from the presearch phase is also useful in performing a presearch. Moreover, presearches can be completed faster since the presearch may be stopped for a search pattern as soon as the first differential characteristic of the search pattern is determined.

4.2   Using Presearch Based on Algorithm 1

This section proposes further improvements on the search algorithm. Search patterns are useful tools for discarding unnecessary search candidates. However, Algorithm 2 repeats the search process for each search pattern, and sometimes repeats similar computations. The search algorithm described in this section can reduce the complexity based on the following.

1. Improving the right side of the inequalities in Algorithm 1.
2. Combining the same computation repeated several times into one.

In the previous section, we found that the number of search patterns which Algorithm 3 should search is 8, when we searched for the 5-round differential characteristics with a probability of 2^{-4} using presearch information obtained from 3-round differential characteristics with a probability of 2^{-1} and 4-round differential characteristics with a probability of 2^{-3}.

Of course, the correct results can be obtained if we search all eight search patterns mentioned above. However, some search patterns have the same probability in the 1-st and 2-nd rounds. For example, search patterns **2**, **6**, **8**, and **9** have the same search pattern from the 1-st round to the 2-nd round $(1, 1)$. This means that the same computation for the search is done repeatedly for the 1-st and the 2-nd rounds which dominates the search complexity.

In Algorithm 1, the inequality shown below determines whether or not the search for more than r-rounds is needed.

$$\prod_{i=1}^{r} p_i \geq \frac{\text{CAND}_N}{\text{BEST}_{N-r}} \tag{2}$$

Since the right side of (2) is constant, it can be determined before Algorithm 1 starts. The greater the right side of (2) is, the smaller the search complexity is. We want the right side of (2) to be as large as possible while still maintaining search exhaustivity.

We define the set comprising the right sides of (2) for r $(1 \leq r \leq N)$ as $(R_1^{(j)}, R_2^{(j)}, \ldots, R_N^{(j)})$[8]. From the definition of $R_r^{(j)}$, the maximum $R_r^{(j)}$ $(1 \leq r \leq N)$ for which Algorithm 1 searches the differential characteristics with a search pattern $(q_1^{(j)}, q_2^{(j)}, \ldots, q_N^{(j)})$ is;

$$R_r^{(j)} = \prod_{i=1}^{r} q_i^{(j)} \ . \tag{3}$$

---

[8] We also define $\text{BEST}_0 = 1$.

Thus, differential characteristics with all search patterns $(q_1^{(j)}, q_2^{(j)}, \ldots, q_N^{(j)})$ will be searched if we set

$$R_r = \min_j R_r^{(j)}$$

where $(R_1^{(j)}, R_2^{(j)}, \ldots, R_N^{(j)})$ is calculated using (3).

The following algorithm was derived from the discussion above.

Algorithm 4.

1. *Derive search patterns* $(q_1^{(j)}, q_2^{(j)}, \ldots, q_N^{(j)})$ *using Algorithm 3.*
2. *For each search pattern, calculate* $(R_1^{(j)}, R_2^{(j)}, \ldots, R_N^{(j)})$

   where $R_r^{(j)} = \prod_{i=1}^{r} q_i^{(j)}$ $(1 \le r \le N)$.
3. *Let* $R_r = \min_j R_r^{(j)}$.

4. *Run Algorithm 1 with the right side of inequalities* $\dfrac{\text{CAND}_N}{\text{BEST}_{N-r}}$ *replaced by*

   R_r $(1 \le r \le N)$.

Some search patterns which Algorithm 3 should search may have considerably smaller probabilities from the 1-st and 2-nd rounds than the others. In this case, some search patterns should be searched directly, and Algorithm 4 is applied to other search patterns.

4.3   Generalization of Algorithm 3

In this section we generalize Algorithm 3 so that we can apply it to other DES-like cryptosystems whose differential characteristic probability also takes variables other than the power of 2. In Algorithm 5, the variable TENT_N is used for treating search patterns effectively.

Algorithm 5 (when p_r is not a power of 2).

1. *Let* $\text{TENT}_N = 2^{\lfloor \log_2 \text{BEST}_{N-1} \rfloor + N}$ *and* $\text{CAND}_N = \text{BEST}_{N-1}$.
2. *Generate all the search patterns* $(q_1, q_2, \ldots, q_N)$ *so that* q_r $(1 \le r \le N)$

   should be powers of 2, and $2 \times \text{TENT}_N \ge \prod_{r=1}^{N} q_r \ge \text{TENT}_N$ *should hold in*

   the similar way as Algorithm 2.
3. *Discard the search patterns similarly as Algorithm 3.*
4. *Perform Algorithm 1 such that at the r-th round search* $(1 \le r \le N)$ *the differential characteristics whose r-round differential characteristic probability p_r satisfies the following conditions:*

$$q_r \ge p_r > 2^{\log_2 q_r - 1} \quad and \quad \prod_{i=1}^{r} p_i \ge \text{CAND}_N .$$

5. *If* $\mathrm{TENT_N} \geq 2^{\lceil \log_2 \mathrm{CAND_N} \rceil}$ *holds, and* $\mathrm{CAND_N}$ *is revised even once, i.e. a differential characteristic probability that is better than all previous is found, let* $\mathrm{TENT_N} = 2^{-1} \times \mathrm{TENT_N}$ *and go to the 2-nd step.*
   Otherwise, we have $\mathrm{BEST_N} = \mathrm{CAND_N}$.

Though in the algorithm above we use the power of 2 as q_r and $\mathrm{TENT_N}$, appropriate numbers, for example 4, may be used.

5   Small Techniques

All algorithms presented above require differential characteristics of the F-function. It is impractical to calculate differential characteristic of the F-function for FEAL every time it is required because a long time is required. It is also impractical to calculate and store all differential characteristics of the F-function in advance because of time and space requirements.

We regard all S_d functions to be independent[9]. We adopt this method: calculate the differential characteristics of the S_d function in advance, and calculate differential characteristics of the F-function every time it is required with a very low level of complexity.

However, this method still requires a long time and an enormous amount of memory. We decrease the complexity using the following properties of the S_d function [S96].

Note that constants and variables in the following theorems and conjectures are l-bit strings.

Theorem 1 (Equivalence of S_0 and S_1). *For any* (a, b, c)*, equations*

$$x \oplus a + y \oplus b \bmod 2^l = (x + y) \oplus c \bmod 2^l \quad and$$
$$x \oplus a + y \oplus b + 1 \bmod 2^l = (x + y + 1) \oplus c \bmod 2^l$$

hold with the same probability.

Theorem 2 (Distribution of differential probability of addition). *For any* (a, b, c)*, probability with which*

$$x \oplus a + y \oplus b \bmod 2^l = (x + y) \oplus c \bmod 2^l$$

holds is 0 or a power of 2.

---

[9] The best differential characteristic (linear expression) searches previously reported in References [M95, MAO96, TSM95] were based on the same assumption.

Sketch of proof of Theorems 1 and 2: First, prove the propositions of Theorems 1 and 2 in linear cryptanalysis by studying the carry propagation of addition. Second, translate the results to those in differential cryptanalysis using the Walsh transformation [CV95].

Conjecture 1. *If*

$$a \oplus b = a' \oplus b' \quad and$$
$$a + b \equiv a' + b' \pmod{2^l}$$

holds, equations

$$x \oplus a + y \oplus b \bmod 2^l = (x + y) \oplus c \bmod 2^l \quad and$$
$$x \oplus a' + y \oplus b' \bmod 2^l = (x + y) \oplus c \bmod 2^l$$

hold with the same probability.

In case of $l = 8$, a computer exhaustive search proves that this conjecture is correct.

By using Conjecture 1, the data (a, b, c) and (a', b', c') can be made up into the datum $(a \oplus b, a + b \bmod 2^l, c)$ to reduce the amount of required memory.

Conjecture 2 (# of non-0 entries in diff. char. dist. table). *The number of pairs (a, b, c) which satisfy equation*

$$x \oplus a + y \oplus b \bmod 2^l = (x + y) \oplus c \bmod 2^l \quad with \ non\text{-}0 \ probability$$

is $4 \cdot 7^{l-1}$.

In the case of $l \leq 8$, a computer exhaustive search proves that this conjecture is correct.

Conjecture 3. *The number of pairs $(a \oplus b, a + b \bmod 2^l, c)$ which satisfies equation*

$$x \oplus a + y \oplus b \bmod 2^l = (x + y) \oplus c \bmod 2^l \quad with \ non\text{-}0 \ probability$$

is $2 \cdot 5^{l-1}$, which is, surprisingly, the same as the number of non-zero entries in the linear distribution table.

In the case of $l = 8$, a computer exhaustive search proves that this conjecture is correct.

6   Search Experiments and Results

The complexity of Algorithm 1 can be estimated using the number of search candidates for the 1-st and 2-nd rounds [MAO96]. Figure 1 illustrates the number of search candidates for the 1-st and 2-nd rounds of FEAL using Algorithms 1 and 4 where we set $CAND_N$ to the best differential characteristic probability. Note that we ignore the presearch complexity in the case of Algorithm 4. We perform a presearch based on Table 2. A 7-round search for FEAL requires the most search time, about 10 minutes (Pentium/166 MHz) using Algorithm 4.

Figure 2 illustrates the best differential characteristic probability of FEAL.

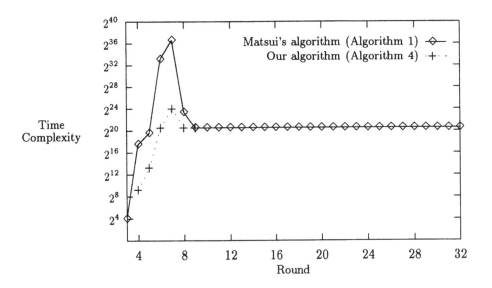

Fig. 1. Complexity of the search for the best differential characteristic probability of FEAL

7   Conclusion

We succeeded in completing the search for the best differential characteristics of FEAL for up to 32 rounds by improving the known search algorithm. We found 6-round differential characteristics with a greater probability, 2^{-11} than that previously discovered, 2^{-12}. All the best 6-round characteristics are illustrated in Fig. 3. We also confirmed that the N-round ($7 \leq N \leq 32$) best differential characteristic probability is 2^{-2N}. These best differential characteristics were previously determined but it was not confirmed whether or not these differential characteristics were the best [BS93]. We conclude that FEAL–32 is secure against differential cryptanalysis in that the best differential characteristic probability is sufficiently small.

Due to the duality of differential cryptanalysis and linear cryptanalysis [M95], we can apply the idea of Algorithms 3, 4, and 5 to the search for the best linear expression.

The remaining problem is to develop an algorithm whose complexity is minimal taking into consideration the presearch complexity in Algorithms 3, 4, and 5.

References

[BS91]   E. Biham and A. Shamir.   Differential Cryptanalysis of DES-like Cryptosystems. *Journal of Cryptology*, Vol. 4, No. 1, pp. 3–72, 1991. (The extended abstract was presented at CRYPTO'90).

Number of rounds N for search	Presearch information	
	Number of rounds	Probability
4	3	$2^{-0}, 2^{-1}, 2^{-2}$
5	3	$2^{-0}, 2^{-1}, 2^{-2}, 2^{-3}$
	4	2^{-3}
6	3	$2^{-0}, 2^{-1}, 2^{-2}, 2^{-3}, 2^{-4}$
	4	$2^{-3}, 2^{-4}, 2^{-5}, 2^{-6}, 2^{-7}, 2^{-8}, 2^{-9}$
	5	$2^{-4}, 2^{-5}, 2^{-6}, 2^{-7}, 2^{-8}, 2^{-9}$
7	3	$2^{-0}, 2^{-1}, 2^{-2}, 2^{-3}, 2^{-4}$
	4	$2^{-3}, 2^{-4}, 2^{-5}, 2^{-6}, 2^{-7}, 2^{-8}, 2^{-9}$
	5	$2^{-4}, 2^{-5}, 2^{-6}, 2^{-7}, 2^{-8}, 2^{-9}$
	6	2^{-11}
8	3	$2^{-0}, 2^{-1}, 2^{-2}, 2^{-3}, 2^{-4}$
	4	$2^{-3}, 2^{-4}, 2^{-5}, 2^{-6}, 2^{-7}, 2^{-8}, 2^{-9}$
	5	$2^{-4}, 2^{-5}, 2^{-6}, 2^{-7}, 2^{-8}, 2^{-9}$
	6	2^{-11}
	7	2^{-14}
9	3	$2^{-0}, 2^{-1}, 2^{-2}, 2^{-3}, 2^{-4}$
	4	$2^{-3}, 2^{-4}, 2^{-5}, 2^{-6}, 2^{-7}, 2^{-8}, 2^{-9}$
	5	$2^{-4}, 2^{-5}, 2^{-6}, 2^{-7}, 2^{-8}, 2^{-9}$
	6	2^{-11}
	7	2^{-14}
	8	2^{-16}

Table 2. Used presearch information

r	ΔY_r	ΔX_r	p_r	r	ΔY_r	ΔX_r	p_{r_1}	r	ΔY_r	ΔX_r	p_{r_1}
1	02000000	80800000	1	1	00000008	00000202	2^{-1}	1	00000202	00800282	2^{-1}
2	80800000	A0008000	2^{-2}	2	00000202	00800282	2^{-1}	2	00800282	80A0A2A2	2^{-3}
3	00000000	00000000	1	3	00000000	00000000	1	3	00000000	00000000	1
4	80800000	A0008000	2^{-2}	4	00000202	00800282	2^{-1}	4	00800282	80A0A2A2	2^{-3}
5	02000000	80800000	1	5	00000008	00000202	2^{-1}	5	00000202	00800282	2^{-1}
6	A8882080	A2008000	2^{-7}	6	8020A2A0	0080028A	2^{-7}	6	0080028A	80A0A0A0	2^{-3}

(The differential characteristics which are the differential characteristics above turned upside down are also the best 6-round differential characteristics.)

Table 3. Best 6-round differential characteristics

[BS93]   E. Biham and A. Shamir. *Differential Cryptanalysis of the Data Encryption Standard.* Springer-Verlag, Berlin, Heidelberg, New York, 1993.

[CV95]   F. Chabaud and S. Vaudenay.   Links Between Differential and Linear Cryptanalysis. In A. D. Santis, editor, *Advances in Cryptology — EURO-CRYPT'94*, Volume 950 of *Lecture Notes in Computer Science*, pp. 356–365. Springer-Verlag, Berlin, Heidelberg, New York, 1995.

[LMM91]   X. Lai, J. L. Massey, and S. Murphy. Markov Ciphers and Differential Cryptanalysis. In D. W. Davies, editor, *Advances in Cryptology — EURO-*

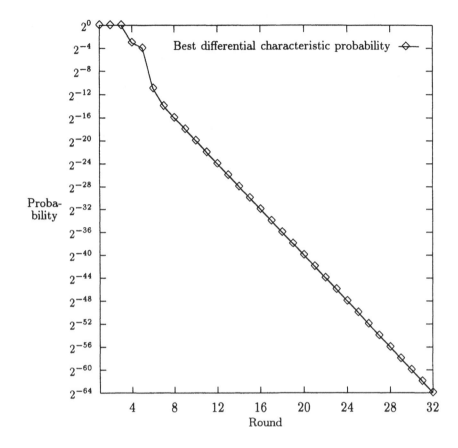

Fig. 2. Best differential characteristic probability of FEAL

CRYPT'91, Volume 547 of *Lecture Notes in Computer Science*, pp. 17–38. Springer-Verlag, Berlin, Heidelberg, New York, 1991.

[LSK95]   S. Lee, S. H. Sung, and K. Kim. An Efficient Method to Find the Linear Expressions for Linear Cryptanalysis. In *1995 Japan-Korea Joint Workshop on Information Security and Cryptology*, pp. 183–190, Inuyama, Aichi, JAPAN, 1995. ISEC Group of IEICE (Japan) and KIISC (Korea).

[M95]   M. Matsui. On Correlation Between the Order of S-boxes and the Strength of DES. In A. D. Santis, editor, *Advances in Cryptology — EUROCRYPT'94*, Volume 950 of *Lecture Notes in Computer Science*, pp. 366–375. Springer-Verlag, Berlin, Heidelberg, New York, 1995.

[MAO96]   S. Moriai, K. Aoki, and K. Ohta. The Best Linear Expression Search of FEAL. *IEICE Transactions Fundamentals of Electronics, Communications and Computer Sciences (Japan)*, Vol. E79-A, No. 1, pp. 2–11, 1996. (The extended abstract was presented at CRYPTO'95).

[MKOM90]   S. Miyaguchi, S. Kurihara, K. Ohta, and H. Morita. Expansion of FEAL Cipher. *Review of Electrical Communication Laboratories*, Vol. 2, No. 6, pp. 117–127, 1990.

[S96]   M. Sugita. Private communications, 1996.

[SM88]   A. Shimizu and S. Miyaguchi. Fast Data Encipherment Algorithm FEAL. In *Advances in Cryptology — EUROCRYPT'87*, Volume 304 of *Lecture Notes in Computer Science*, pp. 267–278. Springer-Verlag, Berlin, Heidelberg, New York, 1988.

[TSM95]   T. Tokita, T. Sorimachi, and M. Matsui. Linear Cryptanalysis of LOKI and s^2-DES. In J. Pieprzyk and R. Safavi-Naini, editors, *Advances in Cryptology — ASIACRYPT'94*, Volume 917 of *Lecture Notes in Computer Science*, pp. 293–303. Springer-Verlag, Berlin, Heidelberg, New York, 1995.

New Block Encryption Algorithm MISTY

Mitsuru Matsui

Information Technology R&D Center
Mitsubishi Electric Corporation
5-1-1, Ofuna, Kamakura, Kanagawa, 247, Japan
matsui@iss.isl.melco.co.jp

Abstract. We propose secret-key cryptosystems MISTY1 and MISTY2, which are block ciphers with a 128-bit key, a 64-bit block and a variable number of rounds. MISTY is a generic name for MISTY1 and MISTY2. They are designed on the basis of the theory of provable security against differential and linear cryptanalysis, and moreover they realize high speed encryption on hardware platforms as well as on software environments. Our software implementation shows that MISTY1 with eight rounds can encrypt a data stream in CBC mode at a speed of 20Mbps and 40Mbps on Pentium/100MHz and PA-7200/120MHz, respectively. For its hardware performance, we have produced a prototype LSI by a process of 0.5μ CMOS gate-array and confirmed a speed of 450Mbps. In this paper, we describe the detailed specifications and design principles of MISTY1 and MISTY2.

1 Fundamental Design Policies of MISTY

Our purpose of designing MISTY is to offer secret-key cryptosystems that are applicable to various practical systems as widely as possible; for example, software stored in IC cards and hardware used in fast ATM networks. To realize this, we began its design with the following three fundamental policies:

1. *MISTY should have a numerical basis for its security,*
2. *MISTY should be reasonably fast in software on any processor,*
3. *MISTY should be sufficiently fast in hardware implementation.*

For the first policy, we have adopted the theory of provable security against differential and linear cryptanalysis [1][2][4], which was originally introduced by Kaisa Nyberg and Lars Knudsen. As far as we know, MISTY is the first block encryption algorithm designed for practical use with provable security against differential and linear cryptanalysis. Although this advantage does not mean information theoretic provable security, we believe that it is a good starting point for discussing secure block ciphers.

Secondly, we have noticed the fact that many recent block ciphers were designed so that they could be fastest and/or smallest on specific targets; for example, 32-bit microprocessors. This often results in slow and/or big implementation on other types of processors. Since we regarded seeking applicability to various systems as more important than pursuing maximum performance on specific targets, we decided to design a cipher that could be reasonably fast and small on any platform, and hence not to adopt software instructions that are effective on special processors only.

For the last policy, we should note that DES is reasonably fast in both software and hardware, while many recent ciphers are seriously slow and/or big when they are implemented in hardware because of their software-oriented structure. On the other hand, since one of our target systems is a fast ATM network of several hundreds Mbps, which cannot be reached in software for the present, we have carefully optimized the look-up tables of MISTY from the viewpoint of its hardware performance. It should be also noted that, in general, a choice of substitution tables does not significantly affect their software execution speed; i.e. memory access time.

2   Discussions on Basic Operations

In this section we classify basic operations that are frequently used in block ciphers into four categories and discuss their applicability to MISTY in terms of compatibility between their security level and software/hardware efficiency.

- **Logical Operations**
  Logical operations such as AND, OR and especially XOR are most common components of secret-key ciphers and are clearly small and fast in any software or hardware system. However we cannot expect much security of them.
- **Arithmetic Operations**
  Arithmetic operations such as additions, subtractions and sometimes multiplications are also commonly used in software-oriented ciphers because they can be carried out by one instruction on many processors and fairly contribute to their security. However, in hardware, their effect on data diffusion is not necessarily high enough, considering their encryption speed, since their delay time due to carry-spreading is often long and expensive.
- **Shift Operations**
  Shift operations, especially rotate-shifting, are frequently used in designing secret-key ciphers. They indirectly improve data diffusion, and in hardware they are obviously cheap and fast if the number of shift counts is fixed. We should note, however, that software performance of shift operations heavily depends on their target size; for instance, when a rotate shift of 32-bit data is executed on 8-bit or 16-bit microprocessors, its speed may be quite slow.
- **Look-up Tables**
  In software, efficiency of loop-up tables strongly depends on memory access speed. In early microprocessors, memory access was much more expensive than register access, while many recent processors can read from and write to memory in one cycle (or often less than one cycle due to parallel processing) under certain conditions. On the other hand, in hardware, the use of ROM is slow in general, but if the tables are optimized for direct construction by

logic gates, their delay time can be drastically reduced. Moreover, as for the security, the look-up table method clearly contributes to data diffusion effectively.

Taking the above discussion into consideration, we have concluded that logical operations and look-up tables arranged in terms of security level and hardware performance meet our design policies and hence they are desirable as basic components of MISTY.

3   Theory of Provable Security

This section briefly summarizes the theory of provable security against differential and linear cryptanalysis. For more detail, see [4]. This theory forms a basis of the security of MISTY.

Definition 1. Let $F_k(x)$ be a function with an n-bit input x and an ℓ-bit parameter k. We define average differential probability DP^F and average linear probability LP^F of the function F as

$$DP^F \overset{def}{=} \frac{1}{2^\ell} \sum_k \max_{\Delta x \neq 0, \Delta y} \frac{\#\{x | F_k(x) \oplus F_k(x \oplus \Delta x) = \Delta y\}}{2^n}, \tag{1}$$

$$LP^F \overset{def}{=} \frac{1}{2^\ell} \sum_k \max_{\Gamma x, \Gamma y \neq 0} \left(2 \frac{\#\{x | x \bullet \Gamma x = F_k(x) \bullet \Gamma y\}}{2^n} - 1 \right)^2, \tag{2}$$

respectively. We also apply this definition to a function $F(x)$ without the parameter k by setting $\ell = 0$.

When $F_k(x)$ is an encryption function with a key k, DP^F and LP^F represent a strict level of security of the function against differential and linear cryptanalysis, respectively. Since we can prove that F is secure against the two attacks when these values are small, we say that F is provably secure if DP^F and LP^F are proved to be sufficiently small.

The following three theorems give relationships between average differential/linear probability of a "small" function and that of a "large" function that is a combination of the small functions. That is to say, using these theorems, we can construct a "large and strong" function from "small and strong" functions. Theorem 2 was first proved for average differential probability by Nyberg and Knudsen [1], and then shown for average linear probability by Nyberg [2].

Theorem 2. *In figure 1, assume that each f_i is bijective and DP^{f_i} (resp. LP^{f_i}) is smaller than p. If the entire function F_k ($k = k_1 || k_2 || k_3 ...$) shown in the figure has at least three rounds, then DP^F (resp. LP^F) is smaller than p^2.*

Note: The authors of [1] originally proved $2p^2$ (not p^2) for a cipher with bijective f_i and at least three rounds, and for a cipher with any f_i and at least four rounds. Recently Aoki and Ohta improved this bound to p^2 when f_i is bijectvie [3].

We proved in [4] that the above theorem is valid for the algorithm shown in figure 2. An essential difference between figures 1 and 2 is that the functions f_i can be processed in parallel in figure 2, and consequently the structure of figure 2 is faster than that of figure 1.

Theorem 3. *In figure 2, assume that each f_i is bijective and DP^{f_i} (resp. LP^{f_i}) is smaller than p. If the entire function F_k ($k = k_1\|k_2\|k_3...$) shown in the figure has at least three rounds, then DP^F (resp. LP^F) is smaller than p^2.*

We found that a similar formula holds even if the input string is divided into two strings of unequal bit length. Specifically, consider the algorithm shown in figure 3, where the input string is divided into n_1 bits and n_2 bits ($n_1 \geq n_2$). Now assuming that in odd rounds the right n_2-bit string is zero-extended to n_1 bits before XOR-ed with the left n_1-bit string, and in even rounds the right n_1-bit string is truncated to n_2 bits before XOR-ed with the left n_2-bit string, we have the following general theorem [4]:

Theorem 4. *In figure 3, assume that each f_i is bijective and DP^{f_i} (resp. LP^{f_i}) is smaller than p. If the entire function F_k ($k = k_1\|k_2\|k_3...$) shown in the figure has at least three rounds, then DP^F (resp. LP^F) is smaller than*

$$max\{p_1p_2,\ p_2p_3,\ 2^{n_1-n_2}p_1p_3\}. \tag{3}$$

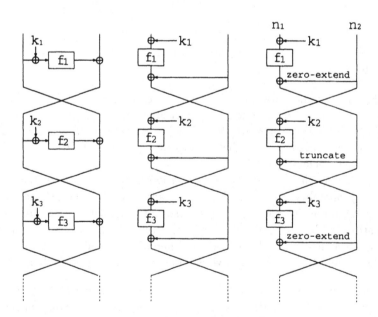

Figure 1.   Figure 2.   Figure 3.

4   Design of the Data Randomizing Part

In this section we discuss the structure of the data randomizing part of MISTY. For a complete description of MISTY1 and MISTY2, see an appendix.

4.1   The Framework

Our basic strategy in designing the data randomizing part of MISTY is to build the entire algorithm from small components using the methods shown in the previous section recursively. This enables us to easily evaluate the security level of the total algorithm by that of the small ones. For instance, let us apply the structure of figure 2 recursively to all f_i functions given in figure 2. In this case, if the average differential/linear probability of the smallest function is less than p, we can prove from theorem 3 that the probability of the entire algorithm is less than p^4.

Now by applying theorem 2 or theorem 3 to a 64-bit block cipher, where theorems 2 and 3 correspond to MISTY1 and MISTY2, respectively, we have a "small" function with 32-bit input/output, which is called an FO function in MISTY (figure 4). Next by applying theorem 2 again to the FO function, we have a "smaller" function with 16-bit input/output, which is referred to as an FI function in MISTY. Since the size of the FI function is still big to use as a look-up table, we have divided the 16-bit string into 9 bits and 7 bits, not 8 bits and 8 bits, using the algorithm given in figure 3.

This unequal division is due to the fact that bijective functions of odd size are generally better than those of even size from the viewpoint of provable security against differential and linear cryptanalysis. More specifically, when the size n of a function is odd, the possible minimal value of its average differential/linear probability is proved to be 2^{-n+1}, but when it is even, it is only conjectured that the possible minimal value is 2^{-n+2} (an open problem). Therefore, if we divide the 16-bit into 8 bits and 8 bits, the average differential/linear probability of the entire 64-bit cipher is proved to be less than $(((2^{-8+2})^2)^2)^2 = 2^{-48}$ (on condition that the above conjecture is correct), while if we divide it into 9 bits and 7 bits, then we can guarantee that the probability is less than $((2^{-9+1}2^{-7+1})^2)^2 = 2^{-56}$ from theorem 4 whenever all subkey bits are independent.

This shows that an unequal division generally has an advantage for security against differential and linear cryptanalysis. On the other hand, it has two penalties in implementation; the first is an obstruction to parallel computation, and the second is a decrease in software performance caused by handling data with an odd number of bits. We have nevertheless adopted the unequal division because of its security. In the following, we refer to the first and third functions of the lowest level as S_9, and the second function as S_7, which are "smallest" components of MISTY. For reducing the size of software, we use the same table in the first and third rounds.

In both MISTY1 and MISTY2, for the sake of flexibility of their security level, the number of rounds n of level 1 (see figure 4) is variable on condition that n is a multiple of four, while that of levels 2 and 3 is fixed to three rounds. Now compare encryption/decryption speed of MISTY1 and MISTY2. If we do not take any parallel processing into consideration, the total complexity of MISTY1 and MISTY2 with the same number of rounds is clearly the same; however if we allow parallel computations, their encryption speed is not the same. This is mainly because MISTY1 can carry out two FI's at a time, while MISTY2 can execute four FI's in parallel.

Table 1 gives encryption/decryption time of MISTY1 and MISTY2, where each entry shows the number of calculations of S_9 assuming the computation time of S_7 is the same as that of S_9. For simplicity we have ignored the time for XOR operations. It is clearly seen from table 1 that MISTY2 is faster than MISTY1 in encryption, but MISTY1 is faster in ECB and CBC decryption. This is because parallel computations are impossible in inverse calculation of MISTY2. MISTY2 is therefore suitable for OFB and CFB modes.

	Encryption ECB,CBC,OFB,CFB	Decryption ECB,CBC	Decryption OFB,CFB
n-round MISTY1	$3n$	$3n$	$3n$
n-round MISTY2	$1.5n$	$9n$	$1.5n$

Table 1. Encryption/Decryption time of MISTY1 and MISTY2 (number of calculations of S_9).

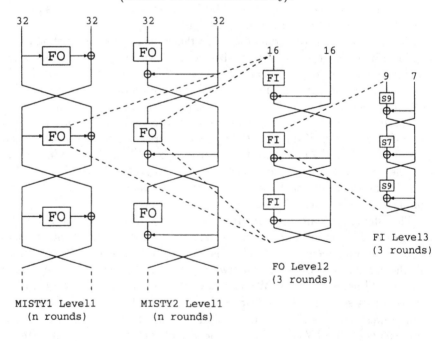

Figure 4: Recursive structure of MISTY

4.2 S_7 and S_9

In selecting S_7 and S_9, we have the following three criteria:

1. *Their average differential/linear probability must be minimal,*
2. *Their delay time in hardware is as short as possible,*
3. *Their algebraic degree is high, if possible.*

For the first criterion, a sequence of power functions over finite fields is known to attain the minimal value (that is, 2^{-6} for S_7 and 2^{-8} for S_9), and as far as we know, this is the only example that can be obtained in a systematic way. Hence we first planned to investigate the hardware delay, whose exact definition we adopted will be given below, for all functions that have the form $S_i(x) = A \circ x^\alpha \circ B$ $(i = 7, 9)$, where A and B are arbitrary bijective linear transformations and α is an integer such that $(2^i - 1, \alpha) = 1$. The last equality is a necessary and sufficient condition that a power function can be bijective.

However, because it was time-consuming for us to calculate the delay for all functions above, we next restricted our search to the functions that have the form $S_i(x) = A \circ x^\alpha$ $(i = 7, 9)$ and have a polynomial basis or a normal basis over $GF(2)$. In other words, we investigated all possible linear transformations for A and a limited number of linear transformations for B. Note that the average differential/linear probability does not depend on a selection of A or B, but the delay does. Now the following is our formal definition of the hardware delay and the algebraic degree of $S_i(x)$:

Definition 5. For a function $y = f(x)$ with an i-bit input $x = (x_0, x_1, x_2, ..., x_{i-1})$ and a j-bit output $y = (y_0, y_1, y_2, ..., y_{j-1})$, we call the following equation an algebraic normal form of the a-th output bit y_a of f:

$$y_a = e^{(a,0)} + \sum_{0 \le k_1 < i} e_{k_1}^{(a,1)} x_{k_1} + \sum_{0 \le k_1 < k_2 < i} e_{k_1,k_2}^{(a,2)} x_{k_1} x_{k_2}$$

$$+ \sum_{0 \le k_1 < k_2 < k_3 < i} e_{k_1,k_2,k_3}^{(a,3)} x_{k_1} x_{k_2} x_{k_3} + ..., \tag{4}$$

where $e^{(a,0)}, e_{k_1}^{(a,1)}, e_{k_1,k_2}^{(a,2)}, e_{k_1,k_2,k_3}^{(a,3)}$ .... are binary values, and the sum $\sum$ denotes an XOR operation.

The hardware length of y_a is defined as the number of non-zero terms of equation 4, and the hardware length of the function f is the maximal hardware length of all output bits of f. Also, the algebraic degree of y_a is defined as the maximal degree of equation 4, and the algebraic degree of the function f is the maximal algebraic degree of all output bits of f.

Note that the hardware length of y_a minus 1 is equivalent to the number of two-input XOR gates required for constructing y_a from x_k $(0 \le k \le i)$ in hardware, and the logarithm of the hardware length of f indicates its hardware delay time. Although we have to count the number of AND gates and fan-outs to see the exact delay time in hardware, we have adopted the above definition for simplicity. Also note that the algebraic degree of $S_i(x) = A \circ x^\alpha \circ B$ agrees with the binary hamming weight of α.

Selection of S_7

For all functions having the form $A \circ x^\alpha$ over $GF(2^7)$ with a polynomial or normal basis and $(2^7 - 1, \alpha) = 1$, we first calculated the algebraic degree and hardware length of each output bit; as a result, we obtained the following:

- If the algebraic degree is at least 4, then the hardware length of any output bit is at least 21.
- If the algebraic degree is equal to 3, then the hardware length of any output bit is at least 10.
- If the algebraic degree is equal to 2, then the hardware length of any output bit is at least 7.

Since we regarded the length as too long when the algebraic degree is four or more, we decided to adopt a function whose algebraic degree is equal to three. Then for all functions whose algebraic degree is three, we calculated their entire hardware length, and found that the minimal length is 13 and the function that attains this length is unique up to the order of output bits. Lastly, by adding a constant value to its output, we determined the final form of S_7, whose concrete logic is as follows:

$$y_0 = x_0 + x_1 x_3 + x_0 x_3 x_4 + x_1 x_5 + x_0 x_2 x_5 + x_4 x_5 + x_0 x_1 x_6 + x_2 x_6 + x_0 x_5 x_6 + x_3 x_5 x_6 + 1$$
$$y_1 = x_0 x_2 + x_0 x_4 + x_3 x_4 + x_1 x_5 + x_2 x_4 x_5 + x_6 + x_0 x_6 + x_3 x_6 + x_2 x_3 x_6 + x_1 x_4 x_6 + x_0 x_5 x_6 + 1$$
$$y_2 = x_1 x_2 + x_0 x_2 x_3 + x_4 + x_1 x_4 + x_0 x_1 x_4 + x_0 x_5 + x_0 x_4 x_5 + x_3 x_4 x_5 + x_1 x_6 + x_3 x_6 +$$
$$\qquad x_0 x_3 x_6 + x_4 x_6 + x_2 x_4 x_6$$
$$y_3 = x_0 + x_1 + x_0 x_1 x_2 + x_0 x_3 + x_2 x_4 + x_1 x_4 x_5 + x_2 x_6 + x_1 x_3 x_6 + x_0 x_4 x_6 + x_5 x_6 + 1$$
$$y_4 = x_2 x_3 + x_0 x_4 + x_1 x_3 x_4 + x_5 + x_2 x_5 + x_1 x_2 x_5 + x_0 x_3 x_5 + x_1 x_6 + x_1 x_5 x_6 + x_4 x_5 x_6 + 1$$
$$y_5 = x_0 + x_1 + x_2 + x_0 x_1 x_2 + x_0 x_3 + x_1 x_2 x_3 + x_1 x_4 + x_0 x_2 x_4 + x_0 x_5 + x_0 x_1 x_5 +$$
$$\qquad x_3 x_5 + x_0 x_6 + x_2 x_5 x_6$$
$$y_6 = x_0 x_1 + x_3 + x_0 x_3 + x_2 x_3 x_4 + x_0 x_5 + x_2 x_5 + x_3 x_5 + x_1 x_3 x_5 + x_1 x_6 + x_1 x_2 x_6 +$$
$$\qquad x_0 x_3 x_6 + x_4 x_6 + x_2 x_5 x_6$$

Selection of S_9

Similarly, for all functions having the form $S_9(x) = A \circ x^\alpha$ over $GF(2^9)$ with a polynomial or normal basis and $(2^9 - 1, \alpha) = 1$, we first calculated the algebraic degree and hardware length of each output bit; as a result, we had the following:

- If the algebraic degree is at least 3, then the hardware length of any output bit is at least 27.
- If the algebraic degree is equal to 2, then the hardware length of any output bit is at least 9.

Since we regarded the length as too long if the algebraic degree is three or more, we decided to adopt a function whose algebraic degree is equal to two. Then for all functions whose algebraic degree is two, we calculated their entire hardware length, and found that the minimal length is 12 and there are nine functions that attain this length up to the order of output bits. Lastly by selecting one of them randomly and adding a constant value to its output, we determined the final form of S_9, whose concrete logic is as follows:

$$y_0 = x_0 x_4 + x_0 x_5 + x_1 x_5 + x_1 x_6 + x_2 x_6 + x_2 x_7 + x_3 x_7 + x_3 x_8 + x_4 x_8 + 1$$
$$y_1 = x_0 x_2 + x_3 + x_1 x_3 + x_2 x_3 + x_3 x_4 + x_4 x_5 + x_0 x_6 + x_2 x_6 + x_7 + x_0 x_8 + x_3 x_8 + x_5 x_8 + 1$$
$$y_2 = x_0 x_1 + x_1 x_3 + x_4 + x_0 x_4 + x_2 x_4 + x_3 x_4 + x_4 x_5 + x_0 x_6 + x_5 x_6 + x_1 x_7 + x_3 x_7 + x_8$$
$$y_3 = x_0 + x_1 x_2 + x_2 x_4 + x_5 + x_1 x_5 + x_3 x_5 + x_4 x_5 + x_5 x_6 + x_1 x_7 + x_6 x_7 + x_2 x_8 + x_4 x_8$$
$$y_4 = x_1 + x_0 x_3 + x_2 x_3 + x_0 x_5 + x_3 x_5 + x_6 + x_2 x_6 + x_4 x_6 + x_5 x_6 + x_6 x_7 + x_2 x_8 + x_7 x_8$$
$$y_5 = x_2 + x_0 x_3 + x_1 x_4 + x_3 x_4 + x_1 x_6 + x_4 x_6 + x_7 + x_3 x_7 + x_5 x_7 + x_6 x_7 + x_0 x_8 + x_7 x_8$$
$$y_6 = x_0 x_1 + x_3 + x_1 x_4 + x_2 x_5 + x_4 x_5 + x_2 x_7 + x_5 x_7 + x_8 + x_0 x_8 + x_4 x_8 + x_6 x_8 + x_7 x_8 + 1$$
$$y_7 = x_1 + x_0 x_1 + x_1 x_2 + x_2 x_3 + x_0 x_4 + x_5 + x_1 x_6 + x_3 x_6 + x_0 x_7 + x_4 x_7 + x_6 x_7 + x_1 x_8 + 1$$
$$y_8 = x_0 + x_0 x_1 + x_1 x_2 + x_4 + x_0 x_5 + x_2 x_5 + x_3 x_6 + x_5 x_6 + x_0 x_7 + x_0 x_8 + x_3 x_8 + x_6 x_8 + 1$$

4.3 The function FL

For the purpose of avoiding possible attacks other than differential and linear cryptanalysis, we have supplemented an additional simple function FL, whose design criteria are (1) to be a linear function for any fixed key and (2) to have a variable form depending on a key value.

Since this function is linear as long as the key is fixed, it does not affect the average differential/linear probability of the entire algorithm. Moreover, this function is obviously fast in both software and hardware since it is constructed by logical operations such as AND, OR and XOR only.

5 Design of the Key Scheduling Part

In designing the key scheduling part of MISTY, we set up the following criteria from the viewpoint of compatibility between its security level and applicability to various systems:

1. *The size of key is 128 bits,*
2. *The size of subkey is 256 bits,*
3. *Every round is affected by all key bits,*
4. *Every round is affected by as many subkey bits as possible.*

For security reasons we have adopted the 128-bit key, and for practical reasons we have limited the size of the subkey to 256 bits. Reducing the size of subkey has two important performance advantages. The first advantage can be obtained in systems whose resources are limited such as in IC cards. In these systems, since RAM size for temporary use is usually strictly limited, it is generally impossible to store all subkey bits in RAM if its size is large; hence we have to carry out the key scheduling part in every data block, which could be a heavy penalty on performance. We decided to choose subkeys of 256 bits, so that all the bits could be stored in RAM even for extremely restricted software environments.

The second advantage comes from the fact that in microprocessors with many integer registers such as RISC processors, the 256-bit subkey can be loaded completely into the registers. In most implementation of block ciphers, all subkey bits are written into memory in key scheduling process, and in encryption process they are read from the memory round by round. Hence if all the subkey bits are kept in the registers during the entire encryption process, the total performance is expected to be significantly improved.

On the other hand, in compensation for this small number of subkey bits and simple key scheduling algorithm, we have established the third and fourth design criteria. In MISTY, an FO function and an FL function use 112 subkey bits and 32 subkey bits, respectively. To generate the 112 subkey bits, all of 128 key bits are required. The number of total independent subkey bits of MISTY1 or MISTY2 with eight rounds, for example, is 1216.

6   Examples of Implementation of MISTY

In this section we show two examples of our software implementation and one example of our hardware implementation of MISTY1 with eight rounds.

6.1   Pentium

Pentium has two independent integer execution units called U-pipe and V-pipe, where the U-pipe is usually used for carrying out instructions. However some instructions can be also executed in the V-pipe while the U-pipe is being occupied by special "pairable" instructions. Though the number of these pairable instructions is small, if we write a program so that these two pipes can be efficiently used, the performance of the software is extremely improved, possibly twice or more due to resolution of register contentions.

We wrote an assembly language program of MISTY1 with eight rounds on Pentium 100MHz, which encrypts an input plaintext stream in CBC mode at a speed of 20Mbps. The program heavily uses V-pipe because of the highly parallel structure of MISTY; it takes approximately 300 cycles to process one block, where the U-pipe has no idle time and the V-pipe is used in more than 95% of the 300 cycles.

6.2   PA-7200

PA-7200 can also execute two integer instructions at a time under various restrictions. Moreover PA-RISC series microprocessors have 32 integer registers, almost all of which can be used freely by users; this means that it is easy to load all 256-bit subkey information of MISTY, even every 16 bits in each register.

PA-7200 has 512KB on-chip cache (256KB for code and 256KB for data), which enables us to reduce computational time of MISTY by having a big predefined table. That is to say, we can make a 128KB table that represents the first two rounds of the FI function in advance. By doing this, calculation of FI is significantly simplified. Note that this technique cannot be used in Pentium because Pentium has only small cache (8KB for data) which generally causes serious penalty cycles due to cache misses.

We wrote an assembly language program of MISTY1 with eight rounds on PA-7200 120MHz using the above techniques. It can encrypt an input plaintext stream in CBC mode at a speed of 40Mbps.

6.3 Hardware

We have also designed a prototype LSI of MISTY1 with eight rounds, which has the following specifications:

Encryption Speed:	*450Mbps (typical)*
Clock:	*14MHz*
I/O:	*32-bit parallel × 3 (plaintext, ciphertext, key)*
Supported Modes:	*ECB, CBC, OFB-64, CFB-64*
Design Process:	*0.5μ CMOS gate-array*
Number of Gates:	*65K gates*
Package:	*208-pin flat package*

This LSI has no repetition structure; that is, it contains the full hardware of eight FO functions and ten FL functions. It takes two cycles to encrypt a 64-bit plaintext. It also has three independent 64-bit registers that store a plaintext, an intermediate text after the fourth round, and a ciphertext, respectively. This structure makes the following pipeline data processing possible:

	plaintext 1	*plaintext 2*	*plaintext 3*	*plaintext 4*
Cycles 1 and 2	*Input*			
Cycles 3 and 4	*Encryption*	*Input*		
Cycles 5 and 6	*Output*	*Encryption*	*Input*	
Cycles 7 and 8		*Output*	*Encryption*	*Input*

7 Conclusions

This paper proposed new secret-key block cryptosystems MISTY1 and MISTY2. At present, the author recommends to use MISTY1 with eight rounds, and to use MISTY2, which has a newer structure, with twelve rounds. The next four pages show a complete and self-contained description of MISTY1 and MISTY2.

References

1. Nyberg, K., Knudsen, L.,: Provable Security against Differential Cryptanalysis. Journal of Cryptology, Vol.8, no.1 (1995)
2. Nyberg, K.,: Linear Approximation of Block Ciphers. Advances in Cryptology – Eurocrypt'94, Lecture Notes in Computer Science **950**, Springer Verlag (1994)
3. Aoki, K., Ohta, K.,: Stricter Evaluation for the Maximum Average of Differential Probability and the Maximum Average of Linear Probability (in Japanese). Proceedings of SCIS'96, SCIS96-4A (1996)
4. Matsui, M.,: New Structure of Block Ciphers with Provable Security against Differential and Linear Cryptanalysis. Proceedings of the third international workshop of fast software encryption, Lecture Notes in Computer Science **1039**, Springer Verlag (1996)

Block Cipher Algorithms MISTY1 and MISTY2

Edition 2.1   December 16 1996

This document shows a complete description of encryption algorithms MISTY1 and MISTY2, which are secret-key ciphers with a 64-bit data block, a 128-bit secret key and a variable number of rounds n, where n is a multiple of four.

Data Randomizing Part

- Figure A and B show the data randomizing part of MISTY1 and MISTY2, respectively: The 64-bit plaintext P is divided into the left 32-bit string and the right 32-bit string, which are transformed into the 64-bit ciphertext C by means of bitwise XOR operations denoted by $\oplus$ and sub-functions FO_i $(1 \leq i \leq n)$ and FL_i $(1 \leq i \leq n+2)$. FO_i uses a 64-bit subkey KO_i and a 48-bit subkey KI_i. FL_i uses a 32-bit subkey KL_i.
- Figure C shows the structure of FO_i: The input is divided into the left 16-bit string and the right 16-bit string, which are transformed into the output by means of bitwise XOR operations and sub-functions FI_{ij} $(1 \leq j \leq 3)$, where KO_{ij} $(1 \leq j \leq 4)$ and KI_{ij} $(1 \leq j \leq 3)$ are the j-th (from left) 16 bits of KO_i and KI_i, respectively.
- Figure D shows the structure of FI_{ij}: The input is divided into the left 9-bit string and the right 7-bit string, which are transformed into the output by means of bitwise XOR operations and substitution tables S_7 and S_9. In the first and third XORs, the 7-bit string is zero-extended to 9 bits, and in the second XOR, the 9-bit string is truncated to 7 bits by discarding its highest two bits. KI_{ij1} and KI_{ij2} are the left 7 bits and the right 9 bits of KI_{ij}, respectively.
- Figure E shows the structure of FL_i. The input is divided into the left 16-bit string and the right 16-bit string, which are transformed into the output by means of bitwise XOR operations, a bitwise AND operation denoted by $\cap$ and a bitwise OR operation denoted by $\cup$, where KL_{ij} $(1 \leq j \leq 2)$ is the j-th (from left) 16 bits of KL_i.
- In the next page, the substitution tables S_7 and S_9 are shown in decimal form.

Key Scheduling Part

- Figure F shows the key scheduling part of MISTY1 and MISTY2: K_i $(1 \leq i \leq 8)$ is the i-th (from left) 16 bits of the secret key K, and K_i' $(1 \leq i \leq 8)$ is the output of FI_{ij} when the input of FI_{ij} is assigned to K_i and the key KI_{ij} is set to K_{i+1}, where K_9 is identified with K_1.
- The correspondence between the round subkeys $KO_{ij}, KI_{ij}, KL_{ij}$ and the actual subkeys K_i, K_i' is as follows, where i is identified with $i-8$ when $i > 8$:

Round	KO_{i1}	KO_{i2}	KO_{i3}	KO_{i4}	KI_{i1}	KI_{i2}	KI_{i3}	KL_{i1}	KL_{i2}
Actual	K_i	K_{i+2}	K_{i+7}	K_{i+4}	K_{i+5}'	K_{i+1}'	K_{i+3}'	$K_{\frac{i+1}{2}}$ (odd i) $K_{\frac{i}{2}+2}'$ (even i)	$K_{\frac{i+1}{2}+6}'$ (odd i) $K_{\frac{i}{2}+4}$ (even i)

Test Data of MISTY1 with eight rounds

Key (K_1 to K_8):   00 11 22 33 44 55 66 77 88 99 aa bb cc dd ee ff
Subkey (K_1' to K_8'): cf 51 8e 7f 5e 29 67 3a cd bc 07 d6 bf 35 5e 11
Plaintext: 01 23 45 67 89 ab cd ef
Ciphertext: 8b 1d a5 f5 6a b3 d0 7c

Table of S7

```
 27,  50,  51,  90,  59,  16,  23,  84,  91,  26,114,115,107,  44,102,  73,
 31,  36,  19,108,  55,  46,  63,  74,  93,  15,  64,  86,  37,  81,  28,   4,
 11,  70,  32,  13,123,  53,  68,  66,  43,  30,  65,  20,  75,121,  21,111,
 14,  85,   9,  54,116,  12,103,  83,  40,  10,126,  56,   2,   7,  96,  41,
 25,  18,101,  47,  48,  57,   8,104,  95,120,  42,  76,100,  69,117,  61,
 89,  72,   3,  87,124,  79,  98,  60,  29,  33,  94,  39,106,112,  77,  58,
  1,109,110,  99,  24,119,  35,   5,  38,118,   0,  49,  45,122,127,  97,
 80,  34,  17,   6,  71,  22,  82,  78,113,  62,105,  67,  52,  92,  88,125
```

Table of S9

```
451,203,339,415,483,233,251, 53,385,185,279,491,307,  9, 45,211,
199,330, 55,126,235,356,403,472,163,286, 85, 44, 29,418,355,280,
331,338,466, 15, 43, 48,314,229,273,312,398, 99,227,200,500, 27,
  1,157,248,416,365,499, 28,326,125,209,130,490,387,301,244,414,
467,221,482,296,480,236, 89,145, 17,303, 38,220,176,396,271,503,
231,364,182,249,216,337,257,332,259,184,340,299,430, 23,113, 12,
 71, 88,127,420,308,297,132,349,413,434,419, 72,124, 81,458, 35,
317,423,357, 59, 66,218,402,206,193,107,159,497,300,388,250,406,
481,361,381, 49,384,266,148,474,390,318,284, 96,373,463,103,281,
101,104,153,336,  8,  7,380,183, 36, 25,222,295,219,228,425, 82,
265,144,412,449, 40,435,309,362,374,223,485,392,197,366,478,433,
195,479, 54,238,494,240,147, 73,154,438,105,129,293, 11, 94,180,
329,455,372, 62,315,439,142,454,174, 16,149,495, 78,242,509,133,
253,246,160,367,131,138,342,155,316,263,359,152,464,489,  3,510,
189,290,137,210,399, 18, 51,106,322,237,368,283,226,335,344,305,
327, 93,275,461,121,353,421,377,158,436,204, 34,306, 26,232,  4,
391,493,407, 57,447,471, 39,395,198,156,208,334,108, 52,498,110,
202, 37,186,401,254, 19,262, 47,429,370,475,192,267,470,245,492,
269,118,276,427,117,268,484,345, 84,287, 75,196,446,247, 41,164,
 14,496,119, 77,378,134,139,179,369,191,270,260,151,347,352,360,
215,187,102,462,252,146,453,111, 22, 74,161,313,175,241,400, 10,
426,323,379, 86,397,358,212,507,333,404,410,135,504,291,167,440,
321, 60,505,320, 42,341,282,417,408,213,294,431, 97,302,343,476,
114,394,170,150,277,239, 69,123,141,325, 83, 95,376,178, 46, 32,
469, 63,457,487,428, 68, 56, 20,177,363,171,181, 90,386,456,468,
 24,375,100,207,109,256,409,304,346,  5,288,443,445,224, 79,214,
319,452,298, 21,  6,255,411,166, 67,136, 80,351,488,289,115,382,
188,194,201,371,393,501,116,460,486,424,405, 31, 65, 13,442, 50,
 61,465,128,168, 87,441,354,328,217,261, 98,122, 33,511,274,264,
448,169,285,432,422,205,243, 92,258, 91,473,324,502,173,165, 58,
459,310,383, 70,225, 30,477,230,311,506,389,140,143, 64,437,190,
120,  0,172,272,350,292,  2,444,162,234,112,508,278,348, 76,450
```

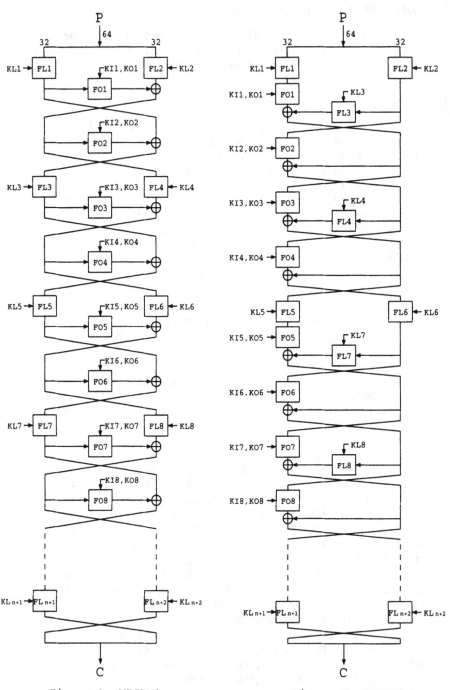

Figure A: MISTY1   Figure B: MISTY2

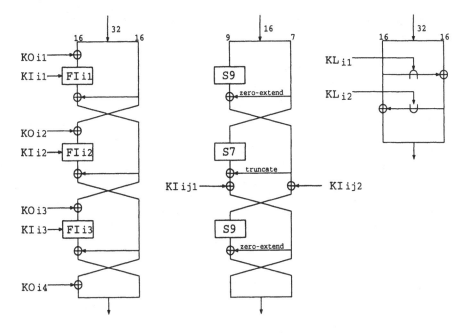

Figure C: FOi   Figure D: FIij   Figure E: FLi

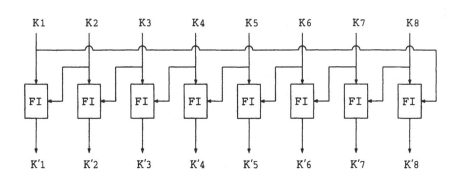

Figure F: Key Scheduling

The Design of the ICE Encryption Algorithm

Matthew Kwan

mkwan@cs.mu.oz.au

Abstract. This paper describes the design and implementation of the ICE cryptosystem, a 64-bit Feistel block cipher. It describes the design process, with the various aims and tradeoffs involved. It also introduces the concept of keyed permutation to improve resistance to differential and linear cryptanalysis, and the use of an extensible key schedule to achieve an explict tradeoff between speed and security.

1   Introduction

The Data Encryption Standard (DES) [8] has been widely used as an international standard since its introduction in 1977. However, in the years since its release, a number of vulnerabilities have come to light.

These include susceptibility to differential cryptanalysis [2], susceptibility to linear cryptanalysis [7], a key/plaintext complementation weakness [4], four weak and twelve semi-weak keys [4], a fixed 56-bit key size, inefficient software performance, and an absence of public design criteria.

While triple-DES [10] provides a larger key size at 112 bits, this is at the expense of a factor of three in encryption speed.

ICE, which stands for *Information Concealment Engine*, was designed to address these issues, while maintaining a compatible interface with DES. This is to allow it to act as a substitute in existing applications.

In addition to the standard ICE algorithm, other variants are described. Thin-ICE is a faster, less secure version, while there are also open-ended variants ICE-n which trade off greater key size for reduced encryption speed.

2   The Structure

ICE is a standard Feistel block cipher, with a structure similar to DES.

It takes a 64-bit plaintext, which is split into two 32-bit halves. In each round of the algorithm the right half and a 60-bit subkey are fed into the function F. The output of F is XORed with the left half, then the halves are swapped. This repeated for all but the final round, where the final swap is left out, as is illustrated in figure 1. At the end of the rounds, the halves are concatenated to form the ciphertext.

Decryption follows the same procedure, except that the subkeys are used in reverse order.

A Feistel structure was chosen for a number of reasons. To begin with, it was guaranteed to carry out one-to-one mappings between plaintext and ciphertext,

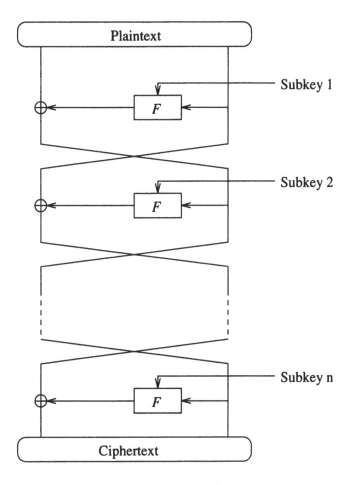

Fig. 1. The structure of n-round ICE

which is necessary for a cipher to be decryptable. This enabled the designer to concentrate on the design of the F function and key schedule, secure in the knowledge that a valid cipher would be produced.

Secondly, Feistel ciphers have been publicly cryptanalysed for more than two decades, and no systematic weakness has been uncovered. In addition, the techniques that have been used to analyse existing Feistel ciphers are generally applicable to new ones. This simplifies the design task, since the designer is not forced to invent as many new forms of cryptanalysis when evaluating a design.

And finally, Feistel ciphers are reasonably fast and simple to implement in software. Speed and simplicity were two important design aims for ICE.

3   The *F* Function

Notation: In this paper, bits will be numbered from right to left, starting at bit zero. So, for example, the rightmost bit of a plaintext half is P_0, while the leftmost bit is P_{31}.

The ICE *F* function is similar in structure to the one used in DES, with the exception of keyed permutation (described below). The function as a whole is illustrated in figure 2.

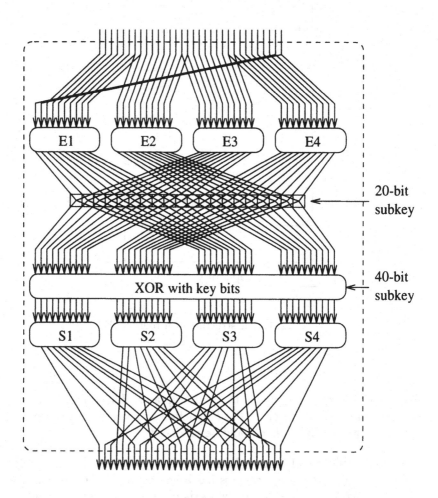

Fig. 2. The ICE *F* function

3.1   The Expansion Function E

The 32-bit plaintext half is expanded to four 10-bit values, E1, E2, E3, E4, in the following manner.

$$E1 = P_1 P_0 P_{31} P_{30} P_{29} P_{28} P_{27} P_{26} P_{25} P_{24}$$
$$E2 = P_{25} P_{24} P_{23} P_{22} P_{21} P_{20} P_{19} P_{18} P_{17} P_{16}$$
$$E3 = P_{17} P_{16} P_{15} P_{14} P_{13} P_{12} P_{11} P_{10} P_9 P_8$$
$$E4 = P_9 P_8 P_7 P_6 P_5 P_4 P_3 P_2 P_1 P_0$$

This expansion function was chosen because four 10-bit values were needed for the S-boxes, and it was reasonably fast to implement in software.

3.2   Keyed Permutation

After expansion, keyed permutation is used. The permutation subkey is 20 bits long, and is used to swap bits between E1 and E3, and between E2 and E4. For example, if bit 19 of the subkey is set, bit 9 of E1 and E3 will be swapped. If bit 0 of the subkey is set, bit 0 of E2 and E4 will be swapped. The cryptographic properties of keyed permutation are described in a later section.

The values E1, E2, E3, and E4, after being permuted, are XORed with 40 bits of subkey, then used as input to the four S-boxes, S1, S2, S3, and S4. Each S-box takes a 10-bit input and produces an 8-bit output. The S-boxes are described in more detail later.

3.3   The Permutation Function P

The four 8-bit S-box outputs are combined via a P-box into a 32-bit value, which becomes the output of the F function. The P-box, which is specified in table 1, was designed to maximize diffusion from each S-box, and to ensure that bits which are separated by 16 places never come from the same S-box, nor from S-boxes separated by two places (eg S1 and S3). Given that these criteria were met, the P-box was also designed to be as regular as possible.

Output bit	31	30	29	28	27	26	25	24	23	22	21	20	19	18	17	16
Source	$S1_7$	$S4_7$	$S3_7$	$S2_7$	$S2_6$	$S3_6$	$S1_6$	$S4_6$	$S3_5$	$S2_5$	$S4_5$	$S1_5$	$S4_4$	$S1_4$	$S2_4$	$S3_4$
Output bit	15	14	13	12	11	10	9	8	7	6	5	4	3	2	1	0
Source	$S2_3$	$S3_3$	$S4_3$	$S1_3$	$S1_2$	$S4_2$	$S2_2$	$S3_2$	$S4_1$	$S1_1$	$S3_1$	$S2_1$	$S3_0$	$S2_0$	$S1_0$	$S4_0$

Table 1. The permutation function P

The 32-bit output is thought of as eight 4-bit blocks, and diffusion is achieved by ensuring that each bit from a particular S-box is permuted to a different 4-bit block in the output. This means that the output of one S-box is guaranteed to affect the inputs to all the four E-boxes in the next round. Whether it affects the inputs to all the S-boxes depends to some extent on the keyed permutation in that round.

Although each S-box output bit goes to a different 4-bit block, there are four positions within each block to choose from. For an even spread, it was decided that each position would be used twice (since there are eight bits coming out of each S-box).

Because of the E-boxes, certain input bits to the F function (bits 0, 1, 8, 9, 16, 17, 24, 25) end up affecting two S-boxes, while the other bits only affect one. The P-box was designed to ensure that only one bit from each S-box will exclusively affect that same S-box in the next round, assuming that the bit isn't redirected by the keyed permutation. However, in the worst case, keyed permutation can result in up to three bits from an S-box exclusively affecting the same S-box in the next round.

4 The S-boxes

The S-boxes in ICE are similar in structure to those used in LOKI [3] in their use of Galois Field exponentiation.

Each S-box takes a 10-bit input X. Bits X_9 and X_0 are concatenated to form the row selector R. Bits $X_8..X_1$ are concatenated to form the 8-bit column selector C. For each row R, there is an XOR offset value O_R, and a Galois Field prime (irreducible polynomial) P_R.

The 8-bit output of an S-box for an input X is given by $(C \oplus O_R)^7 \bmod P_R$, under 8-bit Galois Field arithmetic. The exponent 7 was chosen because it is a one-to-one function, and it produces a reasonably flat XOR profile, useful for resistance to differential cryptanalysis. Regardless of the prime used as the modulus, no input difference produces an output difference with a probability greater than $\frac{6}{256}$.

The XOR offsets for each row in each S-box are given in table 2, while the prime numbers are specified in table 3.

S-box	O_0	O_1	O_2	O_3
S1	83	85	9b	cd
S2	cc	a7	ad	41
S3	4b	2e	d4	33
S4	ea	cd	2e	04

Table 2. The S-box XOR offsets (in hexadecimal)

S-box	P_0	P_1	P_2	P_3
S1	333	313	505	369
S2	379	375	319	391
S3	361	445	451	397
S4	397	425	395	505

Table 3. The S-box Galois Field primes

The choice of these values is described in detail in a later part of this paper. The way the S-boxes are specified, with 16 offsets and 16 primes, was chosen so that the boxes were parameterised. This enabled software to automate the generation and evaluation of millions upon millions of different possible S-boxes.

5   The Key Schedule

ICE has been designed with an extensible key schedule to permit a tradeoff between speed and security.

The standard ICE algorithm takes a 64-bit key and uses 16 subkeys in 16 rounds. There is a fast variant Thin-ICE which uses 8 rounds with a 64-bit key, and there are open-ended variants ICE-n which use $16n$ rounds and $64n$-bit keys. For example ICE-2 uses 32 rounds and a 128-bit key.

5.1   Design criteria

When the key schedule was designed, a number of criteria were used.

- There must be no weak keys. In other words, for an n-round schedule there must be no two keys $K1$ and $K2$ such that the all the subkeys $SK(K1, i) = SK(K2, n - i + 1)$ for $i = 1..n$
- Only a single key complementation weakness, ignoring keyed permutation. In other words, assuming that bits aren't being permuted in the F function, the only values of A, B, C s.t. $ICE(P, K) = ICE(P \oplus A, K \oplus B) \oplus C$ is when A, B, and C have all bits set. This is largely unnecessary, since keyed permutation makes it impossible to exploit complementation weaknesses, but it can't hurt.
- Each subkey bit should only be dependent on only one key bit. This simplifies the proof [5] that the above two conditions are satisfied.
- No meet-in-the-middle attacks. This means that, for any round N, all key bits must be used either in the preceeding rounds, or all must be used in the following rounds.
- Since the F function makes use of 60 key bits per round, each key bit must be used 15 times in the ICE and ICE-n ciphers. Thinking of the 60 key bits as being divided into 15 4-bit blocks, each key bit must use a different 4-bit

block each time it is used. For Thin-ICE, with only eight rounds, every key bit must be used 7 or 8 times, and spread out such that, if the 60 bits are partitioned into three 20-bit blocks, each key bit is used in a 20-bit block 2 or 3 times.

- Immunity to related-key cryptanalysis [1]. This means some sort of irregularity in the key schedule.
- The key scheduling algorithm must be simple to implement in software, yet not too fast, so as to hinder exhaustive key searches.

5.2   Key schedule specification

The Thin-ICE key schedule is simply the first 8 rounds of the standard ICE key schedule. The ICE-n key schedules build on the ICE key schedule using the following algorithm.

```
Start with an empty key schedule.
for i = 1 .. n
  Take the next 64 bits of the key and use them to generate
      a 16-round ICE key schedule.
  Take the schedule so far, split it at the half-way point,
      and insert the new 16-round schedule.
end
```

This ensures that there is no simple meet-in-the-middle attack, and that if ICE has no weak keys, then neither does ICE-n. However, this structure is still susceptible to more sophisticated meet-in-the-middle attacks [11].

In order to satisfy the condition that there be no weak keys, it was firstly necessary to prevent an all-zero key from producing zero subkeys in all rounds. The condition that each subkey bit be dependent on only one key bit means that this can only be achieved by inverting certain bits during the key schedule.

In each round 60 key bits are used. These are typically stored in three 20-bit values, SK1, SK2, and SK3.

SK1 is the value XORed with the inputs to S1 and S2.
SK2 is the value XORed with the inputs to S3 and S4.
SK3 is the value used for key permutation.

The key is first converted into four 16-bit blocks, KB[0 .. 3]. These blocks are used, along with the key rotations KR shown in table 4, to derive the subkeys in each round using the algorithm as follows.

KB[0] $= K_{63}..K_{48}$
KB[1] $= K_{47}..K_{32}$
KB[2] $= K_{31}..K_{16}$
KB[3] $= K_{15}..K_0$

```
for each round n = 1 .. 16
  for SK = SK1, SK2, SK3 in turn, 5 times each
    for i = 0 .. 3
      Set B to bit 0 of KB[(i + KR[n]) mod 4].
      Shift SK left one bit.
      Set bit 0 of SK to B.
      Shift KB[(i + KR[n]) mod 4] right one bit.
      Set bit 19 of KB[(i + KR[n]) mod 4] to the inverse of B.
    end
  end
end
```

Round	1	2	3	4	5	6	7	8	9	10	11	12	13	14	15	16
Rotation	0	1	2	3	2	1	3	0	1	3	2	0	3	1	0	2

Table 4. Key rotations KR

6   Keyed Permutation

The inspiration for the use of keyed permutation comes from the salt value in the Unix password encryption function crypt(3), designed by Ken Thompson and Denis Ritchie at AT&T Bell laboratories. This function uses a fixed, publicly-known, 12-bit value to permute the DES E-box in the manner similar to that described in the ICE F function, except that it permutes only 24 of the 48 E-box output bits.

It was a small step to consider a modification where the salt was increased in size to permute all the bits from the E-box, and where the salt was not publicly known, but rather derived from the secret key in each round via some key schedule. It turns out that this has some useful cryptographic properties.

From an aesthetic point of view it has a certain consistency. The cornerstone of block cipher design has been the use of substitution and permutation networks to encrypt data [9]. Keyed substitution, usually in the form of XORing key bits with an S-box input, has been commonly used, and it is felt that keyed permutation should complement this nicely.

To begin with, keyed permutation gives ICE immunity to complementation weaknesses. These result from situations where key and plaintext bits are inverted, but cancel each other out at the inputs to the S-boxes, causing the S-boxes to produce identical outputs. However, if key bits are inverted in ICE, the S-boxes will receive their inputs from totally different bits, and thus not regularly produce identical outputs.

A great challenge to cryptosystem designers has been resistance to differential cryptanalysis. Although the keyed permutation used in ICE doesn't grant immunity to differential cryptanalysis, it does greatly reduce its effectiveness.

Differential cryptanalysis relies an attacker sending XOR differences to the inputs of S-boxes, where the S-boxes will then produces an XOR difference as output with some probability. But because these input bits are being permuted in ICE, the attacker does not know which S-box will receive the bit.

However, an attacker can exploit the symmetry of the keyed permutation. Think of the 32-bit input to the F function as having left and right 16-bit halves. The permutation in ICE simply swaps bits between the halves. By using input differences where both halves are the same (called *symmetric* inputs), an attacker can be certain that the bits will reach their target S-boxes.

However, since the attacker now has to target at least two S-boxes at a time, the probability of success is typically squared or worse. Detailed values are given in later sections.

Similarly, linear cryptanalysis is forced to use symmetric inputs, again with much lower probabilities.

7   The Design of the S-boxes

The S-boxes in ICE were primarily designed to be resistant to differential cryptanalysis, and in particular to symmetric attacks. In an ideal world, every possible Galois Field prime and XOR offset would be tried, and the resulting S-boxes evaluated. However, with 30 Galois primes to try in 16 positions, and 16 8-bit offsets to choose, this would require more than 10^{71} evaluations. As result, the selection process was done in incremental steps.

1. Of the ten bits input to each S-box, the middle six bits are unique to that S-box, while the remaining four are shared with adjacent S-boxes. Since each row only uses one prime, XOR profiles for all 30 different primes were generated, and groups of four primes were evaluated by adding their profiles together and checking the probabilities of characteristics whose inputs use the middle six bits. 509 groups of four were found whose peak probability was 18/1024.

2. Pairs of these four-prime sets were evaluated, with the proviso that no pair could share a prime in common. The aim was find the pair with the lowest product of probabilities given the same input probability. The pairs chosen appeared to have probability products of 256/1048576, but it later turned out that the analysis software was flawed, and the true probability was 324/1048576. However, by then the design was complete, and the problem was not deemed serious enough to justify a redesign.

3. The XOR offsets in each S-box were selected to minimise the probability of an input difference producing a zero output difference, and consequently to minimise the probability that a symmetric input difference would produce a zero output difference.

4. 4096 sets of values had the same probabilities $(4320/2^{40})$, so the set was chosen which had the lowest probability of a symmetric input difference producing itself as output $(8064/2^{40})$.

With respect to differential cryptanalysis, it makes no difference if all the XOR offsets in a row are XORed with a constant, or if the positions of the primes are XORed with a constant, so there were still 1024 choices available for each S-box. None of these choices would have any effect on differential (and, it turns out, linear) cryptanalysis, so some new, and fairly arbitrary, criteria were used to narrow down the choice of parameters.

1. There should be no x where $F(x) = 0$, assuming a zero subkey. If F in all rounds produces zero outputs, then a plaintext would encrypt to itself.
2. There should be no x where $F(x) = x$.
3. For a given input, no S-box shall produce the same output as another S-box.
4. The sum of the bit count of $F(x) \oplus x$ over all symmetric x values shall be perfectly balanced (i.e. equal to 1048576).
5. Finally there were 16 sets of values to choose from. They were evaluated for the bitcount of $F(x)$ for all 32 single-bit x values. None gave the balanced value of 512, but the closest was 506, which was then chosen as the set of primes and offsets for the S-boxes.

8   Cryptanalysis

During the design process, ICE was subjected to a number of attacks. In addition to the key schedule analysis described previously, it was also subjected to differential cryptanalysis, linear cryptanalysis, and some other specialised attacks.

8.1   Differential Cryptanalysis

There are 22 one-round characteristics with probability 18/1024. However, they require non-symmetric inputs, so are not effective attacks.

The two best symmetric single round characteristics are $F(008c008c) \rightarrow 4042a085$ and $F(00dc00dc) \rightarrow 5920a681$, both with probability 324/1048576. However, because they are not symmetric they cannot be turned into iterative characteristics.

For characteristics with symmetric inputs and outputs, the best examples are $F(b801b801) \rightarrow 02b702b7$ with probability $57600/2^{40}$ and $F(34eb34eb) \rightarrow d82fd82f$ with probability $50176/2^{40}$.

The best symmetric characteristics where the output difference equals the input are $F(80848084) \rightarrow 80848084$ and $F(98619861) \rightarrow 98619861$, both with probability $8064/2^{40}$. Although these can be turned into an iterative attack, the probabilities are too low to be useful.

The best symmetric characteristics that produce zero output differences are $F(b2d6b2d6) \rightarrow 0$ and $F(cad6cad6) \rightarrow 0$, both with probability $4320/2^{40}$

Both of these characteristics can be turned into a five-round characteristic with a probability of $2^{-55.85}$. It is possible that this attack could be used to break Thin-ICE in less time than exhaustive search, although this has not been investigated fully at present.

The full 16-round ICE and all its extended variants appear to be secure against diffferential cryptanalysis.

8.2   Linear Cryptanalysis

Linear cryptanalysis relies on correlations between the XOR sum of certain bits in the input and output of S-boxes. However, the use of keyed permutation means that, although the output of the S-boxes are usable in the same way as the DES attack, the source of the input bits cannot be known for certain unless symmetric input bits are used.

For ICE, the best approximation for a single S-box is $NS_2(457, 136) = 416 = 512 - 96$. However, since this is not symmetric it is not usable as an attack.

Linear approximations that only make use of the two bits shared by adjacent S-boxes do not need knowledge of input bits, and can be combined into iterative expressions. The best approximations of this form are ...

$$X[24, 25] \oplus X[24, 25] \oplus F(X, K)[11, 12, 18, 20, 25, 31] \oplus F(X, K)[4, 9, 17, 22, 27] = K[28, 29] \oplus K[30, 31]$$
$$X[8, 9] \oplus X[8, 9] \oplus F(X, K)[3, 8, 16, 26] \oplus F(X, K)[13, 19, 21, 30] = K[8, 9] \oplus K[10, 11]$$
$$X[8, 9] \oplus X[8, 9] \oplus F(X, K)[3, 14, 23, 26] \oplus F(X, K)[19, 21, 30] = K[8, 9] \oplus K[10, 11]$$

Each of these linear approximations has a probability of $2^{-7.83}$. They can be combined into a symmetric 6-round expression which has a probability of 2^{-42}. If used to attack Thin-ICE it would typically require 2^{82} ciphertexts to achieve a 75% success rate, so it appears that Thin-ICE cannot be broken using this attack.

8.3   Other Attacks

It is possible that the use of keyed permutation introduces weaknesses of its own. One possibility is to somehow trace the path of bits through the cipher, and thus deduce where they were permuted. This would immediately yield key bits. However, early analysis indicates that the avalanche effect of the S and P boxes masks any information of this sort after a few rounds.

For extended-round variants of ICE with keys 128 bits and longer, it must be remembered that the strength of the cipher under a chosen-plaintext attack is only 2^{64} time and memory, since an attacker can theoretically simply store all the plaintext/ciphertext pairs in one massive lookup table, and thus immediately find the plaintext corresponding to a ciphertext. Although this may not be a practical attack, it does represent the theoretical strength of the cipher under chosen-plaintext attacks, and, to a lesser extent, known-plaintext attacks.

9   Software Implementation

Keyed permutation is achieved by simple bitwise operations. This is best demonstrated in the ICE F function source code, written in ANSI C. Note that for speed the S-boxes have been pre-permuted to produce 32-bit outputs. Full source code implementations of ICE, written in ANSI C, C++, and Java, can be found at [6].

```
unsigned long
ice_f (
    unsigned long          p,
    const ICE_SUBKEY       sk
) {
    unsigned long   tl, tr;    /* Expanded 40-bit value */
    unsigned long   al, ar;    /* Salted expanded 40-bit value */

                               /* Left half expansion */
    tl = ((p>>16) & 0x3ff) | (((p>>14) | (p<<18)) & 0xffc00);

                               /* Right half expansion */
    tr = (p & 0x3ff) | ((p<<2) & 0xffc00);

                               /* Perform the keyed permutation */
    al = sk[2] & (tl ^ tr);
    ar = al ^ tr;
    al ^= tl;

    al ^= sk[0];               /* XOR with the subkey */
    ar ^= sk[1];

                               /* S-box lookup and permutation */
    return (ice_sbox[0][al>>10] | ice_sbox[1][al & 0x3ff]
            | ice_sbox[2][ar>>10] | ice_sbox[3][ar & 0x3ff]);
}
```

The C implementations were bechmarked against an optimised version of DES on a 100MHz 486 PC running Linux, as shown in table 5.

The slow speed of key changes was surprising, given the simplicity of the key scheduling algorithm. It turns out this is because the algorithm operates on only one key bit at a time, whereas this DES implementation operates on 28-bit blocks.

10   Summary

The design and analysis of ICE and its variants has been described here. It is hoped that the ciphers will prove secure in the long run, and provide possible

Operation (x 100000)	Time (seconds)
DES encryption	2.37
DES decryption	2.40
DES key change	4.98
ICE encryption	1.63
ICE decryption	1.59
ICE key change	44.79
Thin-ICE encryption	0.88
Thin-ICE decryption	0.87
Thin-ICE key change	22.45
ICE-2 encryption	3.12
ICE-2 decryption	3.04
ICE-2 key change	89.63

Table 5. Benchmark results

Variant	Key	Plaintext	Ciphertext
ICE	deadbeef01234567	fedcba9876543210	7d6ef1ef30d47a96
Thin-ICE	deadbeef01234567	fedcba9876543210	de240d83a00a9cc0
ICE-2	00112233445566778899aabbccddeeff	fedcba9876543210	f94840d86972f21c

Table 6. Certification triplets

alternatives for DES in the future. As a public-domain algorithm, with source code freely available (export restrictions permitting), it should be useful in any number of security and privacy applications.

References

1. E. Biham, New Types of Cryptanalytic Attacks Using Related Keys, *Advances in Cryptology - EUROCRYPT '93 Proceedings*, Springer-Verlag, pp. 386-397, 1994
2. E. Biham and A. Shamir, *Differential Cryptanalysis of the Data Encryption Standard*, Springer-Verlag, 1993
3. L. Brown, J. Pieprzyk and J. Seberry, LOKI: A Cryptographic Primitive for Authentication and Secrecy Applications, *Advances in Cryptology - AUSCRYPT '90 Proceedings*, Springer-Verlag, pp. 229-236, 1990
4. M.E. Hellman, R.C. Merkle, R. Schroeppel, L. Washington, W. Diffie, S. Pohlig and P. Schweitzer, Results of an Initial Attempt to Cryptanalyze the NBS Data Encryption Standard, Technical Report SEL 76-042, Stanford University, September 1976
5. M. Kwan and J. Pieprzyk, A General Purpose Technique for Locating Key Scheduling Weaknesses in DES-Like Cryptosystems, *Advances in Cryptology - ASIACRYPT '91 Proceedings*, Springer-Verlag, pp. 237-246, 1991

6. M. Kwan, The ICE Home Page, http://www.cs.mu.oz.au/~mkwan/ice
7. M. Matsui, Linear Cryptanalysis Method for DES Cipher, *Advances in Cryptology - EUROCRYPT '93 Proceedings*, Springer-Verlag, pp. 386-397, 1994
8. National Bureau of Standards, *Data Encryption Standard*, FIPS PUB 46, U.S. Department of Commerce, 1977
9. C.E. Shannon, Communications Theory of Secrecy Systems, *Bell System Technical Journal*, vol. 28, no. 10, pp. 656-715, October 1949
10. W. Tuchman, Hellman Presents No Shortcut Solutions to DES, *IEEE Spectrum*, v. 16, n. 7, pp. 40-41, July 1979
11. P.C. van Oorschot and M.J. Weiner, A Known-Plaintext Attack on Two-Key Triple Encryption, *Advances in Cryptology - EUROCRYPT '90 Proceedings*, Springer-Verlag, pp. 318-325, 1991

Advanced Encryption Standard

Draft Minimum Requirements and Evaluation Criteria

Abstract. This is the minute of a discussion held at the Fourth Fast Software Encryption Workshop, Haifa, Israel, on Monday January 20, 1997 from 15.30 to 16.30 on the NIST call for comments on the Advanced Encryption Standard proposal. The discussion was held in the presence of over 50 workshop participants from all over the world. These comments were collected during the discussion by Ross Anderson (the discussion chair), Bart Preneel, and Eli Biham, and then circulated by email to the participants who submitted a few further comments. The final draft was prepared by Ross Anderson.

General Comments

1. It was asked whether there should be a standard at all, or whether a diversity of algorithms might be safer and more adapted to applications. (This argument had been advanced by the NSA in opposition to the adoption of triple DES as a standard.) The counterarguments were

    (a) that a standard would be adopted whether we like it or not and we might as well help make it a good one

    (b) for due diligence reasons, many clients would only use an algorithm with a government seal of approval

    (c) that a new standard would give an opportunity for many existing systems to be redeveloped and serious vulnerabilities in protocols etc removed

    (d) that a new standard would concentrate cryptanalytic effort on a single target, which (if unsuccessful) would increase confidence in that target

    (e) that the AES initiative presented an opportunity to establish a standard supported from the outset by government, industry and the academy.

2. Public trust in the algorithm will be harder to build if the rationale behind design decisions is not made fully public, and if the public does not participate in the evaluation process. So the rationale behind all design decisions should be completely explicit.

3. It would be helpful if any S-boxes, constants etc should be chosen by some convincing method (such as at random from a sufficiently large space). There are two reasons for this. Firstly, if all the design choices are made by a single person or organization, then the algorithm will be less likely to be trusted; trapdoors will be suspected. On the other hand, we do not want a "committee" design. A customisable design is probably the best balance between these concerns. Secondly, there are users who will want to customise a standard algorithm (see 11 below).

4. We would favour a process in which the initial submissions are whittled down to a short list of perhaps 3-4 candidates. This would enable the community

to concentrate the analysis and evaluation effort on them rather than dispersing it on dozens of targets. (In this workshop alone about ten ciphers were suggested.)

5. NIST should clarify the role of non-US citizens. Clearly, a new US standard will (like DES) become widely used in other countries. Will non-US submissions be acceptable?

6. There is concern that the proposed timetable does not leave enough time for serious cryptanalysis.

General Requirements

7. It is not clear that one cipher can satisfy the requirements for all applications, and on all kinds of processors (or special hardware). The question arise whether we should have a family of ciphers, appropriate for different environments.

   For example, the majority of fielded DES implementations are on 8-bit processors such as smartcards and microcontrollers, and used in applications such as banking, power metering, pay-TV key management, door locks, road tolls and the like. In such applications, the main 'improvement' sought from a DES successor is a reduction in code size.

   On the other hand, the importance of intellectual property protection is growing and there is wide use of stream ciphers in, for example, pay-TV systems. Here, speed is a definite requirement and code size is relatively unimportant. So NIST should consider whether there should be two standards: a block cipher suitable for 8-bit processors, and a stream cipher optimised for speed.

8. There was wide condemnation of the draft proposal, that C source code be evaluated on a PC. Ideally, a survey of applications, both fielded and planned, should be undertaken so that the relative importance of different performance metrics (speed, code size, etc) could be evaluated and a realistic benchmark suite be specified. At the very least, NIST should be much more explicit about the performance requirements. We expand on this below.

9. NIST should also provide a ranking for the various evaluation criteria to clarify their relative importance.

Technical Requirements

10. There should be procedures agreed in advance for dealing with any weakness of the algorithm that arises later. This might be predictable, such as an advance in chip technology that makes a longer key necessary; unpredictable but minor, such as the discovery of a new but rare class of weak keys; or catastrophic, such as a new shortcut attack that forces a change to a completely different algorithm.

    Several mechanisms are thus likely to be necessary including a review body or process, a 'backup algorithm' and perhaps (as suggested by NIST) a means of increasing the keylength. There was no unanimity on this last point however; an alternative would be to adopt an algorithm with a keysize well beyond possible exhaustive search (e.g., 256 bits) and use part of the keyspace as appropriate.

One possible 'backup algorithm' is using the same algorithm with different parameters, such as with a different set of S boxes. This could provide a rapid and low-cost means of recovering from all but a total break.

11. There are other reasons to support customization by other means than the key. In addition to the building public confidence in the absence of trapdoors, as mentioned above, parametrisation will appeal to those users who want a compromise between a proprietary algorithm and a standard one - such as those who at present use DES with nonstandard S-boxes or other modifications to prevent keysearch. The successor to DES should be chosen so that it is not as difficult to choose strong values of the S-boxes or other constants as it is in the case of DES.

12. An increasing number of applications involve cryptographic authentication protocols (Kerberos being an example). Here, the 64-bit blocksize of DES is a disadvantage; the real requirement is to encrypt variable length blocks. Many implementers use DES-CBC but this can be vulnerable to cut and paste attacks. A block cipher of variable width would be ideal for such applications.

13. Some people felt that a 64 bit blocksize was inadequate for security reasons, as once large volumes of data start to be encrypted the volume limits set by the birthday paradox may be approached.

14. Given that the algorithm may be of variable width and may also have a variable key length, thought needs to be given on how such parameters will be securely expressed. The RC5 approach of packaging the key in a 'control block' with such parameters might provide inspiration here, as could the IBM approach of 'key control vectors' to enforce a functional partition of the keyspace where applications require this. We probably need an algorithm version number as well, and 'fields to be defined later'.

15. In the event that the standardized algorithm is simply another 64-bit block cipher, there is a need for a standard mode of operation that allows a variable length block to be encrypted with error extension in both directions. More generally, it is time to look not just at modes of operation but also at other supporting structures such as APIs and lower level interface definitions.

16. The algorithm should approximate to a random permutation as closely as possible, e.g. there should be no equivalent keys, no complementation properties, no related keys and no weak keys.

17. The bit naming convention should be explicitly defined.

Security Requirements

18. The types of attacks that the cipher must withstand must be made explicit (e.g., known plaintext, chosen plaintext, adaptive chosen plaintext/ciphertext, related-key).

19. The security targets must be quantified, e.g. '2^{10} related key queries, 2^{40} chosen plaintexts, 2^{50} storage, 2^{60} known plaintexts, 2^{80} effort'.
20. There must be minimum values set for security parameters, such as number of rounds, block size and key size, in order to prevent loss of confidence in the standard following a published attack on a legitimate implementation.

Efficiency requirements

21. As noted above, it was widely felt to be unwise to evaluate the candidate algorithms solely on a PC, as the majority of DES implementations are believed to run on 8-bit processors in embedded applications. It appears to be prudent engineering practice to optimise an algorithm for the slowest processor on which it will be widely used - which might mean the 8051 (although 4-bit processors are still used, and GOST appears to have been designed with these in mind). It should also run adequately in Java, as the commercial success of this language cannot be ignored.

    PCs will be important, but we do not know whether the typical PC CPU in five years time will be a RISC processor such as Alpha, a VLIW processor such as Philips' TriMedia, or a combination superscalar/SIMD such as Klamath. Similarly, hardware/firmware implementations (FPGA, ASIC, standard cell,...) should be considered.
22. Some applications, such as B-ISDN require fast key setup. The evaluation criteria should therefore define a maximum key scheduling delay; this might defined relative to encryption as a function of key length. A possible alternative would be ability to cache a number of round keys. However, while 1024 keys might be sufficient for current ATM switches, more keys might be needed by future equipment.
23. There should be targets for code size and memory size, especially for implementations on smartcards and other 8-bit processors. For hardware implementations, there should be a target gate count; and for power-critical applications (such as contactless smartcards) there should be a power target of microjoules per block encrypted.

Evaluation and interface requirements

24. The process of evaluation should involve bounties to attract serious and sustained attack. It is suggested that NIST offer a large sum (say $1m) for a significant shortcut attack. This should ensure that anyone outside the sigint community who discovers such an attack will report it rather than seek to exploit it. The shorter term evaluation procedure should be also clarified: what incentives will there be for outside contributors to invest effort in it?
25. When reducing a large number of candidates to a shortlist, one possible approach to the performance issue would be to define a minimum speed relative to known ciphers such as DES or triple-DES. However, some participants felt

that many people are unaware of, or have no access to, fast DES code for comparison.

26. In any case, a thorough examination of the performance aspects of shortlisted candidates should be carried out. As mentioned above, there would ideally be a study of existing and planned applications leading to the development of a benchmark suite. In the absence of such an exercise, then at the very least the following should be considered for each shortlisted candidate:

    (a) code and memory size, especially on common smartcards and microcontrollers

    (b) speed, not just on currently common chips such as 8051 and Pentium but also RISC and VLIW chips

    (c) gate count for simplest and fully pipelined hardware implementations. Tradeoffs between speed and gate/count should be considered, as well as the minimum number of microjoules per block encrypted

    (d) whether software implementations are significantly different (or more difficult) according to whether the processor is big endian or little endian

    (e) key agility, or round key memory requirements if cacheing is preferred for B-ISDN applications

    (f) whether there is a well understood tradeoff between number of rounds and attack effort

27. NIST should define a standard interface for the algorithms in order to facilitate validation by the wider crypto community.

28. Ease of validation is important. A single test vector is not enough: the algorithm designer should supply a full set of test vectors, plus a validation suite that exercises them via the standard interface mentioned above and performs any other tests required to check all single points of failure and thus ensure that an implementation is correct.

29. Submissions should include not just one or more implementations optimized for speed or memory size on various processors but also an easy-to-read endian-indifferent one, so that correspondence with the description of the cipher can be readily checked.

30. Finally, the evaluation criteria should be more carefully drafted. For example, criteria (b), (c) and (d) overlap, and it is not clear what exactly is meant by 'simplicity' and 'flexibility'.

TWOPRIME: A Fast Stream Ciphering Algorithm

Cunsheng Ding[1], Valtteri Niemi[2], Ari Renvall[3], Arto Salomaa[3]

[1] Turku Centre for Computer Science, Datacity 4th floor, 20520 Turku, Finland
[2] Department of Mathematics & Statiscias, University of Vaasa, FIN-65101 Vaasa, Finland
[3] Department of Mathematics, University of Turku, FIN-20014 Turku, Finland

Abstract. In this paper, we describe an additive stream ciphering algorithm, called "TWOPRIME". It is designed for 32-bit computers, and the key has 128 bits. It is fast in software and analytical in the sense that some security aspects of the algorithm can be controlled. A faster version of TWOPRIME is also presented. We also describe a variant of TWOPRIME, called ONEPRIME, which is for 64-bit machines.

1   Introduction

There are a number of ciphering algorithms, but most of them are limited to a few models such as the Feistel network. The purpose of this paper is to describe a ciphering algorithm that is different to known ones in the public domain with the following properties:

- its structure should be special;
- the key size should be large enough;
- it is for 32-bit computers;
- it should work on blocks of bytes;
- it should be fast in software;
- it should be analytical;
- it should be easily modified for 64-bit computers;
- it is expected to be secure.

TWOPRIME is an additive synchronous stream cipher that is nonproprietary. It has a 128-bit key, and the encryption and decryption algorithms are the same. A C code runs at 1 Mbytes/sec on a Pentium (75 MHz). It should be faster than a number of fast ciphers in the public domain (but looks slower than SEAL [8]). The main feature of our algorithm is that the period, linear complexity, and distributions of some patterns of the keystream sequences are controllable. Compared with other stream ciphers, TWOPRIME seems analytical.

In this paper, we also present a faster version of TWOPRIME. A C code of this faster version runs at 1.3 Mbytes/sec on the same Pentium. As a variant of TWOPRIME, we describe the ONEPRIME ciphering algorithm for 64-bit computers, which should be much faster than TWOPRIME.

2   TWOPRIME: a description

This additive stream cipher works on blocks of 8 bytes. The keystream generator produces an 8-byte keystream block at each time unit, and this keystream block is then bytewise exored with the 8-byte block of input.

The key of the algorithm has 16 bytes, denoted by $k_0 k_1 \cdots k_{15}$, which are divided into four parts. Let

$$K_0 = k_8 + k_9 2^8 + k_{10} 2^{16} + k_{11} 2^{24},$$
$$K_1 = k_{12} + k_{13} 2^8 + k_{14} 2^{16} + k_{15} 2^{24},$$
$$K_2 = (k_0, k_1, k_2, k_3),$$
$$K_3 = (k_4, k_5, k_6, k_7).$$

The algorithm has ten layers. The first layer consists of two (p, a) cyclic counters. A (p, a) *cyclic counter* has an internal register that can store any integer between 0 and $p - 1$, thus the register has $\lceil \log_2 p \rceil$ bits of memory. The initial value of the register is an integer k, where $0 \le k \le p - 1$. The value of the register at time unit i is defined to be

$$r_i = (ai + k \bmod p),$$

where $z \bmod m$ denotes the least nonnegative integer that is congruent to z modulo m, where $\gcd(a, p) = 1$. When $a = 1$, it counts the numbers $k, (k + 1) \bmod p, \cdots, (k + p - 1 \bmod p)$ cyclically. When $a \ne 1 \bmod p$, we call it a cyclic counter with step a and period p, in other words, a (p, a) cyclic counter.

In the two (p_i, a_i) cyclic counters, p_0 and p_1 are two distinct primes having 32 bits, and a_0 and a_1 are two constants between 0 and $p_i - 1$ respectively. The largest two 32-bit primes are $p_1 = 4294967291$ and $p_0 = 4294967279$. Note that

$$p_1 - 1 = 2 \times 5 \times 19 \times 22605091, \quad p_0 - 1 = 2 \times 7 \times 17 \times 18046081.$$

It has been computed that

$$
\begin{aligned}
(2^{18046081} \bmod p_0) &= 1145473156, \\
(2^{7 \times 18046081} \bmod p_0) &= 2366705928, \\
(2^{17 \times 18046081} \bmod p_0) &= 1145324610, \\
(2^{(p_1-1)/2} \bmod p_0) &= 1, \\
(2^{7 \times 17} \bmod p_0) &= 2558525593.
\end{aligned}
$$

Hence the order of 2 modulo p_0 is

$$\mathrm{ord}_{p_0}(2) = (p_0 - 1)/2 = 2147483639 = 2^{31} - 9.$$

We have also computed that

$$
\begin{aligned}
(2^{22605091} \bmod p_1) &= 3079820090, \\
(2^{5 \times 22605091} \bmod p_1) &= 3786472082, \\
(2^{19 \times 22605091} \bmod p_1) &= 1304151046, \\
(2^{(p_1-1)/2} \bmod p_1) &= p_1 - 1, \\
(2^{5 \times 19} \bmod p_1) &= 2147483708.
\end{aligned}
$$

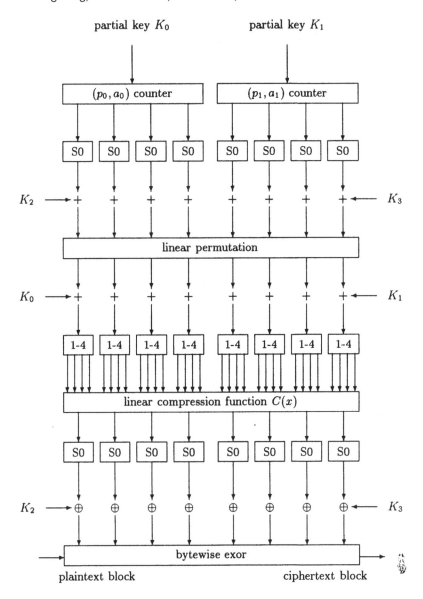

Fig. 1. Structure of the ciphering algorithm.

Hence

$$\mathrm{ord}_{p_1}(2) = p_1 - 1 = 2^{32} - 6.$$

Thus, 2 is a primitive root modulo p_1. We have seen that the two primes are almost equal to $2^{32} - 1$ such that the orders of 2 modulo them are close to 2^{31}.

The two constants a_i are chosen such that

1. $a_i > (p_i - 1)/2$ for $i = 0$ and 1, in order that in every two consecutive two updatings of the two registers of the two cyclic counters in the keystream generator there is one modulo-p_i reduction;
2. they should be different and the difference should be large enough;
3. each a_i should not be too close to p_i;
4. they are primes (this leads to $\gcd(a_1, a_2) = 1$).

Basing on the above considerations we suggest the following two constants:

$$a_0 = 2345986071, \quad a_1 = 3124567807.$$

Of course, there are many such choices for the constants a_i.

The second layer of the algorithm consists of eight S-boxes $S0$ that is a permutation of Z_{256} with good nonlinearity with respect to the addition of the residue class ring Z_{256}. The first, second, third, and fourth bytes of the contents of the two registers of the two cyclic counters are used as the inputs of the eight S-boxes $S0$. This is the first nonlinear layer.

The nonlinear permutation $S0$ is defined by

$$S0(x) = [(x^{255} \bmod 257) \bmod 256], \quad x \in Z_{256}.$$

The permutation $x^{255} = x^{-1}$ has good nonlinearity with respect to the addition of Z_{257} (It is well-known that X^{-1} is a permutation of finite fields with good nonlinearity, and it has been used in the cipher Shark [6]). Computation proves that the above permutation $S0$ has also good nonlinearity with respect to the addition of Z_{256}. The approach to finding a good nonlinear permutation of Z_{256} follows that used by Massey [5].

The third layer is the bytewise addition of the outputs of the second round and the partial keys K_2 and K_3. The outputs of the first four S-boxes $S0$ are added to the four bytes of K_2, and those of the second four S-Boxes S_0 are added to the four bytes of K_3, where all additions are that of Z_{256}, i.e., integer addition modulo 256.

The fourth layer is a linear one that is for diffusion of the key. Let $X_0, \cdots, X_7$ be the eight input bytes of this layer, its eight output bytes are defined by

$$Y_j = \sum_{i=0}^{7} X_i - X_j, \ j = 0, 1, \cdots, 7, \tag{1}$$

where " $+$ " and " $-$ " denote the addition and subtraction of Z_{256}. It is clear that the change of one byte leads to a change of seven of the eight output bytes of this layer. This is a relatively expensive layer, compared with others.

The fifth layer is again a key-addition layer, but this time the partial keys K_0 and K_1 are added. This makes it difficult to find some key-equivalence classes, by which we mean that they determine the same encryption transformation.

The sixth layer is nonlinear and also for data expansion. It has eight-byte inputs, but 32-byte outputs. Each box containing a symbol 1-4 denotes an array of four S-boxes in the order S1, S2, S3, S4. The four S-boxes are defined by

$$S1(x) = [x^3 \bmod 257] \bmod 256,$$
$$S2(x) = [x^{171} \bmod 257] \bmod 256,$$
$$S3(x) = [45^x \bmod 257] \bmod 256,$$
$$S4(x) = \begin{cases} [\log_{45} x \bmod 257] \bmod 256, & \text{if } x \neq 0; \\ 128, & \text{if } x = 0. \end{cases}$$

S3 and S4 are the two S-boxes used in SAFER [5]. As far as nonlinearity is concerned, S3 and S4 are the best nonlinear permutations of Z_{256} with respect to the addition of Z_{256}. S1 and S2 have also good nonlinearity, but not as good as S3 and S4. In fact, S1 and S2 have the same nonlinearity as S0. However, it should be mentioned that the nonlinearity with respect to the bytewise exor of S1 and S2 is much better than that of S3 and S4. One should be careful to use S3 and S4 with respect to bytewise exor operation. This is seen below.

The *nonlinearity* or *differentiality* of a permutation $P(x)$ of Z_{256} with respect to the addition of Z_{256} is measured by the probability

$$\Pr(P(x + a) - P(x) = b).$$

When $a = 0$, this probability is 1 or 0 and it is not interesting in any attack. So we are only interested in the case $a \neq 0$. Note that if $P(x)$ is a permutation, the equation $P(x + a) - P(x) = 0$ has no solution. So for any fixed $a \neq 0$ we have

$$\max_{b \neq 0} \Pr(P(x + a) - P(x) = b) \geq 2/256 = 1/128.$$

Hence

$$\frac{1}{128} \leq \max_{b \neq 0} \Pr(P(x + a) - P(x) = b) \leq \\ \leq \max_{a \neq 0} \max_{b \neq 0} \Pr(P(x + a) - P(x) = b). \qquad (2)$$

For some cryptographic applications the smaller the $\max_{a \neq 0} \max_{b \neq 0} \Pr(P(x + a) - P(x) = b)$ the better the security with respect to some attacks. If the two equalities of (2) hold, we say that the permutation $P(x)$ has the best nonlinearity or differentiality with respect to the addition of Z_{256}.

The two permutations $S3(x)$ and $S4(x)$ have the best nonlinearity or differentiality with respect to the addition of Z_{256}. This has been proved by a C program.

It should be noted that the two permutations $S3(x)$ and $S4(x)$ have bad nonlinearity or differentiality with respect to the bytewise EXOR operation. This is justified by the following formulae:

$$\Pr(S3(x \oplus 128) \oplus S3(x) = b) = \begin{cases} \frac{1}{2}, & b = 253; \\ \frac{1}{4}, & b = 249; \\ \frac{1}{8}, & b = 241; \\ \frac{1}{16}, & b = 225; \\ \frac{1}{32}, & b = 213; \\ \frac{1}{64}, & b = 129; \\ \frac{1}{64}, & b = 1; \\ 0; & \text{otherwise.} \end{cases}$$

Also we have

$$\Pr(S3(x \oplus 64) \oplus S3(x) = b) \in \left\{ \frac{22}{256}, \frac{18}{256}, \frac{10}{256}, \frac{8}{256}, \frac{6}{256}, \frac{4}{256}, \frac{2}{256}, 0 \right\},$$

where "$\oplus$" is the bytewise exor.

These formulae show clearly that the two permutations $S3(x)$ and $S4(x)$ have bad nonlinearity or differentiality with respect to the bytewise exor. Thus, one has to be careful in using the two permutations. However, in our algorithm we have used the two permutations only with respect to the addition of Z_{256}.

The seventh layer is a linear compression one, which has 32-byte inputs and eight-byte outputs. We denote the inputs of this layer from the left to the right by $X_0, X_1, \cdots, X_{31}$, and the outputs from the left to the right by $Y_0, Y_1, \cdots, Y_7$. Then the linear compression function $C(x)$ is defined by

$$\begin{cases} Y_0 = X_0 + X_5 + X_{10} + X_{15} + X_{16} + X_{22} + X_{24} + X_{30}, \\ Y_1 = X_1 + X_6 + X_{11} + X_{12} + X_{17} + X_{23} + X_{25} + X_{31}, \\ Y_2 = X_2 + X_7 + X_8 + X_{13} + X_{18} + X_{20} + X_{26} + X_{28}, \\ Y_3 = X_3 + X_4 + X_9 + X_{14} + X_{19} + X_{21} + X_{27} + X_{29}, \\ Y_4 = X_{16} + X_{21} + X_{26} + X_{31} + X_0 + X_6 + X_8 + X_{14}, \\ Y_5 = X_{17} + X_{22} + X_{27} + X_{28} + X_5 + X_{11} + X_{13} + X_3, \\ Y_6 = X_{18} + X_{23} + X_{24} + X_{29} + X_{10} + X_{12} + X_2 + X_4, \\ Y_7 = X_{19} + X_{20} + X_{25} + X_{30} + X_{15} + X_1 + X_7 + X_9, \end{cases} \tag{3}$$

Thus, every output byte depends on 8 input bytes, and every input byte affects two output bytes. This linear function is mainly for compression, but it also plays an important role in diffusion. The data expansion and compression are designed to prevent one from inverting the whole system backwards. This also makes each output byte dependable on as many S-boxes and inputs of the expansion layer as possible.

The eighth layer is a nonlinear one, where eight S-boxes S0 are applied. The ninth layer is again a key-addition one, but here the addition is bytewise exor. This is also designed to prevent one from going backwards to the front of the keystream generator. The last layer is the bytewise exor of the keystream block and the plaintext block.

3   Theoretical results about the algorithm

Let R be a commutative ring with multiplicative identity 1, and

$$s^N = s_0 s_1 \cdots s_{N-1}$$

be a sequence of length N over R, where $s_i \in R$. If s^N satisfies a linear recurrence relation

$$s_i = a_1 s_{i-1} + a_2 s_{i-2} + \cdots + a_l s_{i-l}, \quad i \geq l, \ a_i \in R,$$

then there exists such a shortest linear recurrence relation, and the shortest l is called the linear complexity or linear span of the sequence and is denoted by $L(s^N)$.

If the linear complexity of a sequence over a field is l, then $2l$ successive characters of the sequence can be used to determine a linear recurrence relation of length l satisfied by the sequence with the Berlekamp-Massey algorithm [4], which has complexity $O(l^2)$. Thus, $2l$ successive characters of the sequence are sufficient to determine the whole sequence. Thus, sequences over fields for additive stream ciphers should have large linear complexity.

For sequences over Z_m, which is the ring $\{0, 1, ..., m-1\}$ with integer addition modulo m and multiplication modulo m, the Berlekamp-Massey algorithm does not work, but the Reeds-Sloane algorithm works. The latter is an analog of the Berlekamp-Massey algorithm, and it is also efficient [7]. Thus, it is necessary to control the linear complexity of sequences over Z_m for additive stream ciphering.

To prove some theoretical results about the algorithm, we need the following lemma [3]:

Lemma 1. *Let $N = p_1^{e_1} \cdots p_t^{e_t}$, where $p_1, \cdots, p_t$ are t pairwise distinct primes, q an integer such that $\gcd(q, N) = 1$. Then for each nonconstant sequence s^∞ of period N over $GF(q)$,*

$$L(s^\infty) \geq \min\{\mathrm{ord}_{p_1}(q), \cdots, \mathrm{ord}_{p_t}(q)\},$$

Where L denotes the linear complexity.

Proposition 2. *Concerning the keystream generator we have the following conclusions:*

1. *Each output sequence of bytes has least period p_0, p_1 or $p_0 p_1$.*
2. *Each output sequence of bytes over the ring Z_{256} has linear complexity at least $\min\{\mathrm{ord}_{p_0}(2), \mathrm{ord}_{p_1}(2)\} = 2^{31} - 9$.*
3. *The elements of Z_{256} are almost equally likely distributed in a cycle of each output sequence of bytes.*
4. *All the above conclusions hold for each output bit sequence.*

Proof: Note that the output sequence of the register of the (p_0, a_0) (resp. (p_1, a_1)) has least period p_0 (resp. p_1). Let X_1, X_2, X_3, X_4 be the four output bytes of the (p_0, a_0) cyclic counter at each time unit, and X_5, X_6, X_7, X_8 be those of the four output bytes of the (p_1, a_1) cyclic counter. It follows that the semi-infinite sequences X_i^∞ have least period p_0 for $i = 1, 2, 3, 4$, and p_1 for $i = 5, 6, 7, 8$.

Let $Y_1 = X_2 + X_3 + X_4 + X_5 + X_6 + X_7 + X_8 \bmod 256$. Then the semi-infinite sequence Y_1^∞ has period $p_0 p_1$. It follows that its least period must be one of $1, p_0, p_1, p_0 p_1$. Obviously, Y_1^∞ is not a constant sequence. Thus, its least period cannot be 1. Suppose that the least period of Y_1^∞ is p_0. Then the semi-infinite sequence $(X_5 + X_6 + X_7 + X_8)^\infty$ must have a period p_0, but it has a period p_1. This is impossible since p_0 and p_1 are distinct primes. Hence, the semi-infinite sequence Y_1^∞ must have least period $p_0 p_1$. The same conclusion holds for Y_i^∞, where $2 \leq i \leq 8$. Since each output byte (bit) sequence cannot be a constant sequence, the least period of each output sequence should be one of $\{p_0, p_1, p_0p_1\}$.

We have already proven that each output bit sequence has a period (not necessary the least one) p_0p_1. By Lemma 1, the linear complexity of each output bit sequence is at least $\min\{\text{ord}_{p_0}(2), \text{ord}_{p_1}(2)\}$.

Let $z_1^\infty = Z_1^\infty \bmod 2$, where Z_1^∞ is the output byte sequence of the first output byte position of the keystream generator, it is easily seen that the linear complexity of the semi-infinite sequence z_1^∞ over Z_{256} is no less than that of Z_1^∞. Thus, we have proved the second claim.

If $p_0 = p_1 = 2^{32}$, then each X_i takes on elements of Z_{256} equally likely, so does each Y_i. However, since $p_0 = 2^{32} - 17$ and $p_1 = 2^{32} - 5$, each output byte Y_i takes on elements of Z_{256} almost equally likely. Since each layer is either a permutation layer or a linear layer, each keystream byte sequence has an almost equally likely distribution of the elements of Z_{256}, so each bit sequence of the keystream block sequence has an almost equally likely distribution of ones and zeroes.   □

Remark: It should be extremely unlikely that the least period of a byte (bit) sequence is p_0 or p_1.

4   Security arguments

A cipher must be secure against ciphertext-only attacks if it is secure against known-plaintext attacks. So in the sequel we shall argue some security aspects of the algorithm only with respect to some known plaintext attacks. When doing so, we assume the cryptanalyst has sufficiently many keystream blocks. Similar to other practical ciphers, it is hard to prove the security of a ciphering algorithm since we cannot sort out all possible attacks on a cipher.

With respect to brute-force attack

An attack that applies to every cipher is the brute-force attack by trying all possible keys. Since the number of possible keys of our ciphering algorithm is 2^{128}, this attack should not work. On the other hand, it might be possible that a number of keys determine the same encryption transformation, but we do not see a way to prove their existence, let alone to determine them if they exist.

With respect to linear complexity attacks

Since this is an additive synchronous stream cipher, it is necessary to control the least period (cycle length) of the keystream sequences and its component bit sequences. As proved before, the least period of the output sequence and its component bit sequences all have least period $> \min\{p_0, p_1\}$, and the linear complexities of the output sequence and its component bit sequences are at least

$$\min\{\text{ord}_{p_0}(2), \text{ord}_{p_1}(2)\} = 2^{31} - 9.$$

Thus, any attack based on the Berlekamp-Massey algorithm [4] or Reeds-Sloane algorithm [7] should not work. We can also prove that the linear complexities of the output sequence and its component bit sequences have ideal stability, thus, it is hard to construct an LFSR to approximate the output sequence of the generator [2].

With respect to inverting attacks

One basic question is whether this key-stream generator is invertible. Except the data-expansion layer and the data-compression layer, all other layers are permutation layers, when the key is fixed. But without the key it could be impossible to invert the keystream generator.

Let $W = (W_0, W_1, \cdots, W_7)$ be a keystream block and let $Y = (Y_0, Y_1, \cdots, Y_7)$ be the corresponding output of the linear compression layer. Then we have

$$W_i = k_i \oplus S0[Y_i], \quad i = 0, 1, \cdots, 7,$$

where the addition is the bytewise exor. Assuming that the key is randomly chosen, the information about W provided by the keystream block is zero. Thus, it is impossible to use one keystream block to go backwards.

Note that the last layer of the keystream generator is linear with respect to bytewise exor, one may consider the difference of two keystream blocks, in order to get rid of the partial keys $K_2 K_3$ added. We now analyze what we can get from this idea.

Let $W = (W_0, \cdots, W_7)$ and $W' = (W'_0, \cdots, W'_7)$ be two keystream blocks at time t and time t', and let $(Z_0, \cdots, Z_7)$ and $(Z'_0, \cdots, Z'_7)$ be the corresponding inputs of the data-expansion layer. Then

$$
\begin{aligned}
W_0 \oplus W'_0 = {}& S0[S1(Z_0) + S2(Z_1) + S3(Z_2) + S4(Z_3) + \\
& S1(Z_4) + S3(Z_5) + S1(Z_6) + S3(Z_7)] \oplus \\
& S0[S1(Z'_0) + S2(Z'_1) + S3(Z'_2) + S4(Z'_3) + \\
& S1(Z'_4) + S3(Z'_5) + S1(Z'_6) + S3(Z'_7)], \\
W_1 \oplus W'_1 = {}& S0[S2(Z_0) + S3(Z_1) + S4(Z_2) + S1(Z_3) + \\
& S2(Z_4) + S4(Z_5) + S2(Z_6) + S4(Z_7)] \oplus \\
& S0[S2(Z'_0) + S3(Z'_1) + S4(Z'_2) + S1(Z'_3) + \\
& S2(Z'_4) + S4(Z'_5) + S2(Z'_6) + S4(Z'_7)], \\
W_2 \oplus W'_2 = {}& S0[S3(Z_0) + S4(Z_1) + S1(Z_2) + S2(Z_3) + \\
& S3(Z_4) + S1(Z_5) + S3(Z_6) + S1(Z_7)] \oplus \\
& S0[S3(Z'_0) + S4(Z'_1) + S1(Z'_2) + S2(Z'_3) + \\
& S3(Z'_4) + S1(Z'_5) + S3(Z'_6) + S1(Z'_7)], \\
W_3 \oplus W'_3 = {}& S0[S4(Z_0) + S1(Z_1) + S2(Z_2) + S3(Z_3) + \\
& S4(Z_4) + S2(Z_5) + S4(Z_6) + S2(Z_7)] \oplus \\
& S0[S4(Z'_0) + S1(Z'_1) + S2(Z'_2) + S3(Z'_3) + \\
& S4(Z'_4) + S2(Z'_5) + S4(Z'_6) + S2(Z'_7)],
\end{aligned}
$$

$$W_4 \oplus W_4' = S0[S1(Z_4) + S2(Z_5) + S3(Z_6) + S4(Z_7)+$$
$$S1(Z_0) + S3(Z_1) + S1(Z_2) + S3(Z_3)]\oplus$$
$$S0[S1(Z_4') + S2(Z_5') + S3(Z_6') + S4(Z_7')+$$
$$S1(Z_0') + S3(Z_1') + S1(Z_2') + S3(Z_3')]],$$
$$W_5 \oplus W_5' = S0[S2(Z_4) + S3(Z_5) + S4(Z_5) + S1(Z_5)+$$
$$S2(Z_1) + S4(Z_2) + S2(Z_3) + S4(Z_0)]\oplus$$
$$S0[S2(Z_4') + S3(Z_5') + S4(Z_6') + S1(Z_7')+$$
$$S2(Z_1') + S4(Z_2') + S2(Z_3') + S4(Z_0')],$$
$$W_6 \oplus W_6' = S0[S3(Z_4) + S4(Z_5) + S1(Z_6) + S2(Z_7)+$$
$$S3(Z_2) + S1(Z_3) + S3(Z_0) + S1(Z_1)]\oplus$$
$$S0[S3(Z_4') + S4(Z_5') + S1(Z_6') + S2(Z_7')+$$
$$S3(Z_2') + S1(Z_3') + S3(Z_0') + S1(Z_1')],$$
$$W_7 \oplus W_7' = S0[S4(Z_4) + S1(Z_5) + S2(Z_6) + S3(Z_7)+$$
$$S4(Z_3) + S2(Z_0) + S4(Z_1) + S1(Z_2)]\oplus$$
$$S0[S4(Z_4') + S1(Z_5') + S2(Z_6') + S3(Z_7')+$$
$$S4(Z_3') + S2(Z_0') + S4(Z_1') + S1(Z_2')]$$

What we can hope is to find the eight Z_i and eight Z_i'. To this end, we may have to use the above set of equations. Due to the first nonlinear layer and the modulo p_0 and p_1 operations in the two cyclic counters, the relation between Z_i and Z_i' should be nonlinear, involving also 16 unknowns k_i. The above set of 8 equations has 16 unknowns Z_i and Z_i', and they are highly nonlinear. We do not see an easy way to solve this set of equations. The brute-force method needs 2^{128} tries. On the other hand, even if one could find the solutions, the number of solutions could be too large. Thus, the difference approach can get rid of the last key-addition layer, but may not go further.

With respect to correlation attacks

There are different kinds of correlation attacks, such as those in [1, 9], but they are only for special stream ciphers. It is impossible to sort out all correlation attacks on a system, but the essential idea of a correlation attack would be to find a relation between output keystream characters and some part or whole of the key, or a relation between some intermediate variables. The purpose of finding such a relation is to get information about the key from known keystream blocks. A way to protect a stream cipher from correlation attacks is to use correlation-immune functions in the system in a proper way.

There are mainly two layers that could protect the system from such an attack. The first is the linear permutation layer described by (1), where each function is correlation-immune of order 6 [10]. The second is the linear compression layer described by (3), where each function is correlation immune of order 7. These correlation-immune functions and layers are expected to protect the cipher from correlation-immune attacks.

With respect to affine approximation attacks

The idea of an affine approximation attack would be to use the best affine approximation (BAA) of some nonlinear components of the system to replace the nonlinear parts, in order to construct a pseudo-keystream generator which produces an output sequence that matches the original keystream sequence with high probability or to recover the key of the original keystream generator. Such an attack, carried out for two kinds of stream ciphers in [2], should not work on our algorithm, due to the high nonlinearity of the S-boxes and the two diffusion layers.

The most reasonable affine approximation of the nonlinear S-boxes is to use affine functions $ax + b$ over Z_{256} to approximate the five S-boxes. However, with a simple C program we have obtained the following result.

Proposition 3. *Let* Pr *denote the probability. Then*

$$\max_{0 \leq a \leq 255} \max_{0 \leq b \leq 255} \Pr(S0(x) = ax + b) = \frac{2}{256}$$

$$\max_{0 \leq a \leq 255} \max_{0 \leq b \leq 255} \Pr(S1(x) = ax + b) = \frac{3}{256}$$

$$\max_{0 \leq a \leq 255} \max_{0 \leq b \leq 255} \Pr(S2(x) = ax + b) = \frac{3}{256}$$

$$\max_{0 \leq a \leq 255} \max_{0 \leq b \leq 255} \Pr(S3(x) = ax + b) = \frac{3}{256}$$

$$\max_{0 \leq a \leq 255} \max_{0 \leq b \leq 255} \Pr(S4(x) = ax + b) = \frac{4}{256}.$$

This result shows that affine approximations of the S-boxes with $ax + b$ over Z_{256} are very poor. Clearly, every permutation $P(x)$ of Z_{256} can be identified as a permutation $P'(y)$ of Z_2^8, and one might therefore be concerned with the affine approximation of $P'(y)$ with respect to Boolean affine functions over Z_2^8. However, as our operations in the ciphering algorithm are almost totally based on those of Z_{256}, such an affine approximation might not work.

5   Performance of the algorithm

On a Pentium (75 MHz) an initial C code (Borland C++ compiler, version 1991) of the TWOPRIME runs at 1 Mbyte/sec. The test is done with a self-feeding 4 Mbyte input data. An optimized code should run faster.

6   TWOPRIME-1: A faster version

It is noted that good linear layers within this ciphering structure is relatively expensive. One way to get a faster version of this algorithm is to use the following simpler linear compression function $C(x)$, but keep other parts the same as before:

$$\begin{cases}
Y_0 = X_0 + X_5 + X_{10} + X_{15}, \\
Y_1 = X_1 + X_6 + X_{11} + X_{12}, \\
Y_2 = X_2 + X_7 + X_8 + X_{13}, \\
Y_3 = X_3 + X_4 + X_9 + X_{14}, \\
Y_4 = X_{16} + X_{21} + X_{26} + X_{31}, \\
Y_5 = X_{17} + X_{22} + X_{27} + X_{28}, \\
Y_6 = X_{18} + X_{23} + X_{24} + X_{29}, \\
Y_7 = X_{19} + X_{20} + X_{25} + X_{30}.
\end{cases}$$

This linear compression function is less powerful than the original one, and its component functions are correlation-immune of order 3. We would refer to this faster version as "TWOPRIME-1". A C code of TWOPRIME-1 runs at 1.3 Mbytes/sec on the same machine.

We do not see that TWOPRIME-1 is significantly weaker than TWOPRIME. In fact, we have not found a weakness of TWOPRIME-1, although it might have some.

7   ONEPRIME: A variant for 64-bit machines

A variant of TWOPRIME is the ONEPRIME for 64-bit machines. ONEPRIME has only one (p, a) cyclic counter, in which p is the closest 64-bit prime to $2^{64} - 1$ such that $\text{ord}_p(2) \geq 2^{32}$, and a is any prime that is approximately $3p/4$. Thus, only the first layer of TWOPRIME is modified and others remain the same. The content of the register of the (p, a) cyclic is similarly divided into 8 bytes which are used as the input of the next layer.

A possible choice of the prime p is

$$p = 18446744073709551557 = 2^{64} - 59.$$

Thus,

$$p - 1 = 2^2 \times 11 \times 137 \times 547 \times 5594472617641.$$

The order of 2 modulo p can be computed with a 64-bit computer, basing on this factorization.

We have about the same theoretical results for ONEPRIME, but have not tested ONEPRIME for performance. However it is clearly much faster than TWOPRIME, as the first layer of ONEPRIME is much faster than that of TWOPRIME.

The security of ONEPRIME and TWOPRIME should be at the same level. The choice between the two algorithms depends on the machines used. For 64-machines we would recommend ONEPRIME, as it is faster. We have emphasized on TWOPRIME, as 32-bit machines are mostly used for the time being.

8   Other versions

In some applications, the available memory is very limited. ONEPRIME, TWOPRIME, and their variants can be modified by using only one or two nonlinear S-boxes, instead of five. Accordingly, the linear expansion and compression layers should be modified too.

Acknowledgements: The authors would like to thank the three anonymous referees and Adi Shamir for their helpful comments.

References

1. R. J. Anderson, *Searching for optimum correlation attacks*, in Fast Software Encryption (B. Preneel, Ed.), LNCS 1008, 137-143.
2. C. Ding, G. Xiao and W. Shan, *The Stability Theory of Stream Ciphers*, LNCS 561, Springer-Verlag, 1991.
3. C. Ding, *Binary cyclotomic generators*, in Fast Software Encryption (B. Preneel, Ed.), LNCS 1008, 29-60.
4. J. L. Massey, *Shift-register synthesis and BCH decoding*, IEEE Trans. Inform. Theory, Vol. IT-15 (1969), 122-127.
5. J. L. Massey, *SAFER K-64: A byte-oriented block-ciphering algorithm*, in Fast Software Encryption (R. Anderson, Ed.), LNCS 809, Springer-Verlag, 1-17.
6. V. Rijmen, J. Daemen, B. Preneel, and E. De Win, *The cipher SHARK*, in Fast Software Encryption (D. Gollmann Ed.), LNCS 1039, Springer-Verlag, 99-112.
7. J. A. Reeds, N. J. A. Sloane, Shift-register synthesis (modulo m), SIAM J. Comput. 14, No. 3 (1985), 505-513.
8. P. Rogaway and D. Coppersmith, *A Software-Optimized Encryption Algorithm*, in Fast Software Encryption (R. Anderson, Ed.), LNCS 809, Springer-Verlag, 1994, 56-63.
9. T. Siegenthaler, *Decrypting a class of stream ciphers using ciphertext only*, IEEE Trans. Computers, Vol. C-034 (1984), 81-85.
10. T. Siegenthaler, *Correlation-immunity of nonlinear combining functions for cryptographic applications*, IEEE Trans. Information Theory, Vol. IT-30 (1984), 776-780.

Appendix A: Test values

key	=	0	0	0	0	0	0	0	0
		0	0	0	0	0	0	0	0
plaintext	=	0	0	0	0	0	0	0	0
ciphertext	=	67	67	42	228	183	104	38	130

key	=	1	1	1	1	1	1	1	1
		1	1	1	1	1	1	1	1
plaintext	=	0	0	0	0	0	0	0	0
ciphertext	=	21	166	110	21	135	244	34	80

key	=	1	2	3	4	5	6	7	8
		8	7	6	5	4	3	2	1
plaintext	=	0	0	0	0	0	0	0	0
ciphertext	=	100	145	30	150	24	92	140	185

Appendix B: C code of the algorithm

The following is a C code of the algorithm for testing the performance. The input is a 4-Mbyte self-feeding data, i.e., only 8-byte inputs are given and the next input block is the output of the previous block.

```
#include <stdio.h>
#include <time.h>
#define length 500000
#define m 255
#define F(x,y,z,w) (((x|(y≪8))|(z≪16))|(w≪24))
#define B(x,y,z) ((((x*y)%257)*z)%257)
#define G(x,y) ((x*y)%257)

main ()
{
register unsigned long int j, time, k0, k1, z1, z2;
register unsigned int i,z,u,
I[8]={0,0,0,0,0,0,0,0}; /* one plaintext block */
unsigned int S0[256],S1[256],S2[256],S3[256],S4[256],h[8],
k[16]={0,0,0,0,0,0,0,0,0,0,0,0,0,0,0,0}; /* key */
const unsigned long int p0=4294967279;
const unsigned long int p1=4294967291;
const unsigned long int a0=2234567891;
const unsigned long int a1=3211223331;
/* computing the seeds for the two registers of the cyclic counters */
k0=F(k[8],k[9],k[10],k[11]); k1=F(k[12],k[13],k[14],k[15]);

/* computation of the four S-boxes */
for(i =0; i < 256; ++i)
{
u=B(i,i,i); S1[i]=u&m; /* S1-box */
z=u=B(u,u,i); u=B(u,u,i); z=B(u,u,z);
u=B(u,u,i); u=B(u,u,i); u=B(u,u,i);
z=G(u,z); S0[i]=B(u,u,i)&m; /* S0-box */
u=B(i,i,i); u=B(u,u,i); S2[i]=G(u,z)&m; /* S2-box */
}
S3[0]=1; S4[1]=0;
for(i =1; i < 256; ++i)
```

```
{
S3[i]=(45*S3[i-1])%257; S4[S3[i]]=i;
}
S3[128]=0; S4[0]=128; S3[0]=1;

for(j=0; j < length; ++j) /* encrypt input "length" times for test */
{
k0+=a0; if(k0>=p0) k0-=p0;
k1+=a1; if(k1>=p1) k1-=p1;
h[0]=k0&m; h[1]=(k0≫8)&m; h[2]=(k0≫16)&m; h[3]=k0≫24;
h[4]=k1&m; h[5]=(k1≫8)&m; h[6]=(k1≫16)&m; h[7]=k1≫24;
u=S0[h[0]]+S0[h[1]]+S0[h[2]]+S0[h[3]]+ S0[h[4]]+S0[h[5]]+S0[h[6]]+S0[h[7]];
u=u+k[0]+k[1]+k[2]+k[3]+k[4]+k[5]+k[6]+k[7];
h[0]=(u-S0[h[0]]-k[0]+k[8])&m;
h[1]=(u-S0[h[1]]-k[1]+k[9])&m;
h[2]=(u-S0[h[2]]-k[2]+k[10])&m;
h[3]=(u-S0[h[3]]-k[3]+k[11])&m;
h[4]=(u-S0[h[4]]-k[4]+k[12])&m;
h[5]=(u-S0[h[5]]-k[5]+k[13])&m;
h[6]=(u-S0[h[6]]-k[6]+k[14])&m;
h[7]=(u-S0[h[7]]-k[7]+k[15])&m;
l[0]=S0[((S1[h[0]]+S2[h[1]]+S3[h[2]]+S4[h[3]]+
S1[h[4]]+S3[h[5]]+S1[h[6]]+S3[h[7]])&m]^k[0]^ l[0];
l[1]=S0[((S2[h[0]]+S3[h[1]]+S4[h[2]]+S1[h[3]]+
S2[h[4]]+S4[h[5]]+S2[h[6]]+S4[h[7]])&m]^k[1]^ l[1];
l[2]=S0[((S3[h[0]]+S4[h[1]]+S1[h[2]]+S2[h[3]]+
S3[h[4]]+S1[h[5]]+S3[h[6]]+S1[h[7]])&m]^k[2]^ l[2];
l[3]=S0[((S4[h[0]]+S1[h[1]]+S2[h[2]]+S3[h[3]]+
S4[h[4]]+S2[h[5]]+S4[h[6]]+S2[h[7]])&m]^k[3]^ l[3];
l[4]=S0[((S1[h[4]]+S2[h[5]]+S3[h[6]]+S4[h[7]]+
S1[h[0]]+S3[h[1]]+S1[h[2]]+S3[h[3]])&m]^k[4]^ l[4];
l[5]=S0[((S2[h[4]]+S3[h[5]]+S4[h[6]]+S1[h[7]]+
S2[h[1]]+S4[h[2]]+S2[h[3]]+S4[h[0]])&m]^k[5]^ l[5];
l[6]=S0[((S3[h[4]]+S4[h[5]]+S1[h[6]]+S2[h[7]]+
S3[h[2]]+S1[h[3]]+S3[h[0]]+S1[h[1]])&m]^k[6]^ l[6];
l[7]=S0[((S4[h[4]]+S1[h[5]]+S2[h[6]]+S3[h[7]]+
S4[h[3]]+S2[h[0]]+S4[h[1]]+S1[h[2]])&m]^k[7]^ l[7];
}
time=clock()/CLOCKS-PER-SEC;
printf("time= %6lu \n", time);
}
```

On Nonlinear Filter Generators

Markus Dichtl

Siemens Corporate Technology
Email: Markus.Dichtl@mchp.siemens.de

Abstract. In this paper the bits in a linear feedback shift register are treated as if they were independent random variables. A necessary condition for filter functions which result in independent random output bits is given. An example shows that the sufficient condition given by Golić in [2] is not necessary.

1 Nonlinear Filter Generators

It is a common technique to compute the output bits of a stream cipher by a nonlinear function applied to some stages of a linear feedback shift register. The nonlinear function is called filter function, and the generators are called nonlinear filter generators.

In [4] Rueppel gives an overview of known results on such generators, which mainly concern the linear complexity of the resulting bit stream. But surprisingly the consequences of the fact that the same bit is used in the input of the filter function repeatedly started to be discussed in the open literature only very recently. In 1994 Anderson ([1]) introduced a new attack on nonlinear filter generators and showed that it worked for some nonlinear filter functions used in practical applications. He suggested that the filter functions needed further study. This suggestion was taken up by Golić who treated the criteria nonlinear filter functions should meet extensively in [2]. Whereas Anderson had only discussed the nonlinear filter functions, Golić also pointed out the importance of the positions of the tabs for the input of the filter function. Nonlinear Filter Generators have also been discussed from another point of view by Lai and Massey in [3]. In their paper the term "Binary Filter with Input Memory" is used. The authors consider the question of delay-d invertibility, that is, whether it is possible to determine the input bits from the output bits with a delay of at most d clocks.

2 The Model

In this paper, we treat the bits of the shift register as if they were independent random variables with probability 1/2 of being 0 or 1. This assumption holds for the rest of this paper, even when it is not explicitly mentioned. This model was introduced in [1], but there the emphasis was on information leakage. In [2] the requirement for the nonlinear filter function was given, that under the assumptions of random bits in the shift register, the output bits must be independent random variables with probabilities of 1/2 of being 0 or 1. Golić also gave a sufficient condition for the filter function to meet this requirement, namely that the filter function is affine in the input from the leftmost or rightmost tab. Furthermore he conjectured that these are the only solutions which are independent of the choice of the tabs.

In this paper, we only consider the question of choosing the filter function in order to get independent random bits in the output. However, it should be noted that this is not sufficient for filtering the output of linear feedback shift registers. [2] gives a list of 9 requirements a nonlinear filter generator should meet.

3   A Necessary Condition

In this sections we give a condition which is necessary to produce independent random bits in the output of the filter function.

Remark. In order to get zeros and ones at the output of the filter function with probability $1/2$, the filter function must be balanced.

In spite of the name of the generator, we do not exclude linear filter functions f. The following theorem is also valid for them.

Theorem 1. *Let $f : GF(2)^n \longrightarrow GF(2)$ $(n \in \mathbb{N})$ be the filter function of a nonlinear filter generator. If the output of the filter functions are random independent bits with probability $1/2$ of being 0 and 1, then there exists at most one index j $(1 \leq j \leq n)$ such that the function $f_j : GF(2)^{n-1} \longrightarrow GF(2)$ with $f_j(x_1, \ldots, x_{j-1}, x_{j+1}, \ldots, x_n) = f(x_1, \ldots, x_{j-1}, 0, x_{j+1}, \ldots, x_n)$, that is the function for which the j-th input bit of f is fixed to 0, is not balanced.*

Proof. Let j and k be positive integers less or equal to n with $j \neq k$. Let p be the probability that f_j is zero, and q the probability that f_k is zero.

Each bit of the shift registers goes, at different times, to the inputs x_j and x_k of f. Let y_j be the output of f when such a bit b goes to x_j and y_k the output of f when the same bit b goes to x_k.

When b is zero, the probability of $y_j y_k$ to be 00 is pq. When b is one this probability is $(1 - p)(1 - q)$. The mean of the two probabilities must be $1/4$. This leads to $2p(2q - 1) = 2q - 1$. This implies $p = 1/2$ or $q = 1/2$, f_j or f_k is balanced.   □

4   Looking for Examples

Golić constructed filter functions of the form $f(x_1, \ldots, x_n) = x_1 + g(x_2, \ldots, x_n)$ where x_1 must come from the leftmost or rightmost tab of the shift register, and g is an arbitrary Boolean function. These filter functions give independent random output bits.

But are they the only functions to achieve this?

When we consider functions of three variables, and use tabs at three subsequent stages of the shift register, the answer (found by brute force search) is yes.

When we look at functions of four variables, and use tabs at four subsequent stages of the shift register, the answer is no.

The following table gives a Boolean function f of four variables which is not affine in any of the variables. f_2 is not balanced, the other f_j are, according to Theorem 1.

x_1 x_2 x_3 x_4	$f(x_1, x_2, x_3, x_4)$
0 0 0 0	0
0 0 0 1	0
0 0 1 0	0
0 0 1 1	0
0 1 0 0	1
0 1 0 1	1
0 1 1 0	1
0 1 1 1	1
1 0 0 0	0
1 0 0 1	0
1 0 1 0	1
1 0 1 1	0
1 1 0 0	1
1 1 0 1	1
1 1 1 0	0
1 1 1 1	1

That f with four subsequent tabs produces independent random output bits can be verified by applying Lemma 1 from [2].

There are choices of tabs for which f does not produce independent random output bits, so this example is not in conflict with Golić's conjecture, that his construction may give the only functions which work for all positions of the tabs.

As a matter of fact, for tabs at the stages of the shift register indexed 0, 1, 3, and 7, all filter functions with four inputs which give independent random output bits are according to the Golić construction.

5   Conclusion and Open Questions

In [2] Golić gave a sufficient condition for a filter function to produce independent random output bits. In this paper we gave a necessary condition. Of course it would be interesting to have a necessary and sufficient condition. Our observations from the previous section show that such a condition must involve the position of the tabs of the shift register.

Perhaps the following question is easier to solve: Are all filter functions which give independent random output bits for tabs which form a full positive difference set according to the Golić construction? This would imply that Golić's conjecture is true.

6   Acknowledgement

I would like to thank a referee for bringing [3] to my attention and for sending me a copy of this paper.

References

1. Anderson, R.: Searching for the Optimum Correlation Attack, in Fast Software Encryption, Second International Workshop, Leuven, Belgium, December 1994. Proceedings, Lecture Notes in Computer Science, vol 1008, B. Preneel (Ed.), Springer Verlag 1995, pp. 137–143.
2. Golić, J.: On the Security of Nonlinear Filter Generators, in Fast Software Encryption, Third International Workshop, Cambridge, UK, February 1996. Proceedings, Lecture Notes in Computer Science, vol 1039, D. Gollman (Ed.), Springer Verlag 1996, pp. 173–188.
3. Lai, X. and Massey, J.: Some Connections between Scramblers and Invertible Automata, Proc. 1988 Beijing Int. Workshop on Info. Theory, Beijing, China, July 4-8, 1988, DI-5.1 - DI-5.5
4. Simmons, G. (ed): Contemporary Cryptology: The Science of Information Integrity, IEEE Press 1992

Chameleon — A New Kind of Stream Cipher

Ross Anderson, Charalampos Manifavas

Cambridge University Computer Laboratory
Pembroke Street, Cambridge CB2 3QG, England
(rja14,cm213)@cl.cam.ac.uk

Abstract. Stream cipher systems are used to protect intellectual property in pay-TV and a number of other applications. In some of these, it would be convenient if a single ciphertext could be broadcast, and subscribers given slightly different deciphering keys that had the effect of producing slightly different plaintexts. In this way, a subscriber who illegally resold material licensed to him could be traced. Previously, such tracing could be done using a one-time pad, or with complicated key management schemes. In this paper we show how to endow any stream cipher with this potentially useful property. We also present a simple traitor tracing scheme based on random coding with which it can be used.

1 Introduction

The electronic distribution of intellectual property such as computer programs, clip art, databases, videos and music, often involves encryption followed by broadcast, with decryption keys being supplied out of band to subscribers who have paid for a particular object.

Computer programs and clip art are commonly distributed on CDs that contain extensive libraries, each item being typically encrypted using a different key. Customers purchase items by calling a service bureau and quoting a credit card number; a key is then read out to them over the phone. A number of firms sell encrypted databases: one is a compendium of building projects in certain counties of California, which is sold to building materials salesmen. Videos are broadcast encrypted on a number of satellite channels, and the decryption keys are sold to subscribers on smartcards.

A common problem with such systems is that some subscribers re-sell the information they have licensed. This is against the terms of their licence, and if they are detected they may be sued. Technical measures may also be used, such as failing to renew their encryption keys. However, given that the available technical measures are imperfect, with pay-TV pirates forging each successive generation of subscriber smartcard [5], and given that strong protection mechanisms are often in conflict with exportability and functionality, there is a shift towards combining technical protection with legal sanctions.

In any case, the important question is how cheaters can be detected.

One common approach is to customise the software as it is installed. Common techniques include inserting the licensee's name, giving a banner at the top of the screen stating something like 'This copy no. 123456 licensed to Bloggs the Butcher'. Another is to monitor the PC environment to detect re-installation, and a third is to have a timelock enforcing re-registration. However, all such mechanisms depend on 'security through obscurity' and can be broken by technically sophisticated pirates tampering with the software.

A second approach is to mark the information before it is encrypted. For example, a database supplier may mark each copy database in a unique way. Such 'fingerprints' have been in use for generations, having been used to mark mathematical tables and other early instances of intellectual property. (For a survey of fingerprinting, see [12].)

If manufacturing a unique database for each customer is too expensive, as it might be if the database is shipped initially on a CD-ROM, the supplier can use other techniques. For example, if he sends out a weekly update to subscribers, he can produce two different versions that differ slightly. By sending these two different versions to different partitions of his N subscribers in successive weeks, he can track down the cheater in $\log N$ weeks.

Whatever strategy is used to mark individual copies of the information, an attacker can always purchase a number of copies and compare them. Nonetheless, not all attackers are well organised, and it is often thought worthwhile to have mechanisms that ensure a certain minimum number of copies will have to be purchased. Matters can be arranged so that any captured pirate copy will correctly identify the subscriber who deciphered it, or — if up to a certain number of subscribers collude — it will correctly identify at least one of them, and will not mistakenly identify any innocent subscribers. This is known as 'Traitor Tracing' [6] and we will return to it below.

Several problems remain to be solved. Firstly, broadcasting more than one ciphertext is expensive and in many applications (such as satellite TV) it is impractical. So we may want there to be only one version of the ciphertext. Secondly, if we rely on software to insert the user's identity on decryption, then it is likely to be disassembled and interfered with by pirates. Even if we use 'trusted' hardware, this will be expensive and may be ultimately vulnerable to attack [5].

So we want a scheme that will enable us to give different keys to different subscribers, in such a way that they decrypt a single broadcast ciphertext in different ways.

The approach taken by [6] and a number of subsequent workers is to mark not the plaintext but a 'virtual key'. This decryption key is computed from a number of user keys; each user gets a sufficient but unique set of these keys, and matters are arranged so that a certain minimum number of users need to collude to construct a key that works but identifies none of them. One problem with this approach is that the bandwidth required for the control messages may not always be available.

If we could use a one-time pad, then we could just well each user a slightly different deciphering key, and they would end up with slightly different plaintexts. However, in applications such as the distribution of videos and music — where such a scheme would be most valuable — the amount of key material required would be prohibitive.

So it would be useful to have an encryption algorithm with the property that a slight change in the key will result in a slight change to the plaintext that is deciphered from a given ciphertext.

One might think that this would expose the cipher to divide-and-conquer attacks, as an attacker would be able to tell when a guess of the key was 'almost right'. But we show that this is not necessarily so. Any stream cipher can be modified simply so that a slight change in the key will cause a slight change to the output keystream. Yet, in practical cases of interest, the construction appears to strengthen rather than weaken the cipher.

2   The Construction

Our construction can be concisely described by a concrete example. We take a conventional pseudorandom generator (which in our prototype is the block cipher that forms the core of the 'Tiger' hash function, run in output feedback mode, rather than in feedforward mode as in the hash function) [4]. The particular choice is unimportant for our construction — we could as easily use any block cipher in output feedback mode, or a dedicated stream cipher such as PIKE [3]. The key for this stream cipher we will call key 'A'.

Next, we take a table of 2^{16} 64-bit words — 512 KB of random data — which we call key 'B'.

In order to encipher a 64-bit word of plaintext we take a 64 bit word from the keystream generator and use it to select four words from key 'B', which we exclusive or together. The result is the keystream; it is exclusive or'ed with the plaintext to get the ciphertext (and, when deciphering, with the ciphertext to get the plaintext).

The effect of a one-bit change in key 'B' is to change about 4 bits per 512KB of keystream generated. These changes are at the same locations in the word as those in the key; thus, when enciphering audio signals that have been digitised into 16-bit words, we can arrange that the copyright marks appear in the least significant bits.

3   Tracing Traitors

A common concern with systems that give intellectual property a unique mark for each subscriber is that a pirate may purchase, say, three copies of a work in different false names and then obtain an unmarked copy by using bitwise majority voting.

There are a number of strategies available to make such attacks more difficult. The basic idea is that for some small integer k, a pirate plaintext (or decoding device) should disclose the identity of at least one of up to k copyright violators who pooled their plaintexts (or secret keys), and that it should not be possible for an innocent subscriber to be framed [6].

These techniques give only lightweight protection in that they are effective only for small values of k. Indeed, Shamir has pointed out that these 'traitor tracing' schemes suffer from the problem that as k increases, the defender does exponentially more work in order to cost the attacker linearly more effort [11]. However there is usually little point in trying to guard against a large conspiracy, as an attacker who could organise it could also manage to subscribe in a false name.

So the realistic goal of traitor tracing is to provide a pragmatic defence against unsophisticated attackers, and in this spirit we offer a simpler way of implementing it than [6]. Our technique was inspired by [7].

In the concrete system given in the above section, with four lookups into a table of 4 megabits, assume that there are 4000 marked bits. Thus, as somewhat over the square root of the total number of bits are marked, we expect that any two users will have a marked bit in common, and that these common bits will be unique to each pair of users. Thus if any two subscribers collude, they will succeed in eliminating all but one of the marks from their 'B' keys, but the remaining mark (or its effects on the plaintext) will identify them.

So if three users collude and attempt to produce a clean copy by bitwise majority voting, the resulting text (or B key) will still incriminate each of them, two at a time, with high probability. Even if four users collude, they can identify the incriminating marks, but not figure out how to remove them. Thus our random coding approach gives us a simple traitor tracing scheme with $k = 4$.

How practical is this? Take for example an audio marking scheme. With 16 bit encoded uncompressed audio, we might want to limit the marks to the least significant bit of each 16-bit word. Thus the number of effective bits in the 'B' key is only 256K, so we need mark at most 1000 of them. This leads to the marking of 1.6% of the least significant bits, which is unnoticeable for most modern music. We will discuss an approach for video signals below.

More complicated marking schemes can be devised (e.g. [9, 10]) and used with our scheme. Our construction is independent of whether the marks on the 'B' key are randomly or systematically generated; the changes they induce in the keystream not only preserve the bit position in the word, but also incidence structures, which is what we generally need for traitor tracing schemes to work.

As with the somewhat different construction of [6], there is no need to penetrate the tamper resistance of a captured pirate decoder. Its behaviour is quite sufficient to identify the subscribers whose keys were used to construct it, assuming that this can be done at all.

4   Key Management

The 'A' keys are quite conventional and can be managed using the conventional machinery of crypto protocols. For example, the current mechanism in several pay-TV systems is to compute a working key as a MAC of all the control packets that have been transmitted in the previous time period. This is so that once a traitor (such as a cloned subscriber card) has been identified, a packet can be sent in each time period instructing that card to commit suicide. If a user blocks this instruction to prevent it reaching his smartcard, then this card cannot calculate the current key and the cloned card is thereby rendered useless. Such key management techniques can be adopted unchanged in the system proposed here.

Managing the 'B' key is more difficult. One might simply treat it as a long term key installed by out-of-band means; if it is used, together with a suitable 'A' key, to generate a lower level 'B' key, then this will have about four times as many marks in it as the long term key did. The possible advantage of having master and session 'B' keys is that re-keying might help discriminate between candidate conspiracies with a higher probability than otherwise. The exact probabilities, and thus the advantage if any, would depend on the parameters of a given application.

5   Performance

The performance degradation is not large, so long as the 'B' key remains in memory. This is the most critical parameter and it can be tuned to the equipment in use.

If the underlying pseudorandom generator is triple DES, then it is unlikely that our construction will add a significant penalty. Even if the generator is a high speed software algorithm, the penalty is not enormous. For example, when we use Tiger, running in output feedback mode on a 275MHz Alpha workstation, we can generate raw pseudorandom bits at 67 Mbps; and when using four lookups to a 512KB table, we still get 42 Mbps. We expect that this can be improved by careful optimisation.

In audio applications, performance is unlikely to be a problem; we can decrypt a minute's worth of music in about a second. Performance is only likely to be an issue in applications such as video, and especially where MPEG decoding places a high load on the processor. In such applications, a bit error rate lower than 0.1% may also be required; so the pragmatic approach is to mark only a subset of the content. One might for example process only one block in a hundred using our construction, and select this block using the native mode stream cipher (which would also be used to encipher the rest of the content).

By using higher density marks, one can construct schemes that are 6-resilient, 8-resilient and so on. The higher density of marking can be offset by marking a smaller subset of the content; however, the comments of section 3 still apply, and there is the further problem that if the marks are made too dense in any subset of the content whose selection is independent of the 'B' key, then an attacker might replace this subset with completely random noise.

6 Security

Despite the poor diffusion of the extra key material, our construction appears to make it more difficult to attack the underlying generator. We can distinguish two cases: the outside attacker (who does not know the value of the 'B' key at all), and the recently revoked insider (who knows all or most of it).

Where the 'B' key is unknown, then it seems that even a very weak generator may resist attack. For example, if we use the multiplexer generator or the nonlinear filter generator, the known attacks [1, 2] do not work.

The more realistic attack scenario is that the attacker knows the 'B' key — or most of it. In this case, attacks are still harder, as there is often equivocation about the pseudorandom input to the tables that generate a given keystream output (and the mapping between the pseudorandom generator output and the keystream is a bit too large to store in any case). The details will be a function of the table size and the number of values that are taken from it; but in general, the effect of the table lookup is similar to that of applying a known pseudorandom function to the generator's output.

Our construction may induce some degenerate behaviour that did not exist before. For example, when we use four lookups to a 512K table, we will get a zero keystream whenever the input pseudorandom value is of the form *abab*, *aabb* or *baab*. This weakness does not arise when using three lookups into an 8 MB table, but in that case we are using only 60 bits of pseudorandomness to generate 64 bits of keystream. But, as far as we can see, these weaknesses are of no practical use to an attacker in the kind of applications in which we envisage our construction being used.

A further security advantage of our construction is that the keys are very much larger than in conventional cryptosystems — hundreds of kilobytes, or megabytes, rather than tens of bytes. In fact, the work that led to this construction was inspired by a realisation that in the modern world, many of the most potent threats to cryptographic security involve either malicious code or attacks over networks; in this environment, big keys are good because they are harder for a virus or network intruder to steal without being detected. Another inspiration for this work was [8], which also uses table lookup and xor to construct a stream cipher, but for a completely different purpose.

Finally, the following attack was suggested from the floor at the workshop: colluders, having removed almost all of the marks, could then insert a number of random marks to provide camouflage and hopefully frame other users. Thus, with the concrete example given above, the three attackers might add another 4000 bits to the three remaining genuine marks. However, the three genuine marks can be detected as they form a 'triangle' joining the three conspirators together.

This attack does force the publisher to examine 4,000 marks rather than three; but, as we have emphasised, a serious attacker would simply organise a sufficient coalition — or subscribe in a false name.

7 Conclusion

We have shown how to do two seemingly contradictory things to an arbitrary stream cipher — to strengthen it, and to endow it with the property that small changes in the key cause only small changes in the keystream. We name this construction 'Chameleon'.

Like other traitor tracing schemes, it has only limited collusion resistance. However, we believe that there are applications in which it could be useful. In addition, as it embodies a new kind of cryptographic mechanism, we hope that it will inspire other new work — whether better fingerprinting schemes, or new applications entirely.

Acknowledgement: This work was carried out as part of the joint EPSRC/DTI funded project 'NetCard', which supported the second author. We also acknowledge the referees whose comments enabled us to improve the presentation.

References

1. "Fast Attack on Certain Stream Ciphers", *Electronics Letters* vol 29 (22 July 93) pp 1322–1323
2. "Searching for the Optimum Correlation Attack", in *'Fast Software Encryption'* (1994), Springer LNCS vol 1008 pp 137–143
3. "On Fibonacci Keystream Generators", RJ Anderson, in *Fast Software Encryption* (1994) Springer LNCS vol 1008 pp 346–352
4. "Tiger: A Fast New Hash Function", RJ Anderson, E Biham, in *Fast Software Encryption* (1996), Springer LNCS vol 1039 pp 89–97
5. RJ Anderson, MG Kuhn, "Tamper Resistance — A Cautionary Note", in *Proceedings of the Second Usenix Electronic Commerce Workshop* (Nov 1996) pp 1–21
6. "Tracing Traitors", B Chor, A Fiat, M Naor, in *Advances in Cryptology — Crypto 94*, Springer LNCS vol 839 pp 257–270
7. "On Key Storage in Secure Networks", M Dyer, T Fenner, A Frieze, A Thomason, in *Journal of Cryptology* v 8 no 4 (Autumn 95) pp 189–200
8. "Conditionally-perfect secrecy and a provably-secure randomized cipher", U Maurer, in *Journal of Cryptology* v 5. no 1 pp 53–66
9. "Asymmetric Fingerprinting", B Pfitzmann, M Schunter, in *Advances in Cryptology — Eurocrypt 96*, Springer LNCS vol 1070 pp 84–95
10. *'Anonymous Fingerprinting'*, B Pfitzmann, M Waidner, IBM Research Report RZ 2881 (#90829) 11/18/96, IBM Research Division, Zurich
11. A Shamir, *comment made from the floor of the conference*
12. "Fingerprinting", in *Proceedings of the 1983 IEEE Symposium on Security and Privacy* pp 18–22

Improving Linear Cryptanalysis of LOKI91
by *Probabilistic Counting* Method

(Extended Abstract)

Kouichi SAKURAI   Souichi FURUYA

Department of Computer Science and Communication Engineering,
Kyushu University, Hakozaki, Higashi-ku, Fukuoka 812-81, Japan
Email: sakurai@csce.kyushu-u.ac.jp

Abstract. We improve linear cryptanalysis by introducing a technique of probabilistic counting into the maximum likelihood stage.

In the original linear cryptanalysis based on maximum likelihood method with deterministic counting, the number of effective key and text bits is a multiple of the number of bit involved in the input to some S-box. Then, when larger S-boxes are used, 2R-method and even the 1R-methods can become impractical just because the number of effective text and key bits become excessive. Though 2R-method is practical for attacking DES, existing examples of ciphers where 2R-method is impractical include LOKI91.

We overcome this problem by selecting a part of the effective key bits and investigating the probabilistic behavior of the remained effective key bits. The previous attacks discusses deterministic evaluation of the given approximated formula only when all values of the effective text/key bits are known, while we compute the probability that the approximated formula with unknown inputs equals to zero.

This extension of linear cryptanalysis make useful for 2R-attack on LOKI91, then improves the performance of previous attacks. Furthermore, we implemented some experiments of attacks on 4-round LOKI91, and confirmed the effectiveness of our method.

1   Introduction

1R- and 2R-methods of Linear cryptanalysis: In a linear cryptanalysis developed by Matsui [Mat93, Mat94], the attacker identifies a (linear) relation between some bits of the plaintext, some bits of the ciphertext, and some bits of the user-provided key. Matsui showed that, if the relation does not hold exactly half the time, key information can be extracted by using maximum likelihood method with a large enough set of known plaintext and ciphertext pairs.

The fundamental method of linear cryptanalysis finds only one bit of the key, which is a parity of a subset of the key bits. Additional techniques of reducing the number of rounds of the approximations, by eliminating the first and/or last rounds, and counting on all the key bits affecting the data at the rounds not in the approximation can reduce the number of required plaintexts, and increase the number of key bits that the attack finds.

The block ciphers that we are now concerned with are iterative and repeatedly use a round transformation during encryption. In the 1R-method, the cryptanalyst guess the value of part of the use-provided key in either the first rounds or the last rounds. In the 2R-method, the guess is for the user-provided key from both the first and the last rounds simultaneously.

In practical implementation, we recover key bits using linear techniques by first counting the number of plaintext/ciphertext pairs that fall into a variety of classes. These classes are defined according to the text involved in the linear approximation (the effective text bits). We then process this data by guessing each possible value for the key bits involved in the linear approximation (the effective key bits) and combine this guess with the effective text, In this way scores can be kept for the number of times the bit identified by the linear approximation to the rest of the cipher is either zero or one. A guess can be made for the value of the effective key bits depending on these final scores.

Limitation of 2R-Method: We should note that the number of effective key and text bits is a multiple of the number of bit involved in the input to some S-box. Then, when larger S-boxes are used, 2R-method and even the 1R-methods can become impractical just because the number of effective text and key bits become excessive. Though 2R-method is practical for attacking DES, existing examples of ciphers where 2R-method is impractical include FEAL and LOKI91.

Our approach: To overcome the difficulty of applying 2R-method to ciphers with large S-boxes, instead of evaluating all effective text/key bits in the approximated equation in deterministic manner, we consider the probabilistic evaluation of the equation with unknown inputs. Namely, even though information on a part of inputs of the equation is not available, we can discuss the probability that the equation equal to zero, where the probability considers all possible patterns of unavailable input bits. Then, our probabilistic counting adds each probability, and finally judges the correct key among all key candidates according to maximum likelihood method. Thus, the probabilistic counting algorithm decreases the number of essentially evaluated effective (key) bits, which makes applicable the approximated formula for 2R-method into to ciphers with larger S-boxes,

Related works: A probabilistic approach in counting algorithm of linear cryptanalysis is initiated in Matsui's only-ciphertext attack of DES [Mat93]. Matsui eliminated the plaintext parts from the best linear expression for DES in a probabilistic manner under the assumption that plaintexts consists of natural English sentences represented by ASCII codes. Thus, Matsui's only-ciphertext attack considers the distribution of bits in plaintexts derived from the specific encoding, whereas we try to eliminate effective key bits by investigating the structure of S-boxes.

Aoki and Ohta [AO94] also considered the level in the role of effective bits in linear cryptanalysis of FEAL, in which even the direct 1R-method is impractical. Then, they applied a similar technique as our probabilistic method in reducing the required memories for implementing the attack on FEAL, though they counted not exact probabilities but only approximated probabilities by $\{0, 1\}$-values.

We should remark that Morris [Mor78] have considered how to reduce the required registers for counting events by probabilistic arguments.

Another solution to discount excessive cost of 2R-method is proposed in [KR96]. They introduce newly discovered non-linear approximations into (multiple) linear cryptanalysis, which also achieve better performance than the previous 1R-attack on LOKI'91 [TSM94], while our probabilistic method uses the previously known (best) linear expressions. Then, this paper compares, via experimental performance, our method to the multiple linear cryptanalysis with non linear approximation [KR96].

Applying our idea into LOKI91:   We investigates how the probabilistic counting method can be applied to linear cryptanalysis for LOKI91. Our theoretical estimation implies that

- For breaking 10-round LOKI91, our method needs 1.78×2^{54} known plaintexts with 2^{21} counter, while the multiple non-linear attack [KR96] requires 1.72×2^{56} known plaintexts with 2^{20} counters.
- For breaking 12-round LOKI91, our method with 2^{21} counters requires 1.88×2^{63} known plaintext, whereas the direct 1R-method [TSM94] with 2^{13} counters requires 1.97×2^{67} known plaintexts.

Experiments of our attacks: We implemented some experiments of attacks on 4 rounds LOKI91 for confirming our theoretical estimation on the number N of known-plaintexts required for successful attack. In particular, the success rate parameter $c = N/(p - 1/2)^2$, where p is the probability related to the used approximated formula, is an important for the practical performance on attacks

Then, we implemented similar experiments as one did in [KR96], to predict the success rate of our method. Namely, instead of direct implementing by using 2^{21} counters, we executed the counting algorithm by assuming that a part of target key (e.g. the effective key of the first round) is known, which decreases the number of implemented counter and makes feasible for implementing.

In our probabilistic method with 2^{21} counters, 4-round LOKI91 is breakable with 1.42×2^{18} known-plaintexts, which is theoretically estimated as $c = 8$. Under the condition that the 12 effective key bits of the first round is given to the cryptanalysis, we have implemented this attack with only $2^{21-12} = 2^9$ counters, and the experience results imply that 209,306 plaintexts ($c = 4.5$) is sufficient for achieving 96% success rate and $c = 6$ is sufficient for achieving 100% success rate, while the multiple non-linear attack [KR96] in the similar experiment with $2^{20-12} = 2^8$ counters is reported to require 1,442,632 plaintexts ($c = 8$) for achieving 96% success rate.

2   Preliminaries

2.1   Notation

Throughout this paper, we use the following notations.

P: The 64-bit data of the message

C: The 64-bit data of the encrypted message

$P_H(resp.P_L)$: The upper (resp. lower) 32-bit data of P, $P = (P_H, P_L)$

$C_H(resp.C_L)$: The upper (resp. lower) 32-bit data of C, $C = (C_H, C_L)$

$F_r(X_r, K_r)$: The r-th round F-function with input X_r and subkey K_r

Y^j: 6 input bits of the S_j box

$S_j(Y^j)$: the output of S_j box with the input Y^j

$A[i]$: The i-th bit of a binary vector A

Γ_X : the masked value of data X

$X[\Gamma_X]$: the even parity value of the bitwise AND between X and ΓX

$A[i, j, \cdots, k] := A[i] \oplus A[j] \oplus \cdots \oplus A[k]$

Note that we refer to the right most bit as the zero-th bit.

2.2 Description of LOKI91

LOKI91 is a 64-bit key/64-bit block cryptosystem similar to DES. This paper omits the details of the key-scheduling part because our analysis is independent of its algorithm. (See [BKPS91] for more precise description.) However, the F-function, which we describe below, plays an important role in our analysis.

The procedure of the LOKI91's i-round function F_i with the 32-bit input X_i and with 32-bit subkey K_i is defined as follows:

$$B' = X_i \oplus K_i, \qquad B = E(B')$$
$$Y^0 = B_{11}B_{10} \cdots B_0, \qquad Y^1 = B_{19}B_{18} \cdots B_8$$
$$Y^2 = B_{27}B_{26} \cdots B_{16}, \qquad Y^3 = B_3B_2 \cdots B_0B_{31} \cdots B_{24}$$
$$O = (S_3(Y^3)S_2(Y^2)S_1(Y^1)S_0(Y^0)), \; F_i(X_i, K_i) = P(O), \qquad (1)$$

where E is an expansion permutation with 32-bit input/48-bit output and P is a permutation over 32-bit. The computation $S_i(Y_i)$ in this procedure (1) is given in the following:

$$row = Y_{11}^i Y_{10}^i Y_1^i Y_0^i, col = Y_9^i Y_8^i \cdots Y_3^i Y_2^i$$
$$S_i(Y^i) = (col + ((row \times 17) \oplus ff_{16})\&ff_{16})^{31}(mod\; g_{row}) \qquad (2)$$

where g_{row}, which is selected from the set

$$\{375, 379, 391, 395, 397, 415, 419, 425, 433, 445, 451, 463, 471, 477, 487, 499\},$$

is an irreducible polynomial over $GF(2^8)$.

3 Our proposed method

3.1 Probabilistic 2R-method

We apply the proposed probabilistic counting into 2R-method in linear cryptanalysis.

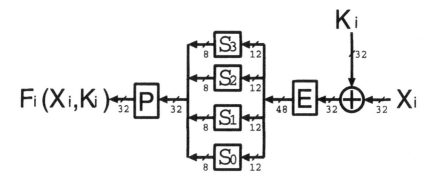

Fig. 1. F function of LOKI91

Consider the following approximated expression for n-round cipher, which is derived from the (best) linear expression for $(n-2)$-round version:

$$P[\Gamma_P] \oplus C[\Gamma_C] \oplus F_1(P_L, K_1)[\Gamma_{P_L}] \oplus F_n(C_L, K_n)[\Gamma_{C_L}] = K[\Gamma_K], \qquad (3)$$

which holds with probability $p \neq \frac{1}{2}$. Suppose that this formula includes t effective text bits and $k = k_1 + k_n$ effective key bits, where k_1 effective key bits are in F_1 and k_n effective key bits are F_n. So, the original counting algorithm requires $2^c + 2^t$ counters [Mat94].

Now we assume that only a part of the effective key bits is available, which we call *visible* effective key bits, and the other effective key bits, which is not available for attacker, are called *invisible*. In the evaluation of the approximated formula with invisible effective key bits, we consider the probability that the approximated formula equals to zero when the effective text bits and the visible effective key bits are deterministically given, where the probability takes the average over all possibilities of the *invisible* effective key bits.

We apply this probabilistic argument into $F_n(C_L, K_n)[\Gamma_{C_L}]$, which is assumed to include k_n^v visible effective key bits, while we use the deterministic evaluation for $F_1(P_L, K_1)[\Gamma_{P_L}]$ Then, the number of INvisible effective key bits is $k_n^{\bar{v}} = k_n - k_n^v$. Under this condition, the maximum likelihood method with probabilistic counting is implemented as follows.

[Probabilistic Counting Algorithm]

Step 1 Prepare 2^t counter $U_i (0 \leq i \leq 2^t)$, where i corresponds to each value on the t effective text bits of Equation (3).

Step 2 For each plaintext and the corresponding ciphertext, compute the value "i" of Step-1 and count up the counter U_i by one.

Step 3 Prepare $2^{k_1+k_n^v}$ counter $T_j (0 \leq j \leq 2^{k_v})$, where j corresponds to each value on the k_1 effective key bits of F_1 and the k_n^v VISIBLE effective key bits of F_n in Equation (3).

Step 4 For each "i" and "j", computes the probability p_{ij} that the left-side of Equation evaluated with "i" and "j" is zero, where the probability takes over all the "invisible" effective key bits. Then, set $T_j = \sum_i p_{ij} \times U_i$

Step 5 Let T_{exact} be the value which maximizes $|T_k - \frac{N}{2}|$, i.e., $|T_{exact} - \frac{N}{2}| = max_k(|T_k - \frac{N}{2}|)$. Then adopt the key candidate corresponding to T_{exact}.

Step 4 If $|T_{exact} - \frac{N}{2}| > \frac{N}{2}$, then guess that the right-side = 0. If $|T_{exact} - \frac{N}{2}| < \frac{N}{2}$, then guess that the right-side = 1.

Thus, the number of the counters required for implementing this algorithm is $2^t + 2^{k_1+k_n^v} = 2^t + 2^{k-k_n^{\overline{v}}}$, which corresponds to the number of counters, U_i and T_j, and the computational complexity for executing this algorithm is $O(N) + O(2^{t+k_1+k_n^v})$, where N is the number of the used known plaintexts.

Remark. Computing the probability p_{ij} for each (i, j) can be done by using a (common) precomputed table with $2^{k_n^v}$-entries, which is given in Appendix B. We describe the construction of this table in Subsection 3.2.

3.2 How to evaluate approximated formulas in a probabilistic manner

We discuss how to evaluate a given approximated formula with some unknown inputs in a probabilistic manner.

In the following, we consider an example of the parity $F_4(C_L, K_4)[18, 22, 26]$, which we use in our practical attacks on 4-round LOKI91 of Section 4 and is also useful for $N(> 4)$-round attack of Section 7. Note that $F_4(C_L, K_4)[18, 22, 26]$ is derived from the only single S-box S_2, namely, $S_2(Y^2)[70_{16}]$. Further, we assume that 4 bits $K_4[23], K_4[19], K_4[18], K_4[17]$ are invisible among 12 effective key bits of $K_4[27] \sim K_4[16]$, under which we attack 4-round LOKI91 in Section 4. Then, the input Y^2 of the corresponding S-box is $abcd?efg???h$, where $a, b, c, d, e, f, g, h \in \{0, 1\}$ and "?" denotes an unknown input bit for the cryptanalysis.

Consider 1001?100???1 as the input Y^2. Then, the possible inputs consist of the following 16 pattern:

100101000001 , 100101000011 , 100101000101 , 100101000111,
100101001001 , 100101001011 , 100101001101 , 100101001111,
100111000001 , 100111000011 , 100111000101 , 100111000111,
100111001001 , 100111001011 , 100111001101 , 100111001111

The each corresponding output of the S-box is the following:

01000110 , 01111111 , 00111000 , 10100010,
00100100 , 00101001 , 10100111 , 11101000,
10100001 , 00100001 , 01001111 , 00100111,
01001000 , 00101100 , 10011011 , 10101011

Then, the each corresponding parity of the output with 01110000 masked is the following:

$$
\begin{array}{cccc}
1, & 1, & 0, & 1, \\
1, & 1, & 1, & 0, \\
1, & 1, & 1, & 1, \\
1, & 1, & 1, & 1
\end{array}
$$

This implies that only 2 times of the zero parity actually occur among 16 types of inputs. Thus, we observe that the parity of the output masked by 01110000 for input 1001?100???1 takes zero with probability 2/16 (under the assumption that all unknown input bits are selected in uniform and random), which is written as

$$
p_s(1001?100???1, 01110000) = \frac{2}{16}.
$$

The probability $p_s(abcd?efg???h, 01110000)$ for each $a, b, c, d, e, f, g, h \in \{0, 1\}$ is listed in Appendix B, which is easily tabulated by computer. Such a is obtained for a given pair of the pattern of the unknown input bits and the mask value of the output for a S-box. Therefor, in the probabilistic counting algorithm presented in Subsection 3.1, a common table is used for computing p_{ij} for any (i, j) at Step 4.

3.3   Analyzing our proposed probabilistic counting

The number of known-plaintexts required for successful linear cryptanalysis, if we use the approximated formula which holds probability p, is generally estimated as $c|p - 1/2|^{-2}$, where c is the success rate parameter that depends on the the the approximated formula [Mat93, Mat94, TSM94]. We should note that, in our probabilistic counting method, the number of known-plaintexts required for successful attacking is depended not only upon the probability with which the approximated formula holds but also upon the probability of the bias of the parity of the output which we apply the probabilistic argument.

We investigate how to effect the probability p_s related to the probabilistic counting on the successful probability of the applied linear cryptanalysis. In the previous subsection, we consider the case of

$$
p_s(1001?100???1, 01110000) = \frac{2}{16}.
$$

as an example. Next we consider the average of the bias of such probabilities over all possible inputs $abcd?efg???h$, i.e.,

$$
\tilde{p}(abcd?efg???h, 01110000) = \frac{1}{2^8} \sum_{i=0000?000???0}^{1111?111???1} |\frac{1}{2} - p_s(i, 01110000)|
$$

Throughout the remained part of this paper, instead of $abcd?efg???h_2$, we use the notation 111101110001, where "1" denotes the positions of the available bits, then

# invisible bits	invisible pattern x	most unbalanced $\tilde{p}(x, 70_{16})$
11	001000000000_2	1.4375000×2^{-7}
10	000001100000_2	1.1250000×2^{-6}
9	010001000001_2	1.6875000×2^{-6}
8	011001000100_2	1.1562500×2^{-5}
7	011001000101_2	1.5468750×2^{-5}
6	011001000111_2	1.0468750×2^{-4}
5	011001100111_2	1.3437500×2^{-4}
4	111101110001_2	1.7265625×2^{-4}
3	011101011111_2	1.1718750×2^{-3}
2	010111111111_2	1.5664063×2^{-3}
1	011111111111_2	1.0175781×2^{-2}
0	111111111111_2	1.0000000×2^{-1}

Table 1. Most unbalanced values of $\tilde{p}(x, 70_{16})$

$\tilde{p}(abcd?efg???h, 01110000)$ is written as $\tilde{p}(111101110001, 01110000)$. The exact value of $\tilde{p}(111101110001, 01110000)$, which can be computed from the table in Appendix B, is $442/4096$. This can be generalized for computing $\tilde{p}(x, \Gamma_y)$.

We have computed which value of x takes the maximum among the set with a common number of invisible inputs for the output mask 70_{16}, which is listed in Table 1. In the linear cryptanalysis with the probabilistic counting, once we decide the number of invisible effective key bits, we choose the position of these invisible bits according to Table 1. Remark that the value of the output mask 70_{16} is decided from the initially given approximated formula.

Thus, as the similar argument as Matsui's *Piling-up Lemma* [Mat93], under the assumption that the all keys are independent, the number of known-plaintexts required for successful probabilistic counting algorithm is theoretically estimated as:

$$N = c \times (2 \times |p_{linear} - 1/2| \times \tilde{p})^{-2}, \tag{4}$$

where p_{linear} is the probability which the given linear approximated formula holds with.

4   Attacking 4-round LOKI91

4.1   Previous deterministic 2R-method

The best linear expression of 2-round LOKI91 is computed in [TSM94] as the following formula (5):

$$P_H[\alpha] \oplus P_L[\alpha] \oplus C_L[\alpha] = K_1[\alpha] \quad (\alpha = 18, 22, 26), \tag{5}$$

which holds with probability $p_2 = \frac{1}{2} - 1.38 \times 2^{-6}$. Then, the 2R-method with this 2-round best linear expression induces the following expression for 4-round LOKI91:

$$P_H[\alpha]P_L[\alpha] \oplus C_H[\alpha] \oplus F_1(P_L, K_1)[\alpha] \oplus F_4(C_L, K_4)[\alpha] = K_2[\alpha] \qquad (6)$$

which holds as the same probability as p_2.

The expression 6 has 24-bits of the keys $K_1[27]K_1[26]\cdots K_1[16]\,K_4[27]\cdots K_4[16]$, which effect the evaluation of its left-side, and 24-bit effective text bits. Then, the original (deterministic) 2R-method requires 2^{25} counters and 2^{48} working complexity for implementing, which requires 1.06×2^{14} known plaintexts.

4.2   Our probabilistic 2R-method

To overcome the time/memory constraint of the previous 2R-method, instead of dealing with all effective key bits, we consider only a part of the effective key bits: we regard 4 bits $K_4[23]$, $K_1[19]$, $K_4[18]$, $K_4[17]$ of the 4th round of F_4 as invisible.

In this case, though the attacker can deterministically decide the parity $F_1(P_L, K_1)[\alpha]$, he cannot evaluate $F_4(C_L, K_4)[\alpha]$ in deterministic manner because he does not get the complete information on inputs K_4. Instead of computing the deterministic parity, the we compute the probability that the parity of $F_4(C_L, K_4)[\alpha]$ takes 0 by using a precomputed table with 2^8 size, which tabulated in Appendix. By using this probability, we can obtain the probability that the left side of Equation (6) equals to zero.

This procedure is executed for all given plain/cipher texts and we get the information of the key by using the maximum likelihood primitive. Thus, the cryptanalysis regards 12 key-bits of F_1 and 8 key-bits of F_4 as the visible effective key bits, and extracts the user-key candidate by using the probabilistic counting algorithm presented in Subsection 3.1.

Now, we discuss the efficiency of our attack on 4-round LOKI91. The probability that the applied approximated formula (5) holds is p_2, and the expected bias of success probability of guessing the parity of the 4th-round function F_4 is $\tilde{p}(f71_{16}, 70_{16}) = 1.73 \times 2^{-4}$. So, *Piling-up Lemma* which we discuss at Subsection 3.3, implies that the success probability of the known plaintext attack is

$$p = \frac{1}{2} + 2 \times (1.38 \times 2^{-6}) \times (1.73 \times 2^{-4}) = \frac{1}{2} + 1.19 \times 2^{-8} \qquad (7)$$

Then, Formula (4) gives a theoretical estimation on the number of the known plaintext required for breaking the cipher that:

$$N_4 = 8.0 \times (1.19 \times 2^{-8})^{-2} = 1.42 \times 2^{18} \qquad (8)$$

Remark. In this theoretical estimation, we assume that the value of the success rate parameter is $c = 8.0$ as the previous attacks in [TSM94, KR96]. In Section 6, we discuss the experimental confirmation on this assumption.

This is 26% of the number 1.38×2^{20} of the required plaintexts in multiple non-linear cryptanalysis with 2^{20} counters and 2^{38} working effort [KR96]. Since this attack uses 20 visible effective key bits and 20 effective text bits, the probabilistic counting can be implemented over 2^{21} counters with 2^{40} time complexity.

Remark. If one counter costs 2 byte memory, then 2^{21} counters corresponds to 4 Mega byte memory, which is available over a recent PC. Furthermore, 2^{40} working complexity is less than the required time for Matsui's experimental attack on 16-round DES [Mat94].

5 More flexibility for attackers

In the original deterministic counting algorithm, the cryptanalyst is restricted to using a number of effective key and text bits which is a multiple of the number of bit involved in the input to some S-box. As we showed in the previous subsection, our new probabilistic method make possible 2R-attack on LOKI'91, which was impractical in the previous deterministic method.

However, the probabilistic method described in Subsection 4.2 requires 2^{21} counters, which is still impractical for certain attackers. This is improved by increasing the number of INVISIBLE effective key bits, which we apply the probabilistic argument. Suppose that the cryptanalyst is restricted to handle with $2^{1}7$ counters because of the memory-constraint on his PC. In this case, the attacker deals with 4 bits, $K_4[26]K_4[25]K_4[22]K_4[18]$ of 12 effective key bits $K_4[27]K_4[26]\cdots K_4[16]$ of the 4-round function in Formula (6) as VISIBLE. Then, he must compute the parity of S_2 with masked by 70_{16} from 4 bits informa-tion $Y^2[10]Y^2[9]Y^2[6]Y^2[2]$ of the 12 bits input Y^2. Note that bias $\tilde{p}(644_{16}, 70_{16}) = 1.16 \times 2^{-5}$, then we can theoretically estimate that the number of plaintexts re-quired for breaking 4-round LOKI91 is

$$N_4' = 8 \times (2 \times 1.16 \times 2^{-5} \times 1.38 \times 2^{-6})^{-2} = 1.58 \times 2^{21} \qquad (9)$$

Though this is 8.92 times of the required number of plaintexts N_4 in 8 of the previous attack with 2^{21} counters, the cryptanalysis can execute this attack with $\frac{1}{256}$ times working complexity as the previous case.

On the other hand, a cryptanalysis could be alive, who has much more memory for counters, though he cannot do complete implementation on the direct 2R-attack which requires 2^{25} counters. Suppose that he can use 2^{23} counters. In this case, the attacker deals with only 1 bits, $K_4[25]K_4[22]K_4[18]$ of 12 effective key bits $K_4[27]K_4[26]\cdots K_4[16]$ of the 4-round function in Formula (6) as *invisible*.

Then, he computes the parity of S_2 with masked by 70_{16} from 11 bits inform-ation expect $Y^2[9]$ of the 12 bits input Y^2. The bias $\tilde{p}(5ff_{16}, 70_{16}) = 1.57 \times 2^{-3}$, then we can theoretically estimate that the number of plaintexts required for breaking 4-round LOKI91 is

$$N_4'' = 8 \times (2 \times 1.57 \times 2^{-3} \times 1.38 \times 2^{-6})^{-2} = 1.72 \times 2^{16} \qquad (10)$$

x	# counters	#$plaintexts$	work effort
	†2^{13}	1.49×2^{22}	2^{24}
060_{16}	2^{15}	1.67×2^{23}	2^{28}
441_{16}	2^{16}	1.49×2^{22}	2^{30}
644_{16}	2^{17}	1.58×2^{21}	2^{32}
645_{16}	2^{18}	1.77×2^{20}	2^{34}
647_{16}	2^{19}	1.93×2^{19}	2^{36}
667_{16}	2^{20}	1.17×2^{19}	2^{38}
$f71_{16}$	2^{21}	1.42×2^{18}	2^{40}
$75f_{16}$	2^{22}	1.54×2^{17}	2^{42}
$5ff_{16}$	2^{23}	1.72×2^{16}	2^{44}
$7ff_{16}$	2^{24}	1.02×2^{16}	2^{46}
fff_{16}	†2^{25}	1.06×2^{14}	2^{48}
	*2^{20}	1.38×2^{20}	2^{38}

† The deterministic 1R attack [TSM94]
‡ The deterministic 2R attack (hypothetical)
* The multiple non-linear attack [KR96]

Table 2. Efficiency of each predict of 4th round F function

Though this requires 16 times working complexity as the attack with 2^{21} counters, the required number of plaintexts decrease into 0.30 times as N_4 of (8) in the previous attack.

Furthermore, the attacker can control the number of the VISIBLE part of the effective key bits in more flexible manner according to his computational resource. The performance of the variant attacks is listed in Table 2. Note that we assume that the success rate $c = 8$ for all cases, the correctness of which we discuss in our experiment of Section 6.

6   Experimental verification of our probabilistic method

We have carried out our experiment on 4-round version of LOKI91 for confirming our theoretical analysis.

The number of known-plaintexts required for successful linear cryptanalysis with the approximated formula, which holds probability p is generally estimated as $c|p - 1/2|^{-2}$, where c is the success rate parameter depends on the the approximated formula. In the case of LOKI91, Tokita et al. [TSM94] estimated $c = 8$ for 1R-attack's achieving 100% success rate. However, if the number of the effective key bits in the used approximated formula increase, the parameter c would become bigger for achieving 100% success rate. So, deciding the exact value of the parameter c is an important for estimating the efficiency of the discussed attack.

To this end, we implemented some experiments, which is similar to one did in [KR96], to predict the success rate of our method. Namely, instead of dir-

# invisible bits	invisible pattern x	c					
		2.0	4.0	6.0	8.0	10.0	12.0
10	060_{16}	39%	48%	51%	56%	55%	65%
9	441_{16}	21%	34%	29%	39%	35%	32%
8	644_{16}	74%	95%	99%	99%	—	—
7	645_{16}	78%	95%	100%	100%	—	—
6	647_{16}	79%	95%	99%	100%	—	—
5	667_{16}	67%	92%	99%	100%	—	—
4	$f71_{16}$	65%	94%	100%	100%	—	—
3	$75f_{16}$	61%	99%	99%	100%	—	—
2	$5ff_{16}$	60%	96%	100%	100%	—	—
1	$7ff_{16}$	61%	98%	100%	100%	—	—

Table 3. Success rate of variant attacks

ect implementing by using 2^{21} counters, we executed the counting algorithm by assuming that a part of target key (e.g. the effective key of the first round) is known, which decreases the number of implemented counter and makes easy for implementing.

The obtained successful rate over 100 trials for the theoretical estimation on Table 2 is given in Table3, which we assume that 12 bits of effective key used in the first round remain fixed and known.

Remark. In Table 3, the (strange) degeneration of the success rate in the experimental result for the pattern $x = fff_{16}$, which corresponds to the deterministic 2R-method, suggests that the performance of the probabilistic counting in our theoretical analysis of Subsection 3.3 should be improved by more refined discussion.

On the other hand, the experimental results for two pattern $x = 060_{16}, 441_{16}$ show that recovery of only a few bits is not reliable, a similar fact is reported in [KR96]. This is because that there are two or more key candidates whose probabilistic behavior is quite similar. In Appendix A, we give a refined analysis with considering the behavior of incorrect-keys, and show other patterns, which achieves better performance than $x = 060_{16}, 441_{16}$ above.

Furthermore, we have implemented with variable counters in the case of the invisible pattern $x = f71_{16}$ with 4 invisible bits for observing how the success rate degrade when the number of recovered key bits increases. The experimented results over 100 trials are tabulated in Table 4, where the number of the known key bits of the first round changes while the number of the recovered key bits is fixed as 8: 4 bits are visible and 4 bits are invisible.

To the end of this section we estimate how success rates degrade. Following Table 5 is the summarized experimental results of variant numbers of the recovered bits for $x = 644_{16}$.

# key bits recoverd		c				
1st round F_1	4th round F_4	2.0	3.0	4.0	6.0	8.0
0	8	65%	84%	94%	100%	100%
1	8	64%	87%	96%	99%	100%
2	8	56%	78%	90%	99%	100%
3	8	48%	72%	94%	100%	100%
4	8	30%	78%	86%	100%	100%

Table 4. Success rate of variant numbers of the recovered bits for $x = f71_{16}$

#key bits recovered		c			
1st round F_1	4th round F_4	$c = 2.0$	$c = 4.0$	$c = 6.0$	$c = 8.0$
0	4	74%	95%	99%	99%
1	4	84%	100%	98%	100%
2	4	48%	94%	98%	100%
3	4	58%	96%	100%	96%
4	4	58%	84%	96%	100%
5	4	38%	88%	100%	98%
6	4	38%	80%	100%	100%
7	4	34%	80%	98%	100%
8	4	34%	72%	92%	100%
9	4	14%	68%	96%	96%
10	4	16%	72%	92%	100%
11	4	-%	-%	86%	97%
12	4	-%	-%	80%	96%

Table 5. Success rate of variant numbers of the recovered bits for $x = 664_{16}$

7   Attacking n-round LOKI91

We apply the probabilistic counting method into $n(> 4)$ round LOKI91.

As the 16-round LOKI91 [TSM94], iterative linear approximations is useful for the best linear approximations of $n = 4, 5, 7, 8, 10$. Then, we can apply the similar argument of the attack on 4-round LOKI91 into these reduced round LOKI91.

We discuss the efficiency of attacks under the assumption that we have only 2^{21} counters for 20 visible effective key bits:

$$K_1[27]K_1[26]K_1[25]K_1[24]K_1[23]K_1[22]K_1[21]K_1[20]K_1[19]K_1[18]K_1[17]K_1[16]$$
$$K_n[27]K_n[26]K_n[25]K_n[24]K_n[22]K_n[21]K_n[20]K_n[16]$$

Then, the remained effective bits $K_n[23], K_n[19]K_n[18]K_n[17]$ are regarded as invisible, over which we argue the probabilistic behavior.

Round	# plaintexts	# counters	work effort
4	1.42×2^{18}	2^{21}	1.00×2^{40}
6	1.50×2^{27}	2^{21}	1.00×2^{40}
7	1.59×2^{36}	2^{21}	1.00×2^{40}
9	1.68×2^{45}	2^{21}	1.68×2^{45}
10	1.78×2^{54}	2^{21}	1.78×2^{54}
12	1.88×2^{63}	2^{21}	1.88×2^{63}

Table 6. Attacks on each round LOKI91 with 2^{21} counters

In the case of 10-round LIKI91, the best linear expression of 8-round LOKI91 implies via 2R-method the following approximated formula (11) for 10-round LOKI91, which hold with probability $\frac{1}{2} - 1.23 \times 2^{-24}$.

$$P_H[\alpha] \oplus P_L[\alpha] \oplus C_H[\alpha] \oplus F_1(P_L, K_1)[\alpha] \oplus F_{10}(C_L, K_{10})[\alpha]$$
$$= K_2[\alpha] \oplus K_4[\alpha] \oplus K_5[\alpha] \oplus K_7[\alpha] \oplus K_8[\alpha] \quad (11)$$

Then, the number of the known plaintext required for breaking 10-round LOKI91 is theoretically given as follows:

$$N_{10} = 8.0 \times (2 \times 1.23 \times 2^{-24} \times 1.73 \times 2^{-4})^{-2} = 1.78 \times 2^{54} \quad (12)$$

We should recall that the multiple non-linear attack requires 1.72×2^{56} known plaintexts with 2^{20} counters for breaking 10-round LOKI91 [KR96].

In the case of 12-round LIKI91, the best linear expression of 10-round LOKI91 implies via 2R-method the following approximated formula (13) for 12-round LOKI91, which hold with probability $\frac{1}{2} + 1.69 \times 2^{-29}$:

$$P_H[\alpha] \oplus C_H[\alpha] \oplus F_1(P_L, K_1)[\alpha] \oplus F_{12}(C_L, K_{12})[\alpha]$$
$$= K_3[\alpha] \oplus K_4[\alpha] \oplus K_6[\alpha] \oplus K_7[\alpha] \oplus K_9[\alpha] \oplus K_{10}[\alpha] \quad (13)$$

Then, a theoretically estimation implies that the number of the known plaintext required for breaking 12-round LOKI91 is

$$N_{12} = 8.0 \times (2 \times 1.69 \times 2^{-29} \times 1.73 \times 2^{-4})^{-2} = 1.88 \times 2^{63} \quad (14)$$

The performance of these attacks is listed in Table 6. Thus, in our theoretical estimation, 12-round LOKI91 is breakable with 1.88×2^{63} known plaintexts faster than an exhaustive search for 64-bits keys, whereas the direct 1R-method [TSM94] with 2^{13} counters requires 1.97×2^{67} known plaintexts with 2^{67} working effort for breaking 12-round LOKI91.

8   Concluding remarks

Yet another extension of linear cryptanalysis has been presented by introducing probabilistic counting method. This new method improves the performance of the linear cryptanalysis of LOKI91, which correctness was confirmed via our implemented experiments. We note that, though we have examined 3R- and 4R-attack on DES, no significant advantage over existing attacks on DES are yet obtained.

A future research topic is to optimize our method. We have used the best linear expression for k-round LOKI91. However, alternative better expression could exist for our probabilistic method, which would improve the perform-ance of our attack. Furthermore, a hybrid attack between our method and the others [LH94, KR94, KR96] shall be investigated for clarifying the limitation of the linear cryptanalysis, which is useful for designing provably secure block ciphers [NK95].

The final remark is that our probabilistic-counting method can be applicable to any statistical attack [Vau96], based on the maximum likelihood principle, which includes differential cryptanalysis [BS91, BS93], though advantage of ap-plied cryptanalysis over existing attacks remains open.

Acknowledgments

The authors would like to thank the referees for helpful comments on the submit-ted version. Also thanks a referee's comment on the strange degeneration of the authors' experimental results and Robshaw's email-answer [Rob96] to the first author's query on the paper [KR96], which lead the authors to more detailed ana-lysis of the proposed attack and improvements described in Appendix A. Final thanks Ronald Rivest for remarking the connection between an initiated research of Morris [Mor78] and our probabilistic counting method.

References

[AO94] K.Aoki, and K.Ohta, "Linear Cryptanalysis of the Fast Data Encipherment Algorithm," *Tech. Rept. of IEICE*, ISEC94-5 (1994).

[BKPS91] L. Brown, M. Kwan, J. Pieprzyk, J. Seberry, "Improving Resistance to Differential Cryptanalysis and the Redesign of LOKI," *Advances in Cryptology, - ASIACRYPT'91, LNCS Vol. 739, Springer-Verlag*, 1991.

[BPS90] L. Brown, J. Pieprzyk, J. Seberry, "LOKI - A Cryptographic Primitive for Au-thentication and Secrecy Applications," *Advances in Cryptology, - AUSCRYPT'90, LNCS Vol. 453, Springer-Verlag*, 1990.

[BS91] E. Biham, A. Shamir, "Differential Cryptanalysis of Snefru, Khafre, REDOC-II, LOKI and Lucifer," *Advances in Cryptology, - CRYPTO'91, LNCS Vol.576, Springer-Verlag*, 1991.

[BS93] E. Biham, A. Shamir, "Differential Cryptanalysis of of the Data Encryption Standard," *Springer-Verlag*, 1993.

[Knu91] L. R. Knudsen, "Cryptanalysis of LOKI," *Advances in Cryptology, - ASIAC-RYPT'91, LNCS Vol. 739, Springer-Verlag*, 1991.

[Knu92] L. R. Knudsen, "Cryptanalysis of LOKI91," *Advances in Cryptology, - AUS-CRYPT'92, LNCS Vol. 718, Springer-Verlag*, 1992.

[KR94] B. S. Kaliski, M. J. B. Robshaw, "Linear Cryptanalysis Using Multiple Approximations," *Advances in Cryptology, - CRYPTO'94, LNCS Vol.839, Springer-Verlag*, 1994.

[KR96] L. R. Knudsen, M. J. B. Robshaw, "Non-linear Approximations in Linear Cryptanalysis," *Advances in Cryptology, - EUROCRYPT'96, LNCS Vol. 1070, Springer-Verlag*, 1996.

[LH94] S. K. Langford, M. E. Hellman, "Differential-Linear Cryptanalysis," *Advances in Cryptology, - CRYPTO'94, LNCS Vol. 839, Springer-Verlag*, 1994.

[Mat93] M. Matsui, "Linear Cryptanalysis Method for DES Cipher," *Advances in Cryptology, - EUROCRYPT'93, LNCS Vol. 765, Springer-Verlag*, 1993.

[Mat94] M. Matsui, "The First Experimental Cryptanalysis of the Data Encryption Standard," *Advances in Cryptology, - CRYPTO'94, LNCS Vol. 839, Springer-Verlag*, 1994.

[Mor78] R. Morris, "Counting large numbers of events in small registers," *Comm. of the ACM*, Vol.21, No.10 (1978).

[Nyb94] K.Nyberg, "Linear approximation of block ciphers," *Advances in Cryptology, - EUROCRYPT'94, LNCS Vol. 950, Springer-Verlag*, 1995.

[NK95] K.Nyberg and L.R.Knudsen, "Provable security against a differential attack." *J. Cryptology*, Vol.8, No.1, pp.27-37 (1995).

[OA94] K. Ohta, K. Aoki, "Linear Cryptanalysis of the Fast Data Encipherment Algorithm," *Advances in Cryptology, - CRYPTO'94, LNCS Vol. 839, Springer-Verlag*, 1994.

[OMA95] K. Ohta, S. Moriai, and K. Aoki, "Improving the search algorithm for best linear expression," *Advances in Cryptology, - CRYPTO'95, LNCS Vol. 963, Springer-Verlag*, 1995.

[Riv97] R. Rivest, "Oral remark on Morris's work [Mor78] at the second author's presentation of Fse4," Jan. 1997

[Rob96] J. B. Robshaw, "Email-answer to the first author's query on the experimental results on [KR96]," Oct. 1996.

[Vau96] S.Vaudenay, "An experiment on DES statistical cryptanalysis," *Proc. of 3rd ACM CCCS* , 1996.

[TSM94] T. Tokita, T. Sorimachi, M. Matsui, "Linear Cryptanalysis of LOKI and s^2DES," *Advances in Cryptology, - ASIACRYPT'94, LNCS Vol.917, Springer-Verlag*, 1994.

A   A refined analysis of the probabilistic method

A.1   Theoretical formula of Success rate

Subsection 3.3 roughly discussed the performance of our probabilistic method by simply applying Matsui's *Piling-up Lemma* [Mat93]. This appendix considers an exact performance by dealing with the behavior of incorrect keys.

First, we consider the following linear approximation to use the cryptanalysis:

$$P_h[\Gamma P_h] \oplus P_l[\Gamma P_l] \oplus C_h[\Gamma C_h] \oplus C_l[\Gamma C_l] \qquad (15)$$
$$\oplus F_1(P_l, K_1)[\Gamma P_h] \oplus F_n(C_l, K_n)[\Gamma C_h] = K[\Gamma K]$$

Now we assume that the parity of the 1st round F-function $F_1(P_l, K_1)[\Gamma P_h]$ and the parity of the nth round F-function $F_n(C_l, K_n)[\Gamma C_h]$ are derived from a single S-box. Then, by using the notation

$$PC[\Gamma PC] = P_h[\Gamma P_h] \oplus P_l[\Gamma P_l] \oplus C_h[\Gamma C_h] \oplus C_l[\Gamma C_l],$$

we can write the above linear expression as follows:

$$PC[\Gamma PC] \oplus S(Y_{1,x})[\Gamma S_1] \oplus S(Y_{n,y})[\Gamma S_n] = K[\Gamma K].$$

In the equation above, the notation $Y_{k,x}$ denotes inputs of the x-th S-box[1] in the k-round F-function, and $\Gamma S_1, \Gamma S_n$ denote each mask obtained from $\Gamma P_h, \Gamma C_h$ via expanding function E of each F-function.

Let $\frac{1}{2} + p'$ $(p' > 0)$ be the probability that this linear expression holds for the correct key $K_1^T K_n^T$. Then, the success rate of the original linear cryptanalysis with N random pairs of plain/cipher-texts is estimated [Mat93] in the following formula:

$$SR' = \int_{\frac{N}{2}}^{\infty} \prod_{\kappa=1}^{m} WE_\kappa(x) \sqrt{\frac{2}{\pi N}} \exp\left\{ -\frac{2(x - \mu_T)^2}{N} \right\} dx$$

Note that, in this formula,

$$\mu_T = N(\frac{1}{2} + p'),$$

m is the number of the key candidates which are not correct, and

$$WE_\kappa(x) = Prob\left(T^\kappa : \frac{N}{2} - x < T^\kappa < \frac{N}{2} + x \right),$$

where T^κ denotes the value of the counter corresponding to the (wrong) key candidates $K^W = K^T \oplus \kappa$.

Next we consider how to modify the formula above in the case with probabilistic counting.

In our attack, assume that there exists 12 effective textbits and 12 effective keybits for the first round S-box with the 12 input bits, and there exists $d(< 12)$ effective textbits and d effective keybits for the n round S-box with the 12 input bits. Then, the number of key candidates is 2^{12+d}, which the correct key exists in.

For simplifying the discussion, we consider the case of $d = 2$. In this case, the attacker guesses the value $S(Y_{n,y})[\Gamma S_n]$ is O with probability $\frac{1}{2} + \varepsilon_{ab}$, which is computed as discussed in Subsection 3.2, and corresponds to such a distribution table given in Appendix B.

---

[1] Note that, in the discussed attack of LOKI91, $x = 2$.

$a\backslash b$	0	1
0	$\frac{1}{2} + \varepsilon_{00}$	$\frac{1}{2} + \varepsilon_{01}$
1	$\frac{1}{2} + \varepsilon_{10}$	$\frac{1}{2} + \varepsilon_{11}$

Now we consider the following assumption.

Assumption A: The distribution of the statistic T_n corresponding to the key candidate $K_n = K^T \oplus \kappa_n$ can be modeled by using a bimonomial distribution.

Theorem: Under Assumption A, the success rate of our probabilistic-counting linear-cryptanalysis with N random pairs of plain/cipher-texts is

$$\int_{\frac{N}{2}}^{\infty} \left\{ \prod_{\kappa_n = 01_2}^{11_2} \int_{\frac{N}{2} - x}^{\frac{N}{2} + x} \sqrt{\frac{2}{\pi N}} \exp \left(-\frac{2}{N} (y - \frac{N}{2} + (\sum_{i=00_2}^{11_2} \varepsilon_{i \oplus \kappa_n} \frac{\varepsilon_i N}{2}))^2 \right) dy \right\}$$

$$\times \sqrt{\frac{2}{\pi N}} \exp \left\{ -\frac{2}{N} \left(x - N(\frac{1}{2} + p') \right)^2 \right\} dx$$

The proof of this theorem is omitted from this extended abstract.

We should remark that this formula can be easily generalized into the case $d(2 \leq d < 12)$ as follows.

$$\int_{\frac{N}{2}}^{\infty} \left\{ \prod_{\kappa_n} \int_{\frac{N}{2} - x}^{\frac{N}{2} + x} \sqrt{\frac{2}{\pi N}} \exp \left(-\frac{2}{N} \times (y - \frac{N}{2} + (\sum_{i=0}^{2^d - 1} \varepsilon_{i \oplus \kappa_n} \frac{\varepsilon_i N}{2}))^2 \right) dy \right\}$$

$$\times \sqrt{\frac{2}{\pi N}} \exp \left\{ -\frac{2}{N} \left(x - N(\frac{1}{2} + p') \right)^2 \right\} dx.$$

We calculated the formula for the corresponding ε_i to the parameters of Table 3 of experimental results.

A.2 Numerical versus Experimental

We compare success rates numerically computed from the theoretical formula to experimentally obtained success rates. Note that, in both case of numerical and experimental, we assume that the attacker knows the key of the first round in advance, and discuss success rate to find the n-th round key.

First, we consider the case when the invisible pattern is $x = 060_{16}$. Though this is the most unbalanced among the patterns with 10 invisible bits (see Table 1), the experimental result in Table 3 shows that this pattern is not so good for attacking. The distribution table $p_s (x = ?????ab?????_2, 01110000)$ is the following:

$$\varepsilon_i = \begin{pmatrix} \frac{22}{1024} & \frac{-20}{1024} \\ \frac{-16}{1024} & \frac{14}{1024} \end{pmatrix}$$

Then, via the theoretical formula with the distribution table above implies the following relation between the parameter c and the success rate.

# invisible bits	invisible pattern x	success rate parameter c					
		2.0	4.0	6.0	8.0	10.0	12.0
10	060_{16}	27%	28%	29%	30%	31%	31%
9	441_{16}	19%	21%	22%	24%	25%	25%
8	644_{16}	93%	99%	100%	100%	—	—
7	645_{16}	99%	100%	100%	100%	—	—
6	647_{16}	99%	100%	100%	100%	—	—
5	667_{16}	95%	99%	100%	100%	—	—
4	$f71_{16}$	87%	97%	99%	100%	—	—
3	$75f_{16}$	74%	89%	95%	98%	—	—
2	$5ff_{16}$	60%	77%	85%	91%	—	—
1	$7ff_{16}$	48%	64%	74%	80%	—	—

Table 7. Calculation of success rate of varianlt attacks

c	2.0	4.0	6.0	8.0	10.0	12.0
Numerical	27%	28%	29%	30%	30%	31%
Experiment	39%	48%	51%	56%	55%	65%

We give some results on other patterns with 10 invisible bits for $c = 8$ in the following table.

x	014_{16}	044_{16}	084_{16}
N	31,551,642	45,434,364	24,929,692
$\begin{pmatrix} \varepsilon_{00_2} & \varepsilon_{01_2} \\ \varepsilon_{10_2} & \varepsilon_{11_2} \end{pmatrix}$	$\begin{pmatrix} \frac{-8}{1024} & \frac{24}{1024} \\ \frac{-12}{1024} & \frac{-4}{1024} \end{pmatrix}$	$\begin{pmatrix} \frac{-2}{1024} & \frac{4}{1024} \\ \frac{-18}{1024} & \frac{16}{1024} \end{pmatrix}$	$\begin{pmatrix} \frac{-27}{1024} & \frac{18}{1024} \\ \frac{7}{1024} & \frac{2}{1024} \end{pmatrix}$
Numerical	85%	51%	71%
Experiment	100%	72%	100%

The comparision on the pattern $x = 441_{16}$, which is the most unbalanced among 9 invisible bits, is the following.

c	2.0	4.0	6.0	8.0	10.0	12.0
Numerical	19%	21%	23%	24%	25%	25%
Experiment	21%	34%	29%	39%	35%	32%

The comparison in the case of other patterns with 9 invisible bits for $c = 8$ is the following.

x	007_{16}	064_{16}	260_{16}	441_{16}	484_{16}	602_{16}	604_{16}
N	6,721,059	7,887,910	8,588,726	6,721,059	8,588,726	6,987,214	6,721,059
Numerical	50%	85%	95%	24%	100%	50%	100%
Experiment	57%	97%	100%	38%	100%	52%	100%

Thus, the refined formula well simulates the experimental results on patterns with a few visible bits, and suggests how to choose the invisible pattern for more strong attacks on LOKI91.

B   The distribution table of p_s

$abcd?e\backslash fg???h$	$00???0_2$	$00???1_2$	$01???0_2$	$01???1_2$	$10???0_2$	$10???1_2$	$11???0_2$	$11???1_2$
$0000?0_2$	9	6	7	3	7	4	6	9
$0000?1_2$	8	9	5	11	7	8	8	8
$0001?0_2$	10	13	7	11	12	7	9	7
$0001?1_2$	3	5	5	6	9	8	11	10
$0010?0_2$	9	9	9	11	9	6	4	8
$0010?1_2$	13	10	8	7	11	8	9	9
$0011?0_2$	8	11	5	5	7	9	5	8
$0011?1_2$	6	9	10	6	12	6	8	9
$0100?0_2$	10	5	8	7	9	8	8	11
$0100?1_2$	5	11	10	8	11	9	10	6
$0101?0_2$	9	8	9	7	8	6	6	5
$0101?1_2$	7	8	6	11	7	11	7	10
$0110?0_2$	6	6	10	7	9	5	10	9
$0110?1_2$	7	8	4	9	5	12	7	8
$0111?0_2$	13	6	11	7	8	6	7	7
$0111?1_2$	7	12	8	8	7	7	7	8
$1000?0_2$	5	6	9	10	6	11	10	7
$1000?1_2$	9	3	5	9	12	6	3	6
$1001?0_2$	7	8	7	11	6	8	9	8
$1001?1_2$	10	2	7	8	8	5	9	9
$1010?0_2$	6	8	11	8	6	12	5	13
$1010?1_2$	11	4	11	9	7	7	6	11
$1011?0_2$	9	7	8	7	10	6	9	8
$1011?1_2$	6	12	9	8	10	11	10	8
$1100?0_2$	9	13	7	8	8	7	11	8
$1100?1_2$	6	7	7	6	8	7	5	9
$1101?0_2$	11	7	9	7	6	10	4	6
$1101?1_2$	6	9	9	9	11	10	5	8
$1110?0_2$	13	10	10	6	11	6	5	9
$1110?1_2$	10	9	9	7	7	5	9	8
$1111?0_2$	8	10	6	11	6	9	10	3
$1111?1_2$	7	7	7	8	9	10	7	7

Table 8. $16 \times p_s(abcd?efg???h, 01110000)$ in S box of LOKI91

Cryptanalysis of Ladder-DES

Eli Biham

Computer Science Department
Technion – Israel Institute of Technology
Haifa 32000, Israel
Email: biham@cs.technion.ac.il
WWW: http://www.cs.technion.ac.il/~biham/

Abstract. Feistel ciphers are very common and very important in
the design and analysis of blockciphers, especially due to four reasons:
(1) Many (DES-like) ciphers are based on Feistel's construction. (2) Luby
and Rackoff proved the security of a four-round Feistel construction when
the round functions are random. (3) Recently several provably secure
ciphers were suggested, which use other (assumed secure) ciphers as the
round function. (4) Other such ciphers use this construction as attempts
to improve the security of other ciphers (e.g., to improve the security of
DES).
In this paper we cryptanalyze Ladder-DES, a four-rounds Feistel cipher
using DES in the round function, and show that its security is smaller
than expected.

1 Introduction

Feistel ciphers are very common and well known. In particular, Feistel's construction is used in the Data Encryption Standard[8], Lucifer[12] and their many
successors (such as Feal[11,7], GDES[10], and many others). This construction
was studied from the theoretical point of view by Luby and Rackoff[4], who concluded that four rounds suffice to prove its security when the round function is
random. Many suggested cryptosystems were designed with this construction,
using another cipher as the round function: some examples are Bear and Lion[2],
Beast[6], and Ladder-DES[9]. Many other works had generalized or adopted the
Feistel/Luby-Rackoff construction (some of them are [5,1]).

In this paper we cryptanalyze Ladder-DES, a four-round Feistel cipher, whose
aim is to increase the security of DES, using DES in the round function. We
describe the attack on Ladder-DES, and show that the security of Ladder-DES
is smaller than expected. This attack can be generalized to many similar Feistel
ciphers whose two parts (halves) are of the same size and whose round functions
are permutations. This attack uses a novel application of the birthday paradox.

2   Description Of Ladder-DES

First we describe ladder-DES. It consists of four Feistel rounds, each applies DES as the round function. The four keys of the four DES applications serve as the key of Ladder-DES. The following figure describes Ladder-DES:

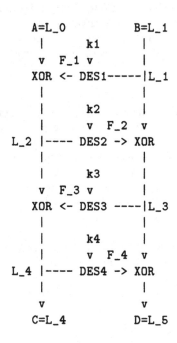

The rounds are numbered from 1 to 4, L_i is the 64-bit input of DES in round i, and F_i is the output of DES in round i. L_0 and L_5 are the left halves of the plaintext and the ciphertext, respectively.

3   A chosen Plaintext Attack

The main tool of the attack is the birthday paradox, which is used in a very unusual way. Usually the birthday paradox is used to find a collision (two equal values) in a set of $\sqrt{n}$ random values. Our attack uses the birthday paradox to identify whether given values are calculated by a pseudo-random function or a pseudo-random permutation. In the first case, the birthday paradox predicts the existence of a collision given $\sqrt{n}$ values. In the later case, collision cannot occur even given all the n values. In our attack the key is found only when we identify that there is no collision. This is the only use of the birthday paradox in this way which we are aware of.

In the attack we choose 2^{36} plaintexts of the form (A,B) where B is your favorite (or random) 64-bit fixed constant, and A gets 2^{36} different 64-bit values.

In this context, L_1 is fixed in all the 2^{36} encryption runs, and L_0 gets 2^{36} different values in the 2^{36} encryption runs. for simplicity, we will call this property of L_0 a *permutation* (i.e., there is no collision; this property holds even in all the 2^{64} possible plaintexts with a fixed B). F_1 is L_1 encrypted under a fixed (but unknown) key, thus it is fixed in all the runs. A permutation XORed with a fixed value is also a permutation, and thus L_2 is a permutation, and F_2 is also a permutation. L_1 is fixed, and thus L_3 and F_3 are permutations as well. L_4 is not a permutation: it is a mix of two permutation, which behaves like a pseudo-random function.

Our aim is to find the permutation in L_3, given the ciphertext $(C, D) = (L_4, L_5)$.

When the 2^{36} ciphertexts are given, we try all the 2^{56} possible keys k4, one by one, using the following algorithm:

```
for each possible key k4 (in range 0 to 2^56 - 1)
    for each ciphertext (Ci, Di) (i = 1, ..., 2^36)
        compute L3^{i,k4} = DES^{-1}_{k4}(Ci) ⊕ Di
        if a collision occurs (i.e., L3^{i,k4} = L3^{j,k4} for some j < i)
            conclude that k4 is wrong, and try next k4
    end for
    – We reach here only when k4 is the right key!!
    conclude that k4 is the key
end for
```

When we decrypt the ciphertext with a wrong candidate for k4, the one-round decryption function (that computes L_3) is expected to behave like a random function. For each candidate key we decrypt the fourth round of all the 2^{36} ciphertexts, or till we get two equal values of L_3. If two equal values of L_3 are found, L_3 is not a permutation, and thus the candidate for k4 is not the key. In average about 2^{32} candidates are required to discard a wrong candidate. The real value of k4 does not imply any collision of L_3 even if all the 2^{64} possible ciphertexts are decrypted by one round, and thus it can be identified.

Later, the values of k3 can be found with the same data, because L_2 is a permutation, but if a wrong value of k3 is used during decryption, the resultant value of L_2 would not be a permutation. A simpler method to find k3 takes two of the plaintexts, compute the difference of the output of F_3 as the XOR of the differences of L_0 and of L_4. Then, it searches exhaustively for the key k3 which satisfies this difference. False alarms can be identified and discarded using a third plaintext.

The remaining key k1 and k2 can then be found by exhaustive search, which would require only one plaintext/ciphertext sample, taken from the data we already have. Each of k1 and k2 would be found with complexity 2^{56}, after k3 and k4 are known.

Some notes on the birthday paradox:

About $\sqrt{2 \cdot \log_e 2 \cdot 2^{64}} = 1.177 \cdot 2^{32} \approx 2^{32}$ random values are required to find two equal 64-bit values with probability $1/2$, and $\sqrt{2 \cdot \log_e 2 \cdot 2^{64} \cdot m} = \sqrt{m} \cdot 1.177 \cdot 2^{32}$ random values are required to find such a pair with probability $1 - 2^{-m}$. In particular, in the interesting case when m=56, and we have an error probability of 2^{-56}, we need only $\sqrt{56} \cdot 1.177 \cdot 2^{32} = 8.1 \cdot 2^{32} = 2^{35}$ values to identify whether they are the result of a random function or a permutation. Thus given 2^{35} ciphertexts we can identify the key almost without mistakes, and with 2^{36} ciphertexts we can be almost ensured to have no mistakes (error probability about $e^{-128} = 2^{-185}$, which causes probability 2^{-129} for a false alarm). In average we need only 2^{32} trys, and only in a few cases we need more than 2^{34} trys for a key (except for the real key).

Complexity:

We try 2^{56} keys, for each we calculate in average 2^{32} single DES's before we discard it. Thus our complexity is about 2^{88} to find k4. The complexity to find k3 is 2^{57}. k1 and k2 can then be found with complexity 2^{56} each. Thus, the total complexity is about 2^{88}. Only 2^{35}–2^{36} chosen plaintexts are required. This complexity is much less than the expected 2^{112} complexity of a meet in the middle attack[3], which was claimed for this cryptosystem.

4   A Known Plaintext Attack

The complexity and number of required plaintexts of this known plaintext attack are about the same as of the chosen plaintext attack (2^{90} complexity, 2^{36} known plaintexts). The amount of required memory is however much larger than the chosen plaintext attack requires.

When the plaintexts/ciphertexts are given, we try all the 2^{56} possible keys k4 one by one. For each k4 we search for collisions in L_3 as in the chosen ciphertext attack, but this time collisions should occur for all the keys. We keep the first two collisions we find (in lexicographic order of the index of the plaintexts) in a table (of size $2 \cdot 2^{56}$, each keeps only the index of the pair). Similarly, we try all the values of k1 and search for collisions in F_3 ($F_3 = A \oplus C \oplus DES_{K1}(B)$). Clearly, a pair collides in L_3 iff it collides in F_3. We then search for pairs in the first table which have the same indices as pairs in the second table: only such pairs can suggest the right k1 and k4. It is expected that only the right k1 and k4 will collide in two same pairs (average of 2^{-16} false alarms; additional safety margins can be added by keeping three colliding pairs in the tables, which reduces the rate of false alarms to 2^{-80}).

The remaining k2 and k3 are easily found later with complexity 2^{57}.

This attack requires 2^{36} known plaintext, 2^{90} work (in average to find the first two colliding pairs for each key) and requires 2^{57} space (about 2^{60}–2^{61} bytes).

5   Acknowledgements

We are very grateful to Don Coppersmith for his various comments which improved the results of this paper. This research was supported by the fund for the promotion of research at the Technion.

References

1. William Aiello, Ramarathnam Venkatesan, *Foiling Birthday Attacks in Length-Doubling Transformations*, Lecture Notes in Computer Science, Advances in Cryptology, proceedings of EUROCRYPT'96, pp. 307–320, 1996.
2. Ross Anderson, Eli Biham, *Two Practical and Provably Secure Block Ciphers: BEAR and LION*, proceedings of Fast Software Encryption, Cambridge, Lecture Notes in Computer Science, pp. 113–120, 1996.
3. W. Diffie, M. E. Hellman, *Exhaustive Cryptanalysis of the NBS Data Encryption Standard*, Computer, Vol. 10, No. 6, pp. 74–84, June 1977.
4. M. Luby, C. Rackoff, *How to construct pseudorandom permutations from pseduorandom functions*, SIAM Journal on Computing, Vol. 17, No. 2, pp. 373–386, 1988.
5. Stefan Lucks, Faster Luby-Rackoff Ciphers, proceedings of Fast Software Encryption, Cambridge, Lecture Notes in Computer Science, pp. 189–203, 1996.
6. Stefan Lucks, BEAST: A Fast Block Cipher for Arbitrary Blocksizes, 1996.
7. Shoji Miyaguchi, Akira Shiraishi, Akihiro Shimizu, *Fast Data Encryption Algorithm FEAL-8*, Review of electrical communications laboratories, Vol. 36, No. 4, pp. 433–437, 1988.
8. National Bureau of Standards, *Data Encryption Standard*, U.S. Department of Commerce, FIPS pub. 46, January 1977.
9. Terry Ritter, *Ladder-DES: A Proposed Candidate to Replace DES*, appeared in the Usenet newsgroup sci.crypt, February 1994.
10. Ingrid Schaumuller-Bichl, *On the Design and Analysis of New Cipher Systems Related to the DES*, technical report, Linz university, 1983.
11. Akihiro Shimizu, Shoji Miyaguchi, *Fast Data Encryption Algorithm FEAL*, Lecture Notes in Computer Science, Advances in Cryptology, proceedings of EUROCRYPT'87, pp. 267–278, 1987.
12. Arthur Sorkin, *Lucifer, a Cryptographic Algorithm*, Cryptologia, Vol. 8, No. 1, pp. 22–41, January 1984.

A Family of Trapdoor Ciphers

Vincent Rijmen*   Bart Preneel**

Katholieke Universiteit Leuven,
Department Electrical Engineering-ESAT/COSIC
K. Mercierlaan 94, B-3001 Heverlee, Belgium
{vincent.rijmen,bart.preneel}@kuleuven.ac.be

Abstract. This paper presents several methods to construct trapdoor block ciphers. A trapdoor cipher contains some hidden structure; knowledge of this structure allows an attacker to obtain information on the key or to decrypt certain ciphertexts. Without this trapdoor information the block cipher seems to be secure. It is demonstrated that for certain block ciphers, trapdoors can be built-in that make the cipher susceptible to linear cryptanalysis; however, finding these trapdoors can be made very hard, even if one knows the general form of the trapdoor. In principle such a trapdoor can be used to design a public key encryption scheme based on a conventional block cipher.

1 Introduction

Researchers have been wary of trapdoors in encryption algorithms, ever since the DES [9] was proposed in the seventies [15]. In spite of this, no one has been able to show how to construct a practical block cipher with a trapdoor. For most current block ciphers it is relatively easy to give strong evidence that there exist no full trapdoors. We define a *full* trapdoor as some secret information which allows an attacker to obtain knowledge of the key by using a very small number of known plaintexts, no matter what these plaintexts are or what the key is.

In this paper we consider *partial* trapdoors, i.e., trapdoors that not necessarily work for all keys, or that give an attacker only partial information on the key. We show that it is possible to construct block ciphers for which there exists a linear relation with a high probability; knowledge of such a relation allows for a linear attack which requires only a very small number of known plaintexts [13, 14]. A trapdoor is said to be *detectable* (*undetectable*) if it is computationally feasible (infeasible) to find it even if one knows the general form of the trapdoor.

The rest of this paper is organized as follows. In §2 we explain how both detectable and undetectable trapdoors can be built into S-boxes. §3 deals with trapdoors in round functions and complete block ciphers. Extensions are discussed in §4, and the conclusions are presented in §6.

---

* F.W.O research assistant, sponsored by the Fund for Scientific Research - Flanders (Belgium).
** F.W.O. postdoctoral researcher, sponsored by the Fund for Scientific Research - Flanders (Belgium).

The inner product of two Boolean vectors x and y with components $x[0]$ through $x[m]$ and $y[0]$ through $y[m]$ will be denoted with

$$x \bullet y = \bigoplus_{i=1}^{m} x[i] \cdot y[i] \,.$$

2   Trapdoor $m \times n$ S-boxes

In this section we discuss the construction and hiding of trapdoors in S-boxes.

2.1   Construction

An $m \times n$ substitution box (or S-box) can be defined by its component functions: a collection of n Boolean functions $f_i(x)$, $i = 1, \ldots, n$, that take as input Boolean vectors x of dimension m. We start with an $m \times (n-1)$ S-box $S(x)$ consisting of $n-1$ functions f_i, $i = 1, \ldots, n$, $i \neq q$ selected randomly according to a uniform distribution (or following an arbitrary design criterion). The trapdoor $m \times n$ S-box $T(x)$ is derived from $S(x)$ by adding an extra function in the following way. We choose an n-bit Boolean vector β with $\beta_q = 1$ for some q with $1 \leq q \leq n$ and ensure that

$$f_q(x) = \bigoplus_{i=1, i \neq q}^{n} \beta[i] \cdot f_i(x) \tag{1}$$

with probability p_T. Then

$$\beta \bullet T(x) = 0 \tag{2}$$

holds with probability $p_T(\beta)$. This is equivalent to a correlation

$$c_T(\beta) = 2 \cdot p_T(\beta) - 1$$

between the constant zero function and $\beta \bullet T(x)$. The trapdoor information is the vector β.

2.2   Hiding the Trapdoor

If the S-box is claimed to be selected randomly according to a uniform distribution from all $m \times n$ S-boxes, it is rather easy to hide a trapdoor in it. Indeed, for large values of n and m, the function $f_q(x)$ is computationally indistinguishable from a randomly selected one. We first prove that this construction in fact introduces only one β-vector with a high correlation value, not accompanied by a range of β-vectors with 'smaller' correlation values. Then we discuss the difficulty of finding this trapdoor vector.

Introducing no more than one β with high correlation: Suppose $S(x)$ is an $m \times (n-1)$ S-box selected such that for all n-bit vectors γ,

$$c_S(\gamma) \leq c_{\max} .$$

Consider now the $m \times n$ S-box $T(x)$ that results from adding $f_q(x)$ to $S(x)$. It still holds that for all γ with $\gamma_q = 0$,

$$c_T(\gamma) = c_S(\gamma) \leq c_{\max} ,$$

so we are left with the cases where $\gamma_q = 1$. If $p_T = 1$, then $\beta \bullet T(x) = 0$ and:

$$\begin{aligned}
\gamma \bullet T(x) &= (\gamma \bullet T(x)) \oplus (\beta \bullet T(x)) \\
&= (\gamma \oplus \beta) \bullet T(x) \\
&= (\gamma \oplus \beta) \bullet S(x) .
\end{aligned} \tag{3}$$

Since $(\gamma \oplus \beta)_q = 0$, for all $\gamma \neq \beta$,

$$c_T(\gamma) = c_S(\gamma \oplus \beta) \leq c_{\max} . \tag{4}$$

Equation (3) holds with probability p_T. If $p_T < 1$ it is possible that (4) does not hold. Consider in this case the S-box $T'(x)$ that results from (1) if $p_T = 1$. All correlations of $T'(x)$ are below $c_{\max}$. Thus $T(x)$ can be considered as being constructed by applying $(1 - p_T) \cdot 2^m$ random changes to one component of $T'(x)$. The probability that these random changes to the random S-box will result in a significant change of $c_{\max}$ is very small.

Recovering β If a cryptanalyst suspects a relation of the form (2), he can decide to examine the $2^n - 1$ non-zero values of β exhaustively. For each value of β, verifying p_T requires the computation of a Walsh-Hadamard transform on an m-bit Boolean function [2], which requires $O(m \cdot 2^m)$ operations. If $(m, n) = (8, 32)$ this is feasible and the trapdoor is detectable, but for $(m, n) = (8, 64)$ this requires about 2^{64} Walsh-Hadamard transformations on 8-bit functions, which is currently quite hard (256 times more difficult than a DES key search). For $(m, n) = (10, 80)$ an exhaustive search is at present not feasible. The search can possibly be sped up by lattice methods (such as LLL [12]) or coding theory techniques, but the applicability of these techniques is still an open problem.

The search for the β-vector that has high correlation is equivalent to the problem of learning a parity function in the presence of noise. The Parity Assumption [4] tells that this problem is probably NP-hard. This classification only deals with the general problem; specific instances might be easier to solve. For instance, if p_T is very close to one, it is possible to use Gaussian elimination to solve the problem.

Define the m Boolean vectors $a^{(j)}$, $j = 1, \ldots, n$ as $a^{(j)}[i] = f_j(i)$, $i = 0, \ldots 2^m - 1$. Equation (1) can then be translated into

$$\bigoplus_{i=1}^{n} \beta[i] \cdot a^{(i)} = \delta . \tag{5}$$

If (1) holds with probability one, or $p_T(\beta) = 1$, then $\delta = 0$. In this case the $a^{(i)}$'s are linearly dependent and the linear relation between the vectors can be recovered in a very efficient way with Gaussian elimination on (5). If the probability of (1) is smaller than one, the vectors $a^{(i)}$ are independent; δ is different from zero and unknown to the cryptanalyst, and the Hamming weight of δ is given by

$$w_h(\delta) = 2^m(1 - p_T).$$

The cryptanalyst can still try to recover β by guessing a (low-weight) value for δ and solving the set of equations (5). Equation (5) will only have a solution when the guess for δ is correct. A more complex strategy for the cryptanalyst is to use the following equations:

$$\bigoplus_{i=1}^{n} \beta[i] \cdot a^{(i)} = \bigoplus_{i=1}^{d} \gamma[i] \cdot \delta^{(i)}. \tag{6}$$

The d vectors $\delta^{(i)}$ are guessed by the cryptanalyst. If the unknown δ can be expressed as a linear combination of the vectors $\delta^{(i)}$, the cryptanalyst can hope to find the trapdoor by solving (6) for $\beta[i]$ and $\gamma[i]$. The probability that δ is a linear combination of the d vectors $\delta^{(i)}$ increases with d.

   If the $\delta^{(i)}$ vectors are linearly independent, they generate a vector space of size 2^d. Note that we are only interested in the vectors with a low Hamming weight, say all vectors with Hamming weight $\leq D$. For simplicity we assume that all the $\delta^{(i)}$ vectors have Hamming weight one. The number of vectors in a d-dimensional space with Hamming weight $\leq D$ is given by

$$\sum_{k=1}^{D} \binom{d}{k}.$$

Table 1 shows the numerical values for several choices of D and d.

D	$d = 64$	$d = 96$	$d = 256$	$d = 1024$
1	2^6	2^7	2^8	2^{10}
10	2^{37}	2^{44}	2^{58}	2^{78}
20	2^{55}	2^{68}	2^{98}	2^{139}
32	2^{63}	2^{86}	2^{136}	2^{202}
40	2^{64}	2^{92}	2^{156}	2^{240}

Table 1. The number of vectors in a d-dimensional space with Hamming weight $\leq D$.

   For example, consider a 10×40 S-box. There are 2^{10} inputs, and for each input the equations may hold or not hold, resulting in a number of $2^{2^{10}}$ possible δ vectors; 2^{202} of them have Hamming weight ≤ 32. If we take $d = 64$, the probability p_{lc} that δ is a linear combination of d randomly chosen $\delta^{(i)}$ vectors is equal to $2^{63}/2^{202}$. The work factor of this algorithm is determined by p_{lc} and by the work to solve (6), which is $\mathcal{O}((2^m + n + d)^3)$ (note that the best asymptotic algorithms reduce the exponent from 3 to 2.376 [6]).

By increasing d we increase p_{lc}. However, if d becomes larger than a certain threshold value, spurious solutions for δ will start to appear that have a large Hamming weight. These unwanted solutions correspond to β vectors with low correlation values. This effect limits the use of Gaussian elimination. This algorithm will be be more useful than exhaustive search for β if D and n are small, and m is large.

2.3   Bent Functions

The construction of §2.1 can be extended to deal with additional constraints imposed on the functions $f_i(x)$. For example, in some block ciphers (such as the CAST family [1]), it is necessary that the component functions $f_i(x)$ are bent functions. The Maiorana construction for bent functions [7] can then be used to obtain an S-box satisfying property (2): an m-bit bent function $f(x)$ (m is even) is obtained from an $m/2$-bit permutation $\pi(y)$ and an arbitrary $m/2$-bit function $g(z)$ as follows:

$$f(x) = f(y, z) = \pi(y) \cdot z \oplus g(z).$$

Here '·' denotes multiplication in $GF(2^{m/2})$. If two component functions $f_i(x)$ and $f_j(x)$ are derived from the same permutation $\pi(y)$, we obtain

$$f_i(y, z) \oplus f_j(y, z) = g_i(z) \oplus g_j(z),$$

which can be chosen arbitrarily close to a constant function. To hide (2) in a bent function based S-box we proceed as follows. we choose a β with even Hamming weight. We divide the set of indices where $\beta_i = 1$ arbitrarily into pairs. For each pair of indices we select a different mapping $x \mapsto (y, z)$ and a different permutation π. We define $m/2$-bit functions $g_i(z)$, and extend them to full m-bit functions by adding zero values. Then

$$\beta \bullet T(x) = \bigoplus_{i=1}^{m} g_i(x) = 0$$

with probability p_T.

This construction shows that is possible to find a set of bent functions that sum to an almost constant function. We believe that is also possible to use other bent functions in a similar construction.

3   Trapdoor Ciphers

In this section we propose several constructions for trapdoors in block ciphers starting from the building blocks, i.e., the round functions.

3.1   Trapdoor Round Functions

We now show that the trapdoors in S-boxes can be extended to trapdoors in the round function of a Feistel cipher [8]. A $2p$-bit Feistel cipher with r rounds operates as follows: plaintext and ciphertext consist of two p-bit halves denoted with L_0, R_0 and L_r, R_r respectively. The key is denoted with K. Each round takes a $2p$-bit message input block L_{i-1}, R_{i-1} and a k-bit key input K_i which is derived from K using the key scheduling algorithm. The output of the ith round is computed as follows:

$$R_i = L_{i-1} \oplus F(K_i \oplus R_{i-1})$$
$$L_i = R_{i-1} \qquad\qquad i = 1, \ldots, r.$$

Here F is the round function of the Feistel cipher. Note that after the last round, the swapping of the halves is undone to make encryption and decryption similar.

In this section we consider the round functions of variants on CAST [10] and LOKI91 [5].

tCAST: The ciphers of the CAST family are 64-bit Feistel ciphers, or $p = 32$. The round function F is based on four 8×32 S-boxes, which have components that are either randomly selected functions or are bent functions [1]. The 32-bit input of the round function is divided into four bytes, each going to one of the four S-boxes; the 32-bit output is obtained as the sum modulo 2 of the outputs of the four S-boxes. Using four S-boxes with the same trapdoor β (but with a different value of c_T, denoted with $c_{T^{(i)}}$), we obtain

$$\beta \bullet F(x^{(1)}, x^{(2)}, x^{(3)}, x^{(4)}) = \bigoplus_{i=1}^{4} \beta \bullet T^{(i)}(x^{(i)}) .$$

Hence the round function correlates with the constant zero function with a correlation equal to

$$c_F = c_{T^{(1)}} c_{T^{(2)}} c_{T^{(3)}} c_{T^{(4)}} .$$

As mentioned before, 8×32 S-boxes can be checked for this type of trapdoors. However, should CAST be extended in a natural way to an 128-bit block cipher by using 8×64-bit S-boxes, finding this trapdoor becomes very difficult. The technique can be extended to CAST variants where the exor operation is replaced by a modular addition or multiplication.

tLOKI: The expansion in the round function of LOKI91 [5] allows for a subtle trapdoor, not visible in the individual S-boxes, but only in the round function.

We denote concatenation with '$||$'. The round function of LOKI91 uses four times the same 12×8 S-box, and is defined as:

$$F(x[1], \ldots, x[32]) = P(S(x[29], x[30], x[31], x[32], x[1], \ldots, x[8]) \;||$$
$$S(x[5], x[6], \ldots, x[16]) \;||\; S(x[13], x[14], \ldots, x[24]) \;||\; S(x[21], x[22], \ldots, x[32]))$$

In this analysis the bit permutation P is not relevant and will be ignored. We can build a trapdoor into this round function as follows. Let $a^{(1)}(x)$, $a^{(2)}(x)$, $a^{(3)}(x)$, and $a^{(4)}(x)$ be four 8-bit Boolean functions and $\beta = \beta^{(1)}||\beta^{(2)}||\beta^{(3)}||\beta^{(4)}$ a 32-bit Boolean vector. Suppose the following nonlinear relations hold with probabilities p_1, p_2, p_3, p_4 respectively.

$$\beta^{(1)} \bullet S(x[1], \ldots x[12]) = a^{(1)}(x[1], x[2], x[3], x[4]) \oplus a^{(2)}(x[9], x[10], x[11], x[12])$$
$$\beta^{(2)} \bullet S(x[1], \ldots x[12]) = a^{(2)}(x[1], x[2], x[3], x[4]) \oplus a^{(3)}(x[9], x[10], x[11], x[12])$$
$$\beta^{(3)} \bullet S(x[1], \ldots x[12]) = a^{(3)}(x[1], x[2], x[3], x[4]) \oplus a^{(4)}(x[9], x[10], x[11], x[12])$$
$$\beta^{(4)} \bullet S(x_1, \ldots x_{12}) = a^{(4)}(x[1], x[2], x[3], x[4]) \oplus a^{(1)}(x[9], x[10], x[11], x[12])$$

The use of nonlinear relations in a linear approximation was already studied by Knudsen and Robshaw [11]. The correlation between

$$\beta \bullet F(x[1], \ldots, x[32]) = \beta^{(1)} \bullet S(x[29], \ldots, x[8]) \oplus \beta^{(2)} \bullet S(x[5], \ldots, x[16])$$
$$\oplus \beta^{(3)} \bullet S(x[13], \ldots, x[24]) \oplus \beta^{(4)} \bullet S(x[21], \ldots, x[32])$$

and the constant zero function is now given by $(2p_1-1)(2p_2-1)(2p_3-1)(2p_4-1)$. For the parameters of LOKI91, this is probably a detectable trapdoor, at least for someone who knows what he is looking for. Again, larger block sizes and S-boxes would make such trapdoors harder to detect.

3.2 Trapdoor Ciphers

The trapdoor round functions defined above can be used to construct a trapdoor cipher. The resulting cipher will have iterative linear relations that approximate the output of every other round. For a cipher with r rounds, one needs $\lfloor r/2 \rfloor - 1$ round approximations.

For example, consider a version of tCAST with 16 rounds, block size 80 bits, and using four 10×40 S-boxes. If $p_T = 1 - 2^{-5}$ we can recover the round key of the first and the last round with Matsui's algorithm 2 [13] using approximately 875 known plaintexts. Since the Hamming weight of δ is 32, the Gaussian elimination technique to find the trapdoor will not work faster than exhaustive search.

4 Extensions

The trapdoors we considered make all use of "type II" linear relations as defined in [14]: correlations that exist between the output bits of the round function. It is also possible to hide "type I" linear relations: correlations between input and output bits of the round function. For example, we can construct S-boxes such that

$$\beta \bullet S[x] = \alpha \bullet x \qquad (A)$$
$$\alpha \bullet S[x] = \beta \bullet x \qquad (B)$$

with high probability. It is easy to see that these relations can be concatenated in the following way: $AB - BA - AB - \ldots$ The main advantage of this type of relations is that there are more of them: 2^{n+m} instead of 2^n. If $(m, n) = (8, 32)$, as in CAST, there are already 2^{40} cases to verify.

When building the trapdoor in the round function of tLOKI, we make use of the fact that in LOKI91 the key is added before the expansion (the input to the round function consists of 32 bits, but some of these are duplicated such that 48 bits are input to the S-boxes). In the DES the key is added after the expansion; in this case one can introduce trapdoors as well. A first approach consists of choosing linear functions $a^{(i)}(x)$. In this way the absolute value of the correlation between bits is independent of the key. However this imposes a severe restriction on the number of possible trapdoors, which makes them easy to detect. (We checked the DES for these trapdoors and have not found any.) Another option is to hide several key dependent trapdoors. The key schedule could be carefully adapted such that only a small number of key bits have an actual influence.

In a similar way one can hide differentials into block ciphers, in order to make them vulnerable to differential cryptanalysis [3]. However, exploitation of such trapdoors requires chosen rather than known plaintexts, which is much less practical.

5   Public Key Encryption

Besides the obvious use by government agencies to catch dangerous terrorists and drug dealers, trapdoor block ciphers can also be used for public key cryptography. For this application on selects a block cipher with variable S-boxes and makes it widely available (it is a system-wide public parameter). Bob generates a set of S-boxes with a secret trapdoor. These S-boxes form his public key. If Alice wants to send a confidential message to Bob, she generates a random session key, encrypts her message and a fixed set of plaintexts and sends the ciphertexts to Bob. The set of plaintexts can be fixed, or can be generated from a short seed using a pseudo-random bit generator. Bob uses the trapdoor and the known plaintexts to recover the session key and decrypts the message.

There seems to be no obvious way to extend this construction to digital signatures.

6   Conclusion

We have shown that is rather easy to hide trapdoors in expanding S-boxes like the 8×32-boxes that are currently used in some ciphers. Extending the S-boxes to 10×80 bits makes the trapdoors undetectable.

The expansion function that is used in LOKI and the DES can be used to combine 'innocent' S-boxes into a trapdoor round function. The fact that key addition in the DES is done after the expansion creates the possibility for key dependent trapdoors.

We conclude that the danger of trapdoors in block ciphers is real. Defending against built-in trapdoors can be done in several ways. For some ciphers it is feasible check for several classes of trapdoors. A pro-active approach is to nourish a healthy distrust for other people's pseudo-random generators. A design that uses random elements should clearly explain the process of the pseudo-random bit generation, and, if applicable, the screening process. For algorithms which are kept secret, such as Skipjack, this is an even more worrying problem.

Acknowledgments

We would like to thank the anonymous reviewers and several attendants of the workshop for the interesting suggestions. Special thanks to Ron Rivest for drawing our attention to some relevant papers on learning theory.

References

1. C.M. Adams, S.E. Tavares, Designing S-boxes for ciphers resistant to differential cryptanalysis, *Proceedings of the 3rd Symposium on State and Progress of Research in Cryptography*, W. Wolfowicz, Ed., Fondazione Ugo Bordoni, 1993, pp. 181–190.
2. K.G. Beauchamp, *Walsh Functions and Their Applications*, Academic Press, New York, 1975.
3. E. Biham, A. Shamir, *Differential Cryptanalysis of the Data Encryption Standard*, Springer-Verlag, 1993.
4. A. Blum, M. Furst, M. Kearns, R.J. Lipton, "Cryptographic primitives based on hard learning problems," *Advances in Cryptology, Proceedings Crypto'93, LNCS 773*, D. Stinson, Ed., Springer-Verlag, 1994, pp. 278–291.
5. L. Brown, M. Kwan, J. Pieprzyk, J. Seberry, "Improving resistance against differential cryptanalysis and the redesign of LOKI," *Advances in Cryptology, Proceedings Asiacrypt'91, LNCS 739*, H. Imai, R.L. Rivest, and T. Matsumoto, Eds., Springer-Verlag, 1993, pp. 36–50.
6. D. Coppersmith, S. Winograd, "Matrix multiplication via arithmetic progressions," *Proceedings of the Nineteenth Annual ACM Symposium on Theory of Computing*, 1987, pp. 1–6.
7. J.F. Dillon, "Elementary Hadamard difference sets," *Proceedings of the Sixth Southeastern Conference on Combinatorics, Graph Theory and Computing, Boca Raton, Florida, Congressum Numerantium No. XIV, Utilitas Math., Winnipeg, Manitoba*, 1975, pp. 237–249.
8. H. Feistel, W.A. Notz, J.L. Smith, "Some cryptographic techniques for machine-to-machine data communications," *Proceedings IEEE*, Vol. 63, No. 11, November 1975, pp. 1545–1554.
9. FIPS 46, *Data Encryption Standard*, NBS, U.S. Department of Commerce, Washington D.C., Jan. 1977.
10. H.M. Heys, S.E. Tavares, On the security of the CAST encryption algorithm, *Canadian Conference on Electrical and Computer Engineering*, pp. 332–335, Sept. 1994, Halifax, Canada.
11. L.R. Knudsen, M.J.B. Robshaw, "Non-linear approximations in linear cryptanalysis," *Advances in Cryptology, Proceedings Eurocrypt'96, LNCS 1070*, U. Maurer, Ed., Springer-Verlag, 1996, pp. 224–236.

12. A.K. Lenstra, H.W. Lenstra, Jr., L. Lovász, "Factoring polynomials with rational coefficients," *Math. Annalen,* No. 261, pp. 513–534, 1982.
13. M. Matsui, "On correlation between the order of S-boxes and the strength of DES," *Advances in Cryptology, Proceedings Eurocrypt'94, LNCS 950,* A. De Santis, Ed., Springer-Verlag, 1995, pp. 366–375.
14. M. Matsui, "The first experimental cryptanalysis of the Data Encryption Standard," *Advances in Cryptology, Proceedings Crypto'94, LNCS 839,* Y. Desmedt, Ed., Springer-Verlag, 1994, pp. 1–11.
15. M.E. Smid, D.K. Branstad, "The Data Encryption Standard. Past and future," in *"Contemporary Cryptology: The Science of Information Integrity,"* G.J. Simmons, Ed., IEEE Press, 1991, pp. 43–64.

The Block Cipher SQUARE

Joan Daemen[1]   Lars Knudsen[2]   Vincent Rijmen[*2]

[1] Banksys
Haachtesteenweg 1442
B-1130 Brussel, Belgium
Daemen.J@banksys.be
[2] Katholieke Universiteit Leuven , ESAT-COSIC
K. Mercierlaan 94, B-3001 Heverlee, Belgium
lars.knudsen@esat.kuleuven.ac.be
vincent.rijmen@esat.kuleuven.ac.be

Abstract. In this paper we present a new 128-bit block cipher called SQUARE. The original design of SQUARE concentrates on the resistance against differential and linear cryptanalysis. However, after the initial design a dedicated attack was mounted that forced us to augment the number of rounds. The goal of this paper is the publication of the resulting cipher for public scrutiny. A C implementation of SQUARE is available that runs at 2.63 MByte/s on a 100 MHz Pentium. Our M68HC05 Smart Card implementation fits in 547 bytes and takes less than 2 msec. (4 MHz Clock). The high degree of parallellism allows hardware implementations in the Gbit/s range today.

1 Introduction

In this paper, we propose the block cipher SQUARE. It has a block length and key length of 128 bits. However, its modular design approach allows extensions to higher block lengths in a straightforward way. The cipher has a new self-reciprocal structure, similar to that of THREEWAY and SHARK [2, 15].

The structure of the cipher, i.e., the types of building blocks and their interaction, has been carefully chosen to allow very efficient implementations on a wide range of processors. The specific choice of the building blocks themselves has been led by the resistance of the cipher against differential and linear cryptanalysis. After treating the structure of the cipher and its consequences for implementations, we explain the strategies followed to thwart linear and differential cryptanalysis. This is followed by a description of an efficient attack that exploits the particular properties of the cipher structure.

We do not encourage anyone to use SQUARE today in any sensitive application. Clearly, confidence in the security of any cryptographic design must be based on the resistance against effective cryptanalysis after intense public scrutiny.

---

* F.W.O research assistant, sponsored by the Fund for Scientific Research – Flanders (Belgium).

A reference implementation of SQUARE is available from the following URL: http://www.esat.kuleuven.ac.be/~rijmen/square.

2 Structure of SQUARE

SQUARE is an iterated block cipher with a block length and a key length of 128 bits each. The round transformation of SQUARE is composed of four distinct transformations. It is however important to note that these four building blocks can be efficiently combined in a single set of table-lookups and exor operations. This will be treated later in the section on implementation aspects.

The basic building blocks of the cipher are five different invertible transformations that operate on a 4×4 array of bytes. The element of a state a in row i and column j is specified as $a_{i,j}$. Both indexes start from 0.

2.1 A Linear Transformation θ

θ is a linear operation that operates separately on each of the four rows of a state. We have

$$\theta : b = \theta(a) \Leftrightarrow b_{i,j} = c_j a_{i,0} \oplus c_{j-1} a_{i,1} \oplus c_{j-2} a_{i,2} \oplus c_{j-3} a_{i,3},$$

where the multiplication is in $GF(2^8)$ and the indices of c must be taken modulo 4. Note that the field $GF(2^n)$ has characteristic 2 [9]. This means that the addition in the field corresponds to the bitwise exor.

The rows of a state can be denoted by polynomials, i.e., the polynomial corresponding to row i of a state a is given by

$$a_i(x) = a_{i,0} \oplus a_{i,1} x \oplus a_{i,2} x^2 a_{i,3} x^3.$$

Using this notation, and defining $c(x) = \bigoplus_j c_j x^j$ we can describe θ as a modular polynomial multiplication:

$$b = \theta(a) \Leftrightarrow b_i(x) = c(x) a_i(x) \bmod 1 \oplus x^4 \qquad \text{for} \qquad 0 \le i < 4.$$

The inverse of θ corresponds to a polynomial $d(x)$ given by

$$d(x) c(x) = 1 \pmod{1 \oplus x^4}.$$

2.2 A Nonlinear Transformation γ

γ is a nonlinear byte substitution, identical for all bytes. We have

$$\gamma : b = \gamma(a) \Leftrightarrow b_{i,j} = S_\gamma(a_{i,j}),$$

with S_γ an invertible 8-bit substitution table or S-box. The inverse of γ consists of the application of the inverse substitution S_γ^{-1} to all bytes of a state.

2.3   A Byte Permutation π

The effect of π is the interchanging of rows and columns of a state. We have

$$\pi : b = \pi(a) \Leftrightarrow b_{i,j} = a_{j,i}.$$

π is an involution, hence $\pi^{-1} = \pi$.

2.4   Bitwise Round Key Addition σ

$\sigma[k^t]$ consists of the bitwise addition of a round key k^t. We have

$$\sigma[k^t] : b = \sigma[k^t](a) \Leftrightarrow b = a \oplus k^t.$$

The inverse of $\sigma[k^t]$ is $\sigma[k^t]$ itself.

2.5   The Round Key Evolution ψ

The round keys k^t are derived from the cipher key K in the following way. k^0 equals the cipher key K. The other round keys are derived iteratively by means of the invertible affine transformation ψ.

$$\psi : k^t = \psi(k^{t-1})$$

2.6   The Cipher SQUARE

The building blocks are composed into the round transformation denoted by $\rho[k^t]$:

$$\rho[k^t] = \sigma[k^t] \circ \pi \circ \gamma \circ \theta \tag{1}$$

SQUARE is defined as eight rounds preceeded by a key addition $\sigma[k^0]$ and by θ^{-1}:

$$\text{SQUARE}[k] = \rho[k^8] \circ \rho[k^7] \circ \rho[k^6] \circ \rho[k^5] \circ \rho[k^4] \circ \rho[k^3] \circ \rho[k^2] \circ \rho[k^1] \circ \sigma[k^0] \circ \theta^{-1} \tag{2}$$

2.7   The Inverse Cipher

As will be shown in Section 9, the structure of SQUARE lends itself to efficient implementations. For a number of modes of operation it is important that this is also the case for the inverse cipher. Therefore, SQUARE has been designed in such a way that the structure of its inverse is equal to that of the cipher itself, with the exception of the key schedule. Note that this identity in *structure* differs from the identity of *components and structure* in IDEA [10].

From (2) it can be seen that

$$\text{SQUARE}^{-1}[k] = \theta \circ \sigma^{-1}[k^0] \circ \rho^{-1}[k^1] \circ \rho^{-1}[k^2] \circ \rho^{-1}[k^3] \circ \rho^{-1}[k^4] \circ$$
$$\rho^{-1}[k^5] \circ \rho^{-1}[k^6] \circ \rho^{-1}[k^7] \circ \rho^{-1}[k^8]$$

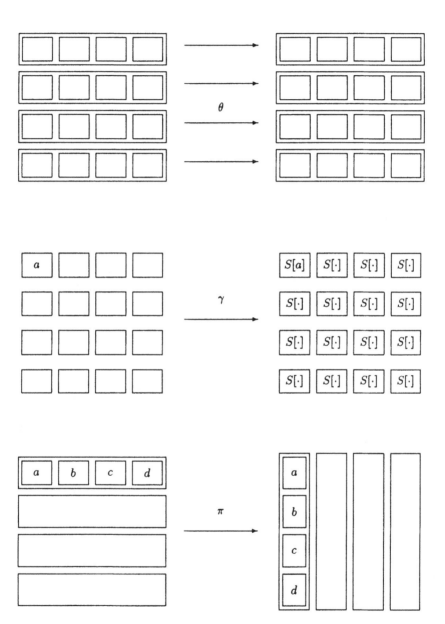

Fig. 1. Geometrical representation of the basic operations of SQUARE. θ consists of 4 parallel linear diffusion mappings. γ consists of 16 separate substitutions. π is a transposition.

with

$$\rho^{-1}[k^t] = \theta^{-1} \circ \gamma^{-1} \circ \pi^{-1} \circ \sigma^{-1}[k^t] = \theta^{-1} \circ \gamma^{-1} \circ \pi \circ \sigma[k^t] \qquad (3)$$

It may seem that the structure of the inverse cipher differs substantially from that of the cipher itself. By exploiting some algebraic properties of the building blocks, we can show this not to be the case. Since π only transposes the bytes $a_{i,j}$ and γ only operates on the individual bytes, independent of their position (i, j), we have

$$\gamma^{-1} \circ \pi = \pi \circ \gamma^{-1}.$$

Moreover, since $\theta^{-1}(a) \oplus k^t = \theta^{-1}(a + \theta(k^t))$, we have

$$\sigma[k^t] \circ \theta^{-1} = \theta^{-1} \circ \sigma[\theta(k^t)],$$

We now define the round transformation of the inverse cipher as

$$\rho'[k^t] = \sigma[k^t] \circ \pi \circ \gamma^{-1} \circ \theta^{-1}, \qquad (4)$$

which has the same structure as ρ itself, except that γ and θ are replaced by γ^{-1} and θ^{-1} respectively. Using the algebraic properties above, we can derive

$$
\begin{aligned}
\theta \circ \sigma[k^0] \circ \rho^{-1}[k^1] &= \theta \circ \sigma[k^0] \circ \theta^{-1} \circ \gamma^{-1} \circ \pi \circ \sigma[k^1] \\
&= \theta \circ \theta^{-1} \circ \sigma[\theta(k^0)] \circ \pi \circ \gamma^{-1} \circ \sigma[k^1] \\
&= \sigma[\theta(k^0)] \circ \pi \circ \gamma^{-1} \circ \sigma[k^1] \\
&= \sigma[\theta(k^0)] \circ \pi \circ \gamma^{-1} \circ \sigma[k^1] \circ \theta^{-1} \circ \theta \\
&= \sigma[\theta(k^0)] \circ \pi \circ \gamma^{-1} \circ \theta^{-1} \circ \sigma[\theta(k^1)] \circ \theta \\
&= \rho'[\theta(k^0)] \circ \sigma[\theta(k^1)] \circ \theta
\end{aligned}
$$

This equation can be generalized in a straightforward way to include more than one round. Now, with $\kappa^t = \theta(k^{8-t})$, we have

SQUARE$^{-1} =$

$$\rho'[\kappa^8] \circ \rho'[\kappa^7] \circ \rho'[\kappa^6] \circ \rho'[\kappa^5] \circ \rho'[\kappa^4] \circ \rho'[\kappa^3] \circ \rho'[\kappa^2] \circ \rho'[\kappa^1] \circ \sigma[\kappa^0] \circ \theta$$

Hence the inverse cipher is equal to the cipher itself with γ replaced by γ^{-1}, with θ by θ^{-1} and different round key values.

2.8 First round

The θ^{-1} before $\sigma[k^0]$ in SQUARE can be incorporated in the first round. We have

$$
\begin{aligned}
\rho[k^1] \circ \sigma[k^0] \circ \theta^{-1} &= \sigma[k^1] \circ \pi \circ \gamma \circ \theta \circ \sigma[k^0] \circ \theta^{-1} \\
&= \sigma[k^1] \circ \pi \circ \gamma \circ \sigma[\theta(k^0)]
\end{aligned}
$$

Hence the initial θ^{-1} can be discarded by omitting θ in the first round and applying $\theta(k^0)$ instead of k^0. The same simplification can be applied to the inverse cipher.

3   Linear and Differential Cryptanalysis

The resistance against linear cryptanalysis [12] and differential cryptanalysis [1] has been the rationale behind the criteria by which the S_γ substitution and the θ multiplication polynomial $c(x)$ have been chosen.

A difference propagation along the rounds of an iterated block cipher is generally called a differential characteristic. A characteristic is specified by a series of difference patterns. The probability associated with a characteristic is the probability that all intermediate difference patterns have the value specified in the above series. We call a differential characteristic a differential *trail*. The probability associated with a differential trail can be approximated by the product of the difference propagations between every pair of subsequent rounds (which can be easily calculated). The probability that a given difference pattern a' at the input of a number of cipher rounds gives rise to a difference pattern b' at the output is equal to the sum of the probabilities of all differential trails starting with a' and ending with b'. In general the propagation from the input difference pattern a' to the output difference pattern b' is called a *differential*.

As can be seen in [12] the correlation between a linear combination of input bits and a linear combination of output bits of an iterated block cipher can be treated in an analogous but slightly different way. A *linear trail* is specified by a series of selection patterns. For a given cipher key, the *correlation coefficient* (positive or negative) corresponding to a linear trail consists of the product of the correlation coefficients between the linear combinations of bits of every pair of subsequent rounds. In [2] it was shown that the correlation between a linear combination of input bits, denoted by selection pattern u, and a linear combination of output bits, denoted by v is equal to the sum of the correlation coefficients of all linear trails starting with u and ending in v. It must be remarked that the correlation coefficients may be positive or negative and that the sign depends on the value of round key bits.

S_γ and $c(x)$ are chosen to minimize the maximum probability of differential trails and the maximum correlation of linear trails over four rounds. This is obtained in the framework of a very specific approach.

3.1   Wide Trail Design Strategy

In [2] the 'wide trail design strategy' was introduced as a means to guarantee low maximum probability of multiple-round differential trails and low maximum correlation of multiple-round linear trails. In this strategy the round transformation is composed of a number of uniform transformations, that are split in the nonlinear blockwise substitution (corresponding to our γ) and the composition of the linear transformations (corresponding to our $\theta \circ \pi$). The round key addition does not play a role in the strategy. It was shown in [2] that the probability of a differential trail is the product of the input-output difference propagation probabilities of the S-boxes with nonzero input difference ('active S-boxes'). The correlation of a linear trail is the product of the input-output correlations of the S-boxes with nonzero output selection patterns ('active S-boxes'). The two mechanisms for eliminating high-probability differential trails and high-correlation linear trails are the following:

- Choose an S-box where the maximum difference propagation probability and the maximum input-output correlation are as small as possible.
- Choose the linear part in such a way that there are no trails with few active S-boxes.

The first mechanism gives us two clear criteria for the selection of the S_γ. The second mechanism gives a hint on how to select the multiplication polynomial $c(x)$. In the following section we will focus on the linear part $\theta \circ \pi$.

4   The Multiplication Polynomial $c(x)$

The transformation θ treats the different rows of a state a completely separately and in the same way. We will now study the difference propagation and correlation properties of θ, concentrating on a single row. Assume an input difference specified by $a'(x) = a(x) \oplus a^*(x)$. The output difference will be given by

$$b'(x) = c(x)a(x) \oplus c(x)a^*(x) \bmod 1 \oplus x^4 = c(x)a'(x) \bmod 1 \oplus x^4.$$

On the other hand, a linear combination of output bits, specified by the selection polynomial $u(x)$ is equal to (i.e., correlated to, with correlation coefficient 1) a linear combination of input bits, specified by the following selection polynomial [2]:

$$v(x) = c(x^{-1})u(x) \bmod 1 + x^4.$$

It is intuitively clear that both linear and differential trails would benefit from a multiplication polynomial that could limit the number of nonzero terms in input and output difference (and selection) polynomials. This is exactly what we want to avoid by choosing a polynomial with a high diffusion power, expressed by the so-called *branch number*.

Let $w_h(a)$ denote the Hamming weight of a vector, i.e., the number of nonzero components in that vector. Applied to a state a, a difference pattern a' or a selection pattern u, this corresponds to the number of non-zero bytes. In [2] the branch number B of an invertible linear mapping was introduced as

$$B(\theta) = \min_{a \neq 0}(w_h(a) + w_h(\theta(a))).$$

This implies that the sum of the Hamming weights of a pair of input and output difference patterns (or selection patterns) to θ is at least B. It can easily be shown that B is a lower bound for the number of active S-boxes in two consecutive rounds of a linear or differential trail. Since θ operates on each row separately, we can have $B = 5$ at most.

In [15] it was shown how a linear mapping over $GF(2^m)^n$ with optimal B ($B = n + 1$) can be constructed from a maximum distance separable code. The MDS-code used is a Reed-Solomon code over $GF(2^m)$: if $G_e = [I_{n \times n} B_{n \times n}]$ is the echelon form of the generation matrix of and $(2n, n, n + 1)$-RS-code, then $\theta : X \mapsto Y = B \cdot X$ defines a linear mapping with optimal branch number.

The polynomial multiplication with $c(x)$ corresponds to a special subset of the MDS-codes, having the additional property that B is a circulant matrix. A circulant matrix is a matrix where every row consists of the same elements, shifted over one position, or $b_{i,j} = b_{0,j-i \bmod n}$. This property is exploited in section 9.2 to produce a memory-efficient implementation of the cipher. In [11] we find the following theorem:

Theorem 1. *An (n, k, d)-code $\mathcal{C}$ with generator matrix $G = [IB]$ is MDS iff every square submatrix of B is nonsingular.*

In a matrix with elements from $\mathrm{GF}(2^m)$ every determinant has a probability of 2^{-m} to evaluate to zero. For increasing size of the matrix the number of determinants increases exponentially, making it infeasible to search randomly for an MDS-code. However, in a circulant matrix the number of distinct determinants is only a fraction of the number for arbitrary matrices (cf. Table 1). By imposing the extra constraint that the matrix should be a circulant, we increase the probability to find an MDS-code by random search.

n	generic	circulant	n	generic	circulant
1	1	1	5	252	41
2	5	3	6	924	111
3	20	7	7	3431	309
4	70	17	8	12869	935

Table 1. The number of square submatrices in a generic matrix of order n, and the number of non-equivalent determinants in a circulant matrix of the same order. The numbers of the last column were obtained by an exhaustive computer search.

$c(x)$ corresponds to a 4×4 matrix, hence if we choose it randomly, the probability that it has $\mathcal{B} = 5$ can be approximated by $(1 - \frac{1}{256})^{17} \approx 0.93$. This gives us a high degree of freedom in the choice of $c(x)$. We choose

$$c(x) = 2_{\mathbf{x}} \oplus 1_{\mathbf{x}} \cdot x \oplus 1_{\mathbf{x}} \cdot x^2 \oplus 3_{\mathbf{x}} \cdot x^3 .$$

This determines $d(x)$ uniquely.

$$d(x) = \mathbf{E}_{\mathbf{x}} \oplus 9_{\mathbf{x}} \cdot x \oplus \mathbf{D}_{\mathbf{x}} \cdot x^2 \oplus \mathbf{B}_{\mathbf{x}} \cdot x^3$$

4.1   Motivation for the Choice of π

Since the branch number of $c(x)$ is 5, the number of active S-boxes in a two-round trail is at least 5. The effect of π, interchanging rows and columns, has the effect that any trail over four consecutive rounds will have at least 25 active S-boxes. A simple and clear proof of this is available and will be published in a more theoretical paper that is being written [3] .

5   The Nonlinear Substitution γ

As explained above, the relevant criteria imposed upon the γ S-box are the highest (in absolute value) occurring correlation between any pair of linear combination of input bits and linear combinations of output bits (denoted by λ) and the highest occurring probability corresponding to any pair of input difference and output difference pattern. This corresponds to the highest value in the so-called exor table of the γ S-box, defined as

$$E_{ij} = \#\{x | S(x) \oplus S(x \oplus i) = j\} .$$

We define $\delta = \max_{i,j}\{E_{ij}\} \cdot 2^{-8}$.

We present three alternative choices for the S-box: explicitly constructed nonlinear algebraic transformations, slightly modified versions of the latter and randomly selected invertible mappings.

5.1   Explicit Construction

In [13] a method is given to construct m-bit S-boxes with $\gamma = 2^{1-m/2}$ and $\delta = 2^{2-m}$, the theoretically minimum possible values. From the proposals in [13] we select the mapping $x \mapsto x^{-1}$ over $\mathrm{GF}(2^8)$, with $\delta = 2^{-6}$ and $\lambda = 2^{-3}$.

The problem with this choice is that the mapping has a very simple description in $\mathrm{GF}(2^8)$. The other components of the round transformation also have a simple description in $\mathrm{GF}(2^8)$. This may enable cryptanalytic attacks based on the algebraic manipulation of equations to derive key information [4].

Note that any m-bit mapping can be represented as a polynomial or a rational form in $\mathrm{GF}(2^m)$. It is however unlikely that this representation can be exploited in cryptanalysis if the polynomial or rational form is of no special, relatively simple, form.

The feasibility of algebraic manipulation can be severely diminished. The elements of $\mathrm{GF}(2^8)$ can be represented with respect to different bases. By choosing a different basis for the definition of θ and γ we can prevent that the round transformation has a simple description in any basis of $\mathrm{GF}(2^8)$.

Still, even specified in another basis, the chosen nonlinear mapping stays an involution and has two fixed points: 0 and 1. By applying an affine transformation on the individual bits of the output these properties can be removed and a simple algebraic expression of the round transformation in any basis of $\mathrm{GF}(2^8)$ can be prevented.

5.2   Modifications

Another method to prevent a simple algebraic description is by choosing a mapping according to the method explained in the previous subsection and subsequently modifying it slightly to destroy the exploitable algebraic structure. It will be seen that the disadvantage of this approach is that δ and/or λ will increase.

We conducted some experiments starting from the mapping multiplicative inverse in $GF(2^8)$ as proposed above ($\delta = 2^{-6}$ and $\lambda = 8 \times 2^{-6}$) and we applied a small number of modifications.

When we consider the mapping as a look-up table and investigate all variants that have a pair of entries swapped, an increase is observed of δ to $6 \cdot 2^{-8}$ and/or λ to 9×2^{-6}. We also tested 300 000 variants in which four or eight entries were swapped. Swapping four entries increases λ to 9×2^{-6}, swapping eight entries increases λ to 10×2^{-6} and δ to $6 \cdot 2^{-8}$.

5.3   Random Search

Algebraically constructed permutations always exhibit some structure that may be exploited in attacks in unanticipated ways, designers often resort to random substitutions: a substitution is selected from a set of substitutions that are generated by the use of a random source and evaluated with respect to (presumably) relevant nonlinearity criteria. In [14] the average differential properties of permutations are investigated and a bound for the expected value of δ is given. For an m-bit permutation

$$\lim_{m \to \infty} \frac{E[\delta 2^m]}{2m} \leq 1 .$$

We verified this experimentally for 1.5 million samples with $m = 8$ and measured at the same time δ and λ. The results are given in table 2. The S-boxes with the highest resistance against both linear and differential cryptanalysis, have $\delta = 10 \cdot 2^{-8}$ and $\lambda = 15 \cdot 2^{-6}$.

λ	δ						
	$8 \cdot 2^{-8}$	$10 \cdot 2^{-8}$	$12 \cdot 2^{-8}$	$14 \cdot 2^{-8}$	$16 \cdot 2^{-8}$	$18 \cdot 2^{-8}$	$20 \cdot 2^{-8}$
15×2^{-6}	0	0.07	0.07	0.006	0.0001	0	0
16×2^{-6}	0.0003	4.77	5.58	0.58	0.04	0.002	0
17×2^{-6}	0.002	15.63	20.55	2.24	0.15	0.007	0.0004
18×2^{-6}	0.0002	12.21	17.17	1.96	0.13	0.007	0.0005
19×2^{-6}	0.0004	4.91	7.31	0.87	0.05	0.003	0
20×2^{-6}	0	1.52	2.34	0.28	0.02	0.001	0
21×2^{-6}	0	0.41	0.64	0.08	0.004	0.001	0

Table 2. Maximum input-output correlation and difference propagation probability of randomly generated nonlinear permutations. The entries denote the percentage of the generated mappings that have the indicated λ and δ.

5.4   Our Choice

Because of its optimal values for λ and δ, we have decided to take for S_γ an S-box that is constructed by taking the mapping $x \mapsto x^{-1}$ and applying an affine transformation (over $GF(2)$) to the output bits. This affine transformation has the property that it has a complicated description in $GF(2^8)$ to thwart interpolation attacks [4].

Our choices force all four-round differential trails to have an associated probability not higher than 2^{-150}, far below the critical noise value of 2^{-127}. Equivalently, four-round linear trails have an associated correlation not over 2^{-75}, far below the critical noise value of 2^{-64}. Hence, for resistance against conventional LC and DC six rounds may seem sufficient. However, the specific blocked structure of the cipher allows for more efficient dedicated differential attacks. This will be explained in the following section.

6   A Dedicated Attack

In this section we describe a dedicated attack that exploits the cipher structure of SQUARE. The attack is a chosen plaintext attack and is independent of the specific choices of S_γ, $c(x)$ and the key schedule. It is faster than an exhaustive key search for SQUARE versions of up to 6 rounds. After describing the basic attack on 4 rounds, we will show how it can be extended to 5 and 6 rounds.

6.1   Preliminaries

Let a Λ-set be a set of 256 states that are all different in some of the (16) state bytes (the *active*) and all equal in the other state bytes (the *passive*). Let λ be the set of indices of the active bytes. We have

$$\forall x, y \in \Lambda : \begin{cases} x_{i,j} \neq y_{i,j} \text{ for } (i,j) \in \lambda \\ x_{i,j} = y_{i,j} \text{ for } (i,j) \notin \lambda \end{cases}$$

In this section we will make use of the geometrical interpretation as presented in Figure 1. Applying the transformations γ and $\sigma[k^t]$ on (the elements of) a Λ-set results in a (generally different) Λ-set with the same λ. Applying π results in a Λ-set in which the active bytes are transposed by π. Applying θ to a Λ-set does not necessarily result in a Λ-set. However, since every output byte of γ is a linear combination (with invertible coefficients) of the four input bytes in the same row, an input row with a single active byte gives rise to an output row with only active bytes.

6.2   Four Rounds

Consider a Λ-set in which only one byte is active. We will now trace the evolution of the positions of the active bytes through 3 rounds. The 1st round contains no θ, hence there is still only one byte active at the beginning of the 2nd round. θ of the

2nd round converts this to a complete row of active bytes, that is subsequently transformed by π to a complete column. θ of the 3rd round converts this to a Λ-set with only active bytes. This is still the case at the input to the 4th round.

Since the bytes of the outputs of the 3rd round (denoted by a) range over all possible values and are therefore balanced over the λ-set, we have

$$\bigoplus_{b=\theta(a), a\in\Lambda} b_{i,j} = \bigoplus_{a\in\Lambda}\bigoplus_k c_{j-k}a_{i,k} = \bigoplus_l c_l \bigoplus_{a\in\Lambda} a_{i,l+j} = \bigoplus_l c_l 0 = 0.$$

Hence, the bytes of the output of θ of the fourth round are balanced. This balancedness is in general destroyed by the subsequent application of γ.

An output byte of the 4th round (denoted by a here) can be expressed as a function of the intermediate state b above

$$a_{i,j} = S_\gamma[b_{j,i}] \oplus k^4_{i,j}.$$

By assuming a value for $k^4_{i,j}$, the value of $b_{j,i}$ for all elements of the Λ-set can be calculated from the ciphertexts. If the values of this byte are not balanced over Λ, the assumed value for the key byte was wrong. This is expected to eliminate all but approximately 1 key value. This can be repeated for the other bytes of k^4.

We implemented the attack and found that two Λ-sets of 256 chosen plaintexts each are sufficient to uniquely determine the cipher key with an overwhelming probability of success.

6.3 Extension by a Round at the End

If an additional round is added, we have to calculate the above value of $b_{j,i}$ from the output of the 5th round instead of the 4th round. This can be done by additionally assuming a value for a set of 4 bytes of the 5th round key. As in the case of the 4-round attack, wrong key assumptions are eliminated by verifying that $b_{j,i}$ is not balanced.

In this 5-round attack 2^{40} key values must be checked, and this must be repeated 4 times. Since by checking a single Λ-set leaves only $1/256$ of the wrong key assumptions as possible candidates, the cipher key can be found with overwhelming probability with only 5 Λ-sets.

6.4 Extension by a Round at the Beginning

The basic idea is to choose a set of plaintexts that results in a Λ-set at the output of the 2nd round with a single active S-box. This requires the assumption of values of four bytes of the round key k^0.

If the intermediate state after θ of the 2nd round has only a single active byte, this is also the case for the output of the 2nd round. This imposes the following conditions on a row of four input bytes of θ of the second round: one particular linear combination of these bytes must range over all 256 possible values (active) while 3 other particular linear combinations must be constant for all 256 states.

This imposes identical conditions on the bytes in the same row in the input to $\sigma[k^1]$, and consequently on a column of bytes in the input to π of the 1st round. If the corresponding column of bytes of k^0 is known, these conditions can be converted to conditions on four plaintext bytes.

Now we consider a set of 2^{32} plaintexts, such that the array of bytes in one column ranges over all possible values and all other bytes are constant.

Now, make an assumption for the value of the 4 bytes of the relevant column of k^0. Select from the set of 2^{32} available plaintexts, a set of 256 plaintexts that obey the conditions indicated above. Now the 4-round attack can be performed. For the given key assumption, the attack can be repeated for a several plaintext sets. If the byte values of k^5 suggested by these attacks are not consistent, the initial assumption must have been wrong. A correct assumption for the bytes of k^0 will result in the swift and consistent recuperation of the last round key.

We implemented this attack where we assumed knowledge of 16 bits of the first-round key. The attack found the other 16 bits of the first-round key and 128 bits of the last-round key using only 2 structures of 256 plaintexts for every key value guessed in the first round.

6.5   Complexity of the Attacks

Combining both extensions results in a 6 round attack. Although infeasible with current technology, this attack is faster than exhaustive key search, and therefore relevant. We have not found extensions to 7 rounds faster than exhaustive key search.

We summarize the attacks in Table 3.

Attack	#Plaintexts	Time	Memory
4-round	2^9	2^9	small
5-round type 1	2^{11}	2^{40}	small
5-round type 2	2^{32}	2^{40}	2^{32}
6-round	2^{32}	2^{72}	2^{32}

Table 3. Complexities of the attack on SQUARE.

7   Number of Rounds

Due to these attacks we have to increase the number of rounds to at least seven. As a safety margin, we fixed the number of rounds to eight.

Conservative users are free to increase the number of rounds. This can be done in a straightforward way and requires no adaptation of the key schedule whatsoever.

8   The Key Evolution ψ

The key schedule specifies the derivation of the round keys in terms of the cipher key. Its function is to provide resistance against the following types of attack:

- Attacks in which part of the cipher key is known to the cryptanalyst, e.g., if the cipher is used with a key shorter than 128 bits.
- Attacks where the key entry to the cipher is known or can be chosen, e.g., if the cipher is used as the compression function of a hash algorithm [7].
- Related-key attacks.

Resistance against the first type of attack can be improved by a key schedule in which the round key undergoes a transformation with high diffusion. For a good scheme, the knowledge of a certain number of bits of one round key fixes very few bits in other round keys. The other two types of attack exploit regularities in the structure of the key schedule by locally compensating round key differences [5, 7].

The key schedule also plays an important role in the elimination of symmetry:

- **Symmetry in the round transformation**: the round transformation treats all bytes of a state in very much the same way. This symmetry can be removed by having round constants in the key schedule.
- **Symmetry between the rounds**: the round transformation is the same for all rounds. This equality can be removed by having round-dependent round constants in the key schedule.

The key schedule is defined in terms of the rows of the key. We can define a left byte-rotation operation $\mathsf{rotr}(a_i)$ on a row as

$$\mathsf{rotl}[a_{i,0}a_{i,1}a_{i,2}a_{i,3}] = [a_{i,1}a_{i,2}a_{i,3}a_{i,0}]$$

and a right byte rotation $\mathsf{rotr}(a_i)$ as its inverse.

The key schedule iteration transformation $k^{t+1} = \psi(k^t)$ and its inverse are defined by

$$
\begin{aligned}
k_0^{t+1} &= k_0^t \oplus \mathsf{rotl}(k_3^t) \oplus C_t & \kappa_3^{t+1} &= \kappa_3^t \oplus \kappa_2^t \\
k_1^{t+1} &= k_1^t \oplus k_0^{t+1} & \kappa_2^{t+1} &= \kappa_2^t \oplus \kappa_1^t \\
k_2^{t+1} &= k_2^t \oplus k_1^{t+1} & \kappa_1^{t+1} &= \kappa_1^t \oplus \kappa_0^t \\
k_3^{t+1} &= k_3^t \oplus k_2^{t+1} & \kappa_0^{t+1} &= \kappa_0^t \oplus \mathsf{rotr}(\kappa_3^t) \oplus C_t'
\end{aligned}
$$

The simplicity of the inverse key schedule is thanks to the fact that θ and ψ commute. The round constants C_t are also defined iteratively. We have $C_0 = 1_\mathbf{x}$ and $C_t = 2_\mathbf{x} \cdot C_{t-1}$.

This choice provides high diffusion and removes the regularities in an efficient way.

9   Implementation Aspects

9.1   8-bit Processor

On an 8-bit processor SQUARE can be programmed by simply implementing the different component transformations. This is straightforward for π, σ and ψ. The transformation γ requires a table of 256 bytes. θ requires multiplication in the field $GF(2^8)$. However, the multiplication polynomial has been chosen to make this very efficient. We have written a program implementing SQUARE in Assembler for the Motorola's M68HC05 microprocessor, typical for Smart Cards. The machine code occupies in total 547 bytes of ROM, needs 36 bytes of RAM and one execution of SQUARE, including the key schedule, takes about 7500 cycles. This corresponds to less than 2 msec with a 4 MHz Clock.

The inverse cipher however is significantly slower than the forward cipher. This is caused by the difference in complexity between θ and θ^{-1}.

9.2   32-bit Processor

In the implementation of the cipher, the succession of steps

$$\theta \circ \sigma[k^t] \circ \pi \circ \gamma = \sigma[k'^t] \circ \theta \circ \pi \circ \gamma$$

with $k'^t = \theta(k^t)$ can be combined in a single set of table lookups. The intermediate state can be represented by four 32-bit words, each containing a row $[a_i]$. Its transpose is denoted by $[a_i]^T$. For $b = \theta(\pi(\gamma(a))) + k'^t$ we have

$$[b_i]^T = \begin{bmatrix} c_0 & c_3 & c_2 & c_1 \\ c_1 & c_0 & c_3 & c_2 \\ c_2 & c_1 & c_0 & c_3 \\ c_3 & c_2 & c_1 & c_0 \end{bmatrix} \cdot \begin{bmatrix} S_\gamma[a_{0,i}] \\ S_\gamma[a_{1,i}] \\ S_\gamma[a_{2,i}] \\ S_\gamma[a_{3,i}] \end{bmatrix} \oplus [k_i'^t]^T$$

$$= \begin{bmatrix} c_0 \\ c_1 \\ c_2 \\ c_3 \end{bmatrix} \cdot S[a_{0i}] \oplus \begin{bmatrix} c_3 \\ c_0 \\ c_1 \\ c_2 \end{bmatrix} \cdot S[a_{1i}] \oplus \begin{bmatrix} c_2 \\ c_3 \\ c_0 \\ c_1 \end{bmatrix} \cdot S[a_{2i}] \oplus \begin{bmatrix} c_1 \\ c_2 \\ c_3 \\ c_0 \end{bmatrix} \cdot S[a_{2i}] \oplus [k_i'^t]^T$$

We define the tables M and T as

$$M[a] = a \cdot [c_0\ c_1\ c_2\ c_3]$$
$$T[a] = M[S[a]].$$

T and M have 256 entries of four bytes each. The table M implements the polynomial multiplication. T combines the nonlinear substitution with this multiplication. Now we have

$$[b_i] = \bigoplus_j \text{rotr}^j (T[a_{ji}]) \oplus [k_i'^t]).$$

We conclude that $\sigma[\theta(k_i^t)] \circ \theta \circ \pi \circ \gamma$ can be done with 16 table lookups, 12 rotations and 16 exors of 32-bit words. This implementation needs the table T, with 256 entries of four bytes, i.e. one kilobyte in total.

Last Round It can be seen that in this implementation, θ of the last round is already executed in the previous set of table-lookups. In the last round the function to be applied is $\sigma[k^8] \circ \pi \circ \gamma$. This can be realised by replacing the table $T[x] = M[S[x]]$ by $S[x]$. Since $c_2 = 1_x$, the unity in $GF(2^8)$, the entries of the small table S can be extracted from T, removing the extra storage requirement for S.

Performance The reference implementation is written in C and runs at 2.63 MByte/s on a 100 MHz Pentium with the Windows95 operating system. The inverse cipher can be implemented in exactly the same way as the cipher itself and has the same performance. The difference is in the tables and the precalculation of the round keys.

10   Acknowledgements

We thank Paulo Barreto, who wrote the optimized reference implementation of SQUARE. Paulo Barreto can be reached at **pbarreto@uninet.com.br**.

References

1. E. Biham and A. Shamir, "Differential cryptanalysis of DES-like cryptosystems," *Journal of Cryptology*, Vol. 4, No. 1, 1991, pp. 3–72.
2. J. Daemen, "Cipher and hash function design strategies based on linear and differential cryptanalysis," *Doctoral Dissertation*, March 1995, K.U.Leuven.
3. J. Daemen and V. Rijmen, "Self-reciprocal cipher structures," *COSIC internal report 96-3*, 1996.
4. T. Jakobsen and L.R. Knudsen, "The interpolation attack on block ciphers," *these proceedings*.
5. J. Kelsey, B. Schneier and D. Wagner, "Key-schedule cryptanalysis of IDEA, G-DES, GOST, SAFER, and Triple-DES," *Advances in Cryptology, Proceedings Crypto'96, LNCS 1109*, N. Koblitz, Ed., Springer-Verlag, 1996, pp. 237–252.
6. L.R. Knudsen, "Truncated and higher order differentials," *Fast Software Encryption, LNCS 1008*, B. Preneel, Ed., Springer-Verlag, 1995, pp. 196–211.
7. L.R. Knudsen, "A key-schedule weakness in SAFER-K64," *Advances in Cryptology, Proceedings Crypto'95, LNCS 963*, D. Coppersmith, Ed., Springer-Verlag, 1995, pp. 274–286.
8. L.R. Knudsen and T.A. Berson, "Truncated differentials of SAFER," *Fast Software Encryption, LNCS 1039*, D. Gollmann, Ed., Springer-Verlag, 1996, pp. 15–26.
9. N. Koblitz, *"A Course in Number Theory and Cryptography,"* Springer-Verlag, New York, 1987.
10. X. Lai, J.L. Massey and S. Murphy, "Markov ciphers and differential cryptanalysis," *Advances in Cryptology, Proceedings Eurocrypt'91, LNCS 547*, D.W. Davies, Ed., Springer-Verlag, 1991, pp. 17–38.
11. F.J. MacWilliams, N.J.A. Sloane, *"The Theory of Error-Correcting Codes,"* North-Holland, Amsterdam, 1977.

12. M. Matsui, "Linear cryptanalysis method for DES cipher," *Advances in Cryptology, Proceedings Eurocrypt'93, LNCS 765*, T. Helleseth, Ed., Springer-Verlag, 1994, pp. 386–397.

13. K. Nyberg, "Differentially uniform mappings for cryptography," *Advances in Cryptology, Proceedings Eurocrypt'93, LNCS 765*, T. Helleseth, Ed., Springer-Verlag, 1994, pp. 55–64.

14. L. O'Connor, "On the distribution of characteristics in bijective mappings," *Journal of Cryptology*, Vol. 8, No. 2, 1995, pp. 67–86..

15. V. Rijmen, J. Daemen et al., "The cipher SHARK," *Fast Software Encryption, LNCS 1039*, D. Gollmann, Ed., Springer-Verlag, 1996, pp. 99–112.

XMX: A Firmware-Oriented Block Cipher Based on Modular Multiplications

David M'Raïhi, David Naccache

Gemplus - Cryptography Department

1, place de la Méditerranée

F-95206, Sarcelles CEDEX, France

100145.2261@compuserve.com

100142.3240@compuserve.com

Jacques Stern, Serge Vaudenay

Ecole Normale Supérieure

45, rue d'Ulm

F-75230, Paris CEDEX 5, France

jacques.stern@ens.fr

serge.vaudenay@ens.fr

Abstract. This paper presents xmx, a new symmetric block cipher optimized for public-key libraries and microcontrollers with arithmetic co-processors. xmx has no S-boxes and uses only modular multiplications and xors. The complete scheme can be described by a couple of compact formulae that offer several interesting time-space trade-offs (number of rounds/key-size for constant security).

In practice, xmx appears to be tiny and fast : 136 code bytes and a 121 kilo-bits/second throughput on a Siemens SLE44CR80s smart-card (5 MHz oscillator).

1   Introduction

Since efficiency and flexibility are probably the most appreciated design criteria, block ciphers were traditionally optimized for either software (typically SAFER [4]) or hardware (DES [2]) implementation. More recently, autonomous agents and object-oriented technologies motivated the design of particularly tiny codes (such as TEA [9], 189 bytes on a 68HC05) and algorithms adapted to particular programming languages such as PERL.

Surprisingly, although an ever-increasing number of applications gain access to arithmetic co-processors [5] and public-key libraries such as BSAFE, MIRACL, BIGNUM [8] or ZEN [1], no block cipher was specifically designed to take advantage of such facilities.

This paper presents xmx (xor-multiply-xor), a new symmetric cipher which uses public-key-like operations as confusion and diffusion means. The scheme does not require S-boxes or permutation tables, there is virtually no key-schedule and the code itself (when relying on a co-processor or a library) is extremely compact and easy to describe.

xmx is firmware-suitable and, as such, was specifically designed to take a (carefully balanced) advantage of hardware and software resources.

2   The algorithm

2.1   Basic operations

xmx is an iterated cipher, where a keyed primitive f is applied r times to an ℓ-bit cleartext m and a key k to produce a ciphertext c.

Definition 1. Let $f_{a,b}(m) = (m \circ a) \cdot b \bmod n$ where :

$$x \circ y = \begin{cases} x \oplus y & \text{if } x \oplus y < n \\ x & \text{otherwise} \end{cases}$$

and n is an odd modulus.

Property 2. $a \circ b$ is equivalent to $a \oplus b$ in most cases (when $n \leq 2^\ell$, and $\{a, b\}$ is uniformly distributed, $\Pr[a \circ b = a \oplus b] = n/2^\ell$).

Property 3. For all a and b, $a \circ b \circ b = a$.

f can therefore be used as a simply invertible building-block ($a < n$ implies $a \circ b < n$) in iterated ciphers :

Definition 4. Let n be an ℓ-bit odd modulus, $m \in \mathbb{Z}_n$ and k be the key-array $k = \{a_1, b_1, \ldots, a_r, b_r, a_{r+1}\}$ where $a_i, b_i \in \mathbb{Z}_n^*$ and $\gcd(b_i, n) = 1$.

The block-cipher xmx is defined by :

$$\mathsf{xmx}(k, m) = (f_{a_r,b_r}(f_{a_{r-1},b_{r-1}}(\ldots(f_{a_1,b_1}(m))\ldots))) \circ (a_{r+1})$$

and :

$$\mathsf{xmx}^{-1}(k, c) = (f_{a_1,b_1}^{-1}(f_{a_2,b_2}^{-1}(\ldots(f_{a_r,b_r}^{-1}(c \circ a_{r+1}))\ldots)))$$

2.2   Symmetry

A crucially practical feature of xmx is the symmetry of encryption and decryption. Using this property, xmx and xmx^{-1} can be computed by the same procedure :

Lemma 5.

$$k^{-1} = \{a_{r+1}, b_r^{-1} \bmod n, a_r, \ldots, b_1^{-1} \bmod n, a_1\} \Rightarrow \mathsf{xmx}^{-1}(k, x) = \mathsf{xmx}(k^{-1}, x) \ .$$

Since the storage of k requires $(2r + 1)\ell$ bits, xmx schedules the encryption and decryption arrays k and k^{-1} from a single ℓ-bit key s :

$$k(s) = \{s, s, \ldots, s, s, s \oplus s^{-1}, s, s^{-1}, \ldots, s, s^{-1}\}$$

where $k^{-1}(s) = k(s^{-1})$.

For a couple of security reasons (explicited *infra*) s must be generated by the following procedure (where $w(s)$ denotes the Hamming weight of s) :

1. Pick a random $s \in \mathbb{Z}_n^*$ such that $\frac{\ell}{2} - \log_2 \ell < w(s) < \frac{\ell}{2} + \log_2 \ell$
2. If $\gcd(s, n) \neq 1$ or $\ell - \log_2 s \geq 2$ go to 1.
3. output the key-array $k(s) = \{s, s, \ldots, s, s, s \oplus s^{-1}, s, s^{-1}, \ldots, s, s^{-1}\}$

Although equally important, the choice of n is much less restrictive and can be conducted along three engineering criteria : prime moduli will greatly simplify key generation ($\gcd(b_i, n) = 1$ for all i), RSA moduli used by existing applications may appear attractive for memory management reasons and dense moduli will increase the probability $\Pr[a \circ b = a \oplus b]$.

As a general guideline, we recommend to keep n secret in all real-life applications but assume its knowledge for the sake of academic research.

3   Security

xmx's security was evaluated by targeting a weaker scheme (wxmx) where $\circ \cong \oplus$ and $k = (s, s, s, \ldots, s, s, \ldots, s, s, s)$.

Using the trick $u \oplus v = u + v - 2(u \wedge v)$ for eliminating xors and defining :

$$h_i(x) = ((\ldots(x \oplus a_1) \cdot b_1 \bmod n \ldots) \oplus a_{i-1}) \cdot b_{i-1} \bmod n$$

we get by induction :

$$\mathsf{wxmx}(k, x) = b_1' \cdot x + a_1 \cdot b_1' \ldots + a_{r+1} - 2(g_1(x) \cdot b_1' + \ldots + g_{r+1}(x)) \bmod n$$

where $b_i' = b_i \cdots b_r \bmod n$ and $g_i(x) = h_i(x) \wedge a_i$.

Consequently,

$$\mathsf{wxmx}(k, x) = b_1' \cdot x + b - 2g(x) \bmod n \quad \text{where} \quad b = a_1 \cdot b_1' + a_2 \cdot b_2' \ldots + a_{r+1}$$

$$\text{and } g(x) = g_1(x) \cdot b_1' + g_2(x) \cdot b_2' + \ldots + g_{r+1}(x) \bmod n .$$

3.1   The number of rounds

When $r = 1$, the previous formulae become $g_2(x) = h_2(x) \wedge s$ and

$$\mathsf{wxmx}(k, x) = ((x \oplus s) \cdot s \bmod n) \oplus s = x s + s^2 + s - 2(g_1(x) s + g_2(x)) \bmod n$$

Assuming that $w(\delta)$ is low, we have (with a significantly high probability) :

$$g_1(x + \delta) = (x + \delta) \wedge s = g_1(x) \bmod n .$$

Therefore, selecting δ such that $s \wedge \delta = 0 \quad \Rightarrow \quad g_1(x \oplus \delta) = g_1(x)$, we get

$$\mathsf{wxmx}(k, x \oplus \delta) - \mathsf{wxmx}(k, x) = (x \oplus \delta - x) \cdot s - 2(s \wedge h_2(x \oplus \delta) - s \wedge h_2(x)) \bmod n .$$

Plugging $\delta = 2$ and an x such that $x \wedge \delta = 0$ into this equation, we get :

$$\text{wxmx}(k, x \oplus \delta) - \text{wxmx}(k, x) = 2\left(s - s \wedge h_2(x+2) + s \wedge h_2(x)\right) \bmod n \ .$$

Since $h_2(x) = s \cdot x + s^2 - 2\,g_1(x) \bmod n$ (where $g_1(x) = x \wedge s$), it follows that $h_2(x)$ and $h_2(x+2)$ differ only by a few bits. Consequently, information about s leaks out and, in particular, long sequences of zeros or ones (with possibly the first and last bits altered) can be inferred from the difference $\text{wxmx}(k, x \oplus \delta) - \text{wxmx}(k, x)$.

In the more general setting $(r > 1)$, we have

$$\text{wxmx}(k, x \oplus \delta) - \text{wxmx}(k, x) = (x \oplus \delta - x)s^r + 2\,e(x, \delta, s) \bmod n$$

where $e(x, \delta, s)$ is a linear form with coefficients of the form $\alpha \wedge s - \beta \wedge s$.

Defining $\Delta = \{\text{wxmx}(k, x \oplus \delta) - \text{wxmx}(k, x)\}$, we get $\|\Delta\| < 2^{rw(s)}$ since Δ is completely characterized by s.

The difference will therefore leak again whenever :

$$2^{rw(s)} < 2^{\ell} \quad \Rightarrow \quad r < \frac{\ell}{w(s)} \ . \tag{1}$$

3.2   Key-generation

The weight of s :   Since $g(x)$ is a polynomial which coefficients (b'_i) are all bitwise smaller than s, the variety of $g(x)$ is small when $w(s)$ is small. In particular, when $w(s) < \frac{80}{r+1}$, less than 2^{80} such polynomials exist.

A 2^{40}-pair known plaintext attack would therefore extract s^r from :

$$\text{wxmx}(k, y) - \text{wxmx}(k, x) = (y - x) \cdot s^r \bmod n$$

using the birthday paradox (the same $g(x)$ should have been used twice). One can even obtain collisions on g with higher probability by simply choosing pairs of similar plaintexts. Using [7] (refined in [6]), these attacks require almost no memory.

Since a similar attack holds for $\bar{s}$ when $w(s)$ is big $(x \oplus y = x + 2\,(\bar{x} \wedge y) - y)$, $w(s)$ must be rather close to $\ell/2$ and (1) implies that r must at least equal three to avoid the attack described in section 3.1.

The size of s :   Chosen plaintext attacks on wxmx are also possible when s is too short : if $s\,m < n$ after r iterations, s can be recovered by encrypting $m = 0_\ell$ since $\text{wxmx}(k, 0_\ell) = b - 2\,g(x)$ and g's coefficients are all bounded by s.

Observing that $0 \leq \text{wxmx}(k, 0_\ell) - s^{r+1} \leq s \cdot 2^r$, we have :

$$0 \leq s - \sqrt[r+1]{\text{wxmx}(k, 0_\ell)} < \frac{1}{r+1} \quad \Rightarrow \quad s = \left\lceil \sqrt[r+1]{\text{wxmx}(k, 0_\ell)} \right\rceil \ .$$

More generally, encrypting short messages with short keys may also reveal s. As an example, let $\ell = 512$, $r = 4$, $s = 0_{432}|s'$ and $m = 0_{432}|m'$ where s' and m' are both 80-bit long. Since $\Pr[x \oplus s = x + s] = (3/4)^{80} \cong 2^{-33}$ when s is 80-bit long, a gcd between ciphertexts will recover s faster than exhaustive search.

3.3 Register size

Since the complexity of section 3.1's attack must be at least 2^{80}, we have :

$$\sqrt{2^{r \cdot w(s)}} > 2^{80}$$

and considering that $w(s) \cong \ell/2$, the product $r\ell$ must be at least 320.

$r = 4$ typically requires $\ell > 80$ (brute force resistance implies $\ell > 80$ anyway) but an inherent $2^{\ell/2}$-complexity attack is still possible since wxmx is a (keyed) permutation over ℓ-bit numbers, which average cycle length is $2^{\ell/2}$ (given an iteration to the order $2^{\ell/2}$ of wxmx(k, x), one can find x with significant probability).

$\ell = 160$ is enough to thwart these attacks.

4 Implementation

Standard implementations should use xmx with $r = 8$, $\ell = 512$, $n = 2^{512} - 1$ and

$$k = \{s, s, s, s, s, s, s, s, s \oplus s^{-1}, s, s^{-1}, s, s^{-1}, s, s^{-1}, s, s^{-1}\}$$

while high and very-high security applications should use $\{r = 12, \ell = 768, n = 2^{786} - 1\}$ and $\{r = 16, \ell = 1024, n = 2^{1024} - 1\}$.

A recent prototype on a Siemens SLE44CR80s results in a tiny (136 bytes) and performant code (121 kilo-bits/second throughput with a 5 MHz oscillator) and uses only a couple of 64-byte buffers.

The algorithm is patent-pending and readers interested in test-patterns or a copy of the patent application should contact the authors.

5 Further research

As most block-ciphers xmx can be adapted, modified or improved in a variety of ways : the round output can be subjected to a constant permutation such as a circular rotation or the chunk permutation $\pi(ABCD) \to BADC$ where each chunk is 128-bit long (since $\pi(\pi(x)) = x$, xmx's symmetry will still be preserved). Other variants replace modular multiplications by point additions on an elliptic curve (ecxmx) or implement protections against [3] (taxmx).

It is also possible to define f on two ℓ-bit registers L and R such that :

$$f(L_1, R_1) = \{L_2, R_2\}$$

where

$$L_2 = R_1 \quad \text{and} \quad R_2 = L_1 \oplus ((R_1 \oplus k_2) \cdot k_1 \bmod n).$$

and the inverse function is :

$$R_1 = L_2, L_1 = R_2 \oplus ((R_1 \oplus k_2) \cdot k_1 \bmod n) = R_2 \oplus ((L_2 \oplus k_2) \cdot k_1 \bmod n)$$

Since such designs modify only one register per round we recommend to increase r to at least twelve and keep generating s with xmx's original key-generation procedure.

6   Challenge

It is a tradition in the cryptographic community to offer cash rewards for successful cryptanalysis. More than a simple motivation means, such rewards also express the designers' confidence in their own schemes. As an incentive to the analysis of the new scheme, we therefore offer (as a souvenir from FSE'97...) 256 Israeli *Shkalim* and 80 *Agorot* (n is the smallest 256-bit prime starting with 80 ones) to the first person who will degrade s's entropy by at least 56 bits in the instance :

$$r = 8, \ell = 256 \text{ and } n = (2^{80} - 1) \cdot 2^{176} + 157$$

but the authors are ready to carefully evaluate and learn from any feedback they get.

References

1. F. Chabaud and R. Lercier, *The ZEN library*, http://lix.polytechnique.fr/~zen/

2. FIPS PUB **46**, 1977, *Data Encryption Standard.*

3. P. Kocher, *Timing attacks in implementations of Diffie-Hellman, RSA, DSS and other systems*, Advances in Cryptology - CRYPTO '96, LNCS **1109**, 1996, pp. 104-113.

4. J. Massey, *SAFER K-64 : a byte oriented block cipher algorithm*, Fast Software Encryption, Cambridge Security Workshop, 1993, LNCS **809**, pp. 1-17.

5. D. Naccache and D. M'Raïhi, *Cryptographic smart cards*, IEEE Micro, June 1996, vol. **16**, no. 3, pp. 14-23.

6. P. van Oorschot and M. J. Wiener, *Parallel collision search with application to hash functions and discrete logarithms*, 2nd ACM Conference on Computer and Communication Security, Fairfax, Virginia, ACM Press, 1994, pp. 210-218.

7. J-J. Quisquater and J-P. Delescaille, *How easy is collision search? Application to DES*, Advances in Cryptology - EUROCRYPT'89, LNCS **434**, 1990, pp. 429-434.

8. B. Serpette, J. Vuillemenin and J. C. Hervé, *BIGNUM : a portable and efficient package for arbitrary-precision arithmetic*, PRL Research Report ♯2, 1989, ftp://ftp.digital.com/pub/DEC/PRL/research-reports/PRL-RR-2.ps.Z.

9. D. J. Wheeler and R. M. Needham, *TEA, a tiny encryption algorithm*, Fast Software Encryption, Leuven, LNCS **1008**, 1994, pp. 363-366.

MMH: Software Message Authentication in the Gbit/Second Rates

Shai Halevi[1] and Hugo Krawczyk[2]

[1] Lab. for Computer Science, MIT, 545 Tech Square, Cambridge, MA 02139, USA. Email: shaih@theory.lcs.mit.edu. Work was done while the author was visiting the IBM Watson Research Center.

[2] IBM T.J. Watson Research Center, PO Box 704, Yorktown Heights, New York 10598, USA. Email: hugo@watson.ibm.com

Abstract. We describe a construction of almost universal hash functions suitable for very fast software implementation and applicable to the hashing of variable size data and fast cryptographic message authentication. Our construction uses fast single precision arithmetic which is increasingly supported by modern processors due to the growing needs for fast arithmetic posed by multimedia applications.

We report on hand-optimized assembly implementations on a 150 MHz PowerPC 604 and a 150 MHz Pentium-Pro, which achieve hashing speeds of 350 to 820 Mbit/sec, depending on the desired level of security (or collision probability), and a rate of more than 1 Gbit/sec on a 200 MHz Pentium-Pro. This represents a significant speed-up over current software implementations of universal hashing and other message authentication techniques (e.g., MD5-based). Moreover, our construction is specifically designed to take advantage of emerging microprocessor technologies (such as Intel's MMX, 64-bit architectures and others) and then best suited to accommodate the growing performance needs of cryptographic (and other universal hashing) applications.

The construction is based on techniques due to Carter and Wegman for universal hashing using modular multilinear functions that we carefully modify to allow for fast software implementation. We prove the resultant construction to retain the necessary mathematical properties required for its use in hashing and message authentication.

1 Introduction

Universal hash functions, which were first introduced by Carter and Wegman in [CW79], have a wide range of applications in many areas of computer science, including compilers, databases, search engines, parallel architectures, complexity theory, cryptography, and many others. The use of universal hashing for message authentication (introduced by Wegman and Carter [WC81] as well) received much attention lately. In particular, many recent works deal with efficient implementation of universal hashing as a tool for achieving fast and secure message authentication (e.g., [St94, Kr94, Kr95, Ro95, AS96, HJ96, Sh96, AGS97]).

This is also the motivation for our work; however, the construction presented here applies to the other (non-cryptographic) uses of universal hashing as well.

Roughly speaking, universal hash functions are collections of hash functions that map strings (or messages) into short outputs such that the probability of any given pair of messages to collide (i.e., have the same hash value) is small. This probability does not depend on any particular distribution of the input data but only on the random choice of the particular function used to hash the data from the set of all hash functions in the universal family. A stronger version of universal hash functions guarantees that elements are mapped into their images in a pairwise independent way.

These properties make universal hashing a prime tool for data storage and retrieval. In addition, as originally observed in [WC81], universal hashing can be used for building secure message authentication schemes where the adversary's ability to forge messages is bounded by the collision probability of the hash family. In such a scheme, a message is authenticated by first hashing it and then encrypting the hash value with a one-time pad (where the pad is of the length of the hash output rather than of the length of the message). The resultant encrypted hash is transmitted together with the message as an authentication tag that is recomputed and validated by the receiver.

In this setting the communicating parties share a secret and random index to a particular function in the hash family as well as the random one-time pads used for encryption. The security of such a message authentication scheme is *unconditional*, namely, *no adversary* (not even a computationally unlimited one) can forge a message with probability better than the collision probability of the universal hash family. In practice, one time pads are usually replaced with pseudorandom pads (or pseudorandom functions) and the security conditioned on the strength of this encryption.

The attractiveness of using universal hashing in the context of message authentication comes then from two sources. First, it allows decoupling the cryptographic work (reduced to the encryption of the hash value) from the bulk work on the data. Second, given the simplicity of the requirements from a universal hash function it allows for potentially efficient constructions.

In this paper we strongly demonstrate the validity of these properties. We show how to build very fast universal hash families with good (and controllable) levels of security. Our emphasis is on high speed implementation using software only. (For hardware optimized universal hashing see [Kr94].) To this end, we exploit todays' microprocessor technology as well as the current trends in microprocessor design.

On a very high level, this construction is obtained by implementing a well-known family of universal hash functions and then modifying the implementation so as to eliminate costly software operations. The result can be thought of as a "buggy implementation" of the original functions, but with a much faster software implementation. Most importantly, we can prove that the obtained construction is "almost as good" (for its collision probability and security) as the original function.

We report on a hand-optimized assembly implementations on a 150 MHz PowerPC 604 and a 150 MHz Pentium-Pro, which achieve hashing speeds of 350 to 820 Mbit/second, depending on the desired collision probability. The same implementation in a 200 MHz Pentium-Pro exceeds the 1 Gbit/second speed (for a 32 bit hash value).

This represents a significant speed-up over current software implementations of universal hashing, or any other secure message authentication technique. An exact comparison is not possible since the data available on the most efficient implementations of other functions are based on different platforms. The reader is referred to [Sh96] for results on the implementation of division hash (or cryptographic CRC [Ra79, Kr94]), and to [BGV96] for results on the implementation of MD5 and SHA-1 which are currently the most popular bases for software implementation of message authentication codes (e.g, [BCK96]). The best reported time on these functions is 114 Mbit/sec for a hand-optimized assembly implementation of MD5 in a Pentium 90 MHz, and half of it for SHA[3]. (See also [To95] for an analysis of the inherent performance limits of MD5 and for motivation of the needs for faster message authentication techniques.)

Very importantly, our construction is specifically designed to take further advantage of emerging microprocessor technologies (such as Intel's MMX, 64 bit architectures and others) in order to accommodate the growing performance needs of cryptographic applications. (In particular, *single precision* scalar-products are increasingly supported in these new architectures as a means to accelerate multimedia and graphic applications; MMH takes direct advantage of that acceleration by using single precision scalar-products as its most basic operation.) It is worth remarking that the need for performance in the Gbit/sec range is not just for Gbit networks; the goal is that machines will spend only a small portion of their power (say less than 10%) in cryptographic operations while they can use most of that power to do other operations (e.g. doing something "useful" with the authenticated data, like playing a multimedia title).

What's in the name. The name MMH stands for Multilinear-Modular-Hashing. It is also intended to hint to MultiMedia applications, which serve both as motivation for the need of fast software message authentication (for example, to verify the integrity of an on-line multimedia title), and as the motivation for the improved support of integer scalar-products in modern microprocessors, which is a crucial factor for MMH high performance.

Organization. In Section 2 we briefly go over the notions of universal hashing and the connections between those and message-authentication, and recall a well-known construction for universal hashing. In Section 3 we describe our modifications of this well-known construction and our implementation of the resulting function, and provide some experimental results. In Section 4 we show that the

---

[3] By extrapolating these figures one could expect a maximal rate of about 300 Mbit/sec for MD5 in a Pentium-Pro 200 MHz, though no such actual implementation is known to us.

resulting function is "almost as good" as the original one, and thus suitable for secure message authentication applications as well as other universal hashing uses. Finally, a related and alternative construction is discussed in Section 5. In Appendix A we present sample C code for the implementation of the core routine in MMH.

2   Preliminaries

2.1   Notation

For integers y, z, p, we write $y = z$ to assert that they are equal (over the integers) and $y \equiv z \pmod{p}$ to assert that they are congruent mod p. By $z \bmod p$ we denote the residue of the division of z by p. We denote vectors by boldface small letters, e.g., $\mathbf{x} = \langle x_1 \cdots x_k \rangle$. Also we identify bit strings with binary-represented integers, and in particular we identify $\{0,1\}^{32}$ with $\{0, 1, \cdots 2^{32} - 1\}$.

2.2   Universal hashing

In the definitions below, H is a family of functions from a domain D to a range R and ϵ is a constant $1/|R| \le \epsilon \le 1$. The probabilities below, denoted by $\Pr_{h \in H}[\cdot]$, are taken over the choice of $h \in H$ according to a given probability distribution on H (usually, the uniform distribution). These definitions and terminology are due to Carter and Wegman [CW79, WC81], Krawczyk [Kr94], Rogaway [Ro95] and Stinson [St95].

Definition 1.   1. H is a *universal family of hash functions* if for all $x \ne y \in D$, $\Pr_{h \in H}[h(x) = h(y)] = \frac{1}{|R|}$

   H is an *ϵ-almost-universal (ϵ-AU) family of hash functions* if for all $x \ne y \in D$, $\Pr_{h \in H}[h(x) = h(y)] \le \epsilon$

   2. Assume that R is an Abelian group and denote by '$-$' the group subtraction operation. H is a *Δ-universal family of hash functions* if for all $x \ne y \in D$ and all $a \in R$, $\Pr_{h \in H}[h(x) - h(y) = a] = \frac{1}{|R|}$

   H is an *ϵ-almost-Δ-universal (ϵ-AΔU) family of hash functions* if for all $x \ne y \in D$ and all $a \in R$, $\Pr_{h \in H}[h(x) - h(y) = a] \le \epsilon$

   (We stress that Δ-universality is relative to a given group operation in the set R.)

   3. H is a *strongly universal family of hash functions* if for all $x \ne y \in D$ and all $a, b \in R$,
   $\Pr_{h \in H}[h(x) = a,\ h(y) = b] = \frac{1}{|R|^2}$

   H is a *ϵ-almost-strongly universal (ϵ-ASU) family of hash functions* if for all $x \ne y \in D$ and all $a, b \in R$, $\Pr_{h \in H}[h(x) = a,\ h(y) = b] \le \frac{\epsilon}{|R|}$

2.3   Universal hashing and message authentication

In this work we consider a typical communication scenario in which two parties communicate over an unreliable link where messages can be maliciously altered by an adversary. To authenticate the communications over this link, the legitimate parties share a secret key which is unknown to the adversary. They use this secret key to compute a *message authentication code* (MAC) on every message they send on the link. A MAC is a function which takes the secret key x and the message m and returns a tag $\mu \leftarrow MAC_x(m)$. The sender sends the pair (m, μ) over the untrusted link. On receipt of (m', μ'), the receiver repeats this computation and verifies that $\mu' = MAC_x(m')$.

We evaluate the security of a MAC function in the usual model which was introduced in [GMR88]. The adversary A, which is not given the shared secret key, has as a goal to *forge* the MAC value for a message not sent between the legitimate parties. In order to do so, A can eavesdrop the communication between these parties, choose the messages to be sent between them (i.e., to see the output of MAC_x computed on messages of its own choice), and can also modify the message and tags in their way between sender and receiver. In the later case A gets to see whether the replaced values are accepted or not by the receiver; we call these attempts "verification queries". If any of these verification queries uses a message not previously sent between the legitimate parties and is accepted by the receiver (i.e., the right MAC_x was computed by the adversary) then the MAC is broken. For a MAC function to be "good", any adversary with "reasonable" resources (time, memory, number of queries, etc.) should have only a negligible probability of breaking the MAC. We refer to [BKR94] for a formal definition of security of MAC functions.

In the Wegman-Carter paradigm [WC81], the secret key shared by the communicating parties consists of a hash function h drawn randomly from a family of hash functions H and a sequence of random pads $d_1, d_2, \ldots$. To authenticate the i-th message m_i, the sender computes the authentication tag $h(m_i) + d_i$, that is, $MAC_{h,d}(m_i) \stackrel{\text{def}}{=} h(m_i) + d_i$. If h is drawn from an ϵ-AΔU-universal family the probability of an adversary (even one with unlimited computational power) to forge a single message is bounded by ϵ [WC81, Kr94]. If the attacker is allowed to perform q verification queries then its probability to successfully forge a MAC is at most $q\epsilon$. (Notice that in this case passive gathering of information does not buy anything to the attacker, only active tampering with messages and authentication tags can help him, thus making the attack harder to mount and easier to detect.) Hence, universal hashing provides with a simple paradigm for achieving secure cryptographic message authentication.

One important thing to notice is that in this approach one first processes the message m using a non-cryptographic operation (universal hashing), and then applies a cryptographic operation (one-time-pad encryption) on $h(m)$, which is typically much shorter than m itself. In practical implementations, the random pad may be replaced by a pseudo-random one, so the parties only need to share the function h and the seed s to the pseudo-random generator (s can also be a key to a pseudorandom function). One can also directly apply a pseudorandom function to the output of the hash function concatenated with a counter. The reader is referred to [WC81, Br82, St94, Kr94, Ro95, Sh96] for more elaborate discussions of these issues.

In the reminder of this paper we concentrate on the construction of an efficient ϵ-$A\Delta U$-universal family of hash functions for small ϵ.

2.4   A Well-Known Construction

The starting point for our construction is a well known construction due to Carter and Wegman [CW79]. This construction works in the finite field Z_p for some prime integer p. The family of hash functions consists of all the multilinear functions over Z_p^k for some integer k. Namely,

Definition 2. Let p be a prime and let k be an integer $k > 0$. Define a family MMH* of functions from Z_p^k to Z_p as follows

$$\text{MMH}^* \stackrel{\text{def}}{=} \{g_{\mathbf{x}} : Z_p^k \to Z_p \mid \mathbf{x} \in Z_p^k\}$$

where the functions $g_{\mathbf{x}}$ are defined for any $\mathbf{x} = \langle x_1, \cdots x_k \rangle, \mathbf{m} = \langle m_1, \cdots, m_k \rangle$, $x_i, m_i \in Z_p$,

$$g_{\mathbf{x}}(\mathbf{m}) \stackrel{\text{def}}{=} \mathbf{m} \cdot \mathbf{x} \bmod p = \sum_{i=1}^{k} m_i x_i \bmod p$$

Theorem 3. *The family* MMH* *is Δ-universal.*

Proof. Fix some $a \in Z_p$, and let $\mathbf{m}, \mathbf{m}'$ be two different messages. Assume w.l.o.g. that $m_1 \neq m_1'$. Then for any choice of $x_2, \cdots x_k$ we have

$$\Pr_{x_1} [g_{\mathbf{x}}(\mathbf{m}) - g_{\mathbf{x}}(\mathbf{m}') \equiv a \pmod{p}]$$
$$= \Pr_{x_1} \left[(m_1 - m_1')x_1 \equiv a - \sum_{i=2}^{k}(m_i - m_i')x_i \pmod{p} \right] = \tfrac{1}{p}$$

□

Reducing the collision probability. Depending on the choice of p and the application at hand, collision probability of $1/p$ may be insufficient (i.e., too large). A simple way to reduce the collision probability is to hash the message twice using two independent keys. This yields a collision probability of $1/p^2$, at the expense of doubling the computational work and length of output. It also requires double-size keys.

The last aspect (key size) can be resolved by recurring to a "Toeplitz matrix construction". That is, instead of choosing two independent k-vectors $\mathbf{x}, \mathbf{x}'$, we choose $k + 1$ scalars $x_1, \cdots x_{k+1}$ and set $\mathbf{x} = \langle x_1, \cdots, x_k \rangle, \mathbf{x}' = \langle x_2, \cdots, x_{k+1} \rangle$. It is a well-known fact that such a choice of keys will still result in a reduced collision probability of $1/p^2$ while increasing the key size by a single scalar x_{k+1}. The same methodology can be applied to further reduce the collision probability to $1/p^n$ for any integer value n. In this case the computational work and output are increased by a factor of n, while keys are only increased with n scalars.

Dealing with arbitrary long messages. The above function can only be applied to a fixed-size messages (namely, to vectors in $Z_p{}^k$). The standard approach for dealing with messages of arbitrary length is to use tree-hashing as already suggested by Carter and Wegman [WC81]. That is, we break the message into blocks of k elements (over Z_p) each and hash each block separately (using the *same* hash function). We then concatenate the hash results of all these blocks and hash them again using an independent key, and so on. (The drawback of this approach is that both the key-size and the collision probability grow linearly with the height of the tree. A suggestion to counter this problem appears in section 3.4.)

3   Multilinear Modular Hashing

The MMH family described here is a variant of the construction MMH* described in section 2.4 designed to achieve fast software implementations while preserving the low collision probability. For the sake of simplicity, we describe the function for a specific set of parameters (particularly suited for 32-bit word architectures); however, the approach is general and can be used with different parameters according to application needs and hardware platform.

The main characteristics of our implementation are

- We work with 32-bit integers (as said the same approach can be used in machines with different word-length).
- We work with the prime integer $p = 2^{32} + 15$, so we can implement a division-less modular reduction.[4]

Therefore our starting point is a specific "very slightly modified" instance of MMH*:

$$\mathrm{MMH}^*_{32} \stackrel{\mathrm{def}}{=} \left\{ g_{\mathbf{x}} : \left(\{0,1\}^{32} \right)^k \to Z_p \ \middle| \ \mathbf{x} \in \left(\{0,1\}^{32} \right)^k \right\}$$

where the functions $g_{\mathbf{x}}$ are defined for any $\mathbf{x} = \langle x_1, \cdots x_k \rangle, \mathbf{m} = \langle m_1, \cdots, m_k \rangle$,

$$g_{\mathbf{x}}(\mathbf{m}) \stackrel{\mathrm{def}}{=} \mathbf{m} \cdot \mathbf{x} \bmod (2^{32} + 15) = \left[\sum_{i=1}^{k} m_i x_i \right] \bmod (2^{32} + 15)$$

Similarly to Theorem 3, we have

Theorem 4. *The family* MMH^*_{32} *is* $\epsilon\text{-}A\Delta U$ *with* $\epsilon = 2^{-32}$.

Notice that this function is only defined for fixed-length messages (namely, messages of $32k$ bits). To handle arbitrary-length messages we use the tree construction from Section 2.4.

---

[4] The idea is adopted from a suggestion by Carter and Wegman [CW] to use the primes $2^{16} + 1$ or $2^{31} - 1$. In our approach, any prime which satisfies $2^{32} < p < 2^{32} + 2^{16}$ will do; $2^{32} + 15$ is the smallest among those primes.

3.1 Implementing the "ideal" function

Modular reduction. The one operation in MMH_{32}^* which is by far the most expensive is the modular reduction. However, since we work with the prime $p = 2^{32} + 15$, we can implement a division-less modular reduction as follows.

Let x be an (unsigned) 64-bit integer, and denote $x = 2^{32}a + b$, where a, b are both unsigned 32-bit integers. Then we note that

$$2^{32}a + b \equiv (2^{32}a + b) - a \cdot (2^{32} + 15) = b - 15a \quad (\text{mod } 2^{32} + 15)$$

Moreover, since $a, b \in [0, 2^{32} - 1]$, then $b - 15a \in [-15 \cdot (2^{32} - 1), \ 2^{32} - 1]$. Thus, if we denote $y = b - 15a$, then $y \equiv x \pmod{2^{32} + 15}$, and y can be represented as a signed 64-bit integer (in two's complement) $y = c \cdot 2^{32} + d$, where $c \in \{-15, \ldots, 0\}$ and d is an unsigned 32-bit integer.

We can now repeat this process once more and compute $z = d - 15c$. Then $z \equiv y \equiv x \pmod{2^{32} + 15}$, and $z \in \{0, \cdots 2^{32} + 15^2\}$. Finally, we test to see if z is still larger than $2^{32} + 15$. If not we return z, otherwise we return $z - (2^{32} + 15)$.

Inner-product. Even the above implementation of a division-less reduction still takes about 10-15 machine instructions, which is very expensive if we need to execute it too often. Therefore, we carry the whole inner-product operation over the integers and then do just one modular reduction at the end. This approach forces us to deal with addition of integers of 64-bits. This, in particular, involves the use of machine instructions for addition-with-carry.

To implement the integer multiplications, we take advantage of machine instructions for multiplying two 32-bit integers and obtaining the 64-bit result. Both addition-with-carry and 32-by-32 multiplication are available in just about all the architectures of todays' computers (in 32-bit word architectures the 64 bit result of the multiplication is returned in two 32-bit registers). However, there is no (official) syntax in high-level programming language to access these operations, so we write our implementation in assembly language.[5]

Fixing the value of **k**. The value of k (the length of the message- and key-vectors) has two effects on the implementation.

- Since we amortize the costly modular reduction over k (cheaper) multiply-and-add operations, increasing k should increase the speed.
- Since the key **x** consists of k 32-bit integers, increasing k results in a longer key.

As a reasonable tradeoff between these conflicting objectives, we work with $k = 32$. For this value of k, the cost of modular reduction amounts to only about 10-15% of the total cost of the implementation.

---

[5] Even if not part of the language definition, some C compilers support 64 bit types (*long long integers*). See Appendix A.

3.2   Modifying the implementation

We make two modifications to the implementation of MMH^*_{32}

- We make the output of the function a 32-bit integer rather than an element in Z_p. This is done by ignoring the most-significant bit in the output of the original function, which is equivalent to reducing it modulo 2^{32}.
- In the inner-product operation, we ignore any carry-bit out of the 64 bit-location. This is equivalent to computing the sum mod 2^{64}.

These two modifications together define the following family of functions.

Definition 5. Set $p = 2^{32} + 15$ and $k = 32$. Define a family MMH of functions from $(\{0,1\}^{32})^k$ to $\{0,1\}^{32}$ as follows

$$MMH_{32} \overset{def}{=} \left\{ h_{\mathbf{x}} : \left(\{0,1\}^{32}\right)^k \to \{0,1\}^{32} \ \middle| \ \mathbf{x} \in \left(\{0,1\}^{32}\right)^k \right\}$$

where the functions $h_{\mathbf{x}}$ are defined for any $\mathbf{x} = \langle x_1, \cdots x_k \rangle, \mathbf{m} = \langle m_1, \cdots, m_k \rangle$,

$$h_{\mathbf{x}}(\mathbf{m}) \overset{def}{=} \left[\left[\left[\sum_{i=1}^{k} m_i x_i \right] \bmod 2^{64} \right] \bmod (2^{32} + 15) \right] \bmod 2^{32}$$

In Section 4 we show that MMH_{32} is ϵ-AΔU with $\epsilon = \frac{1.5}{2^{30}}$.

Instruction count. To give an estimated instruction count for an implementation of MMH, we consider a machine with the following properties

- 32-bit machine integers.
- Arithmetic operations are done in registers.
- A multiplication of two 32-bit integers which yields a 64-bit result takes two machine instructions.

A pseudo-code for MMH on such machine may be as follows

```
MMH(msg, key)
1.    SumHigh = SumLow = 0
2.    For i = 1 to k
3.        load msg[i]
4.        load key[i]
5.        ⟨ProdHigh, ProdLow⟩ = msg[i] * key[i]
6.        SumLow = SumLow + ProdLow
7.        SumHigh = SumHigh + ProdHigh + carry
8.    Reduce ⟨SumHigh, SumLow⟩ mod 2^32 + 15 and then mod 2^32
```

Each multiply-and-add operation takes total of about 7 instructions: 2 for loading the message- and key-words to registers, 2 for the multiplication, 2 for the addition, and 1 more to handle the loop. We repeat these operation $k = 32$ times, and then we need about 10-15 instructions for the modular reduction. This yields an instruction count of about $7k + 15$ to handle a k-word message. That is, we have about about 7.5 instructions per-word, or less than 2 instructions per-byte.

This instruction-count can be further reduced by unrolling the loop a few times and by working on several messages (more precisely, several k-word message blocks) at the same time, so we can load a key word just once and use it on several messages. For example, in our implementation on the PowerPC we have about 6 instructions per-word. On a 64-bit machine we may be able to get as low as 4 instructions per 32 bits of input. (An even faster implementation in a 64-bit machine can be achieved by working on 64-bit words using a prime modulus which is slightly larger than 2^{64} – e.g. $2^{64} + 13$ – as long as the architecture supports the integer multiplication of two 64 bit words.) The implementation of the hashing-tree adds less than 10% to the total work of the function.

Moreover, the structure of the hashing procedure (and in particular the inner-product operation) leaves plenty of room for parallelization. Emerging microprocessor technologies which are aimed at multimedia applications tend to include a good support for inner-product operations (e.g., Intel's MMX, Sun's VIS, etc.) even for standard processors; therefore, we can expect even faster implementations of MMH in the near future.

3.3   Experimental results

Below we describe the results of a few experimental implementations of MMH. We implemented MMH on PowerPC and Intel x86 architectures. The basic MMH function itself was hand-optimized in assembly language on each machine and the tree structure and various initializations were implemented in C. For each of these architectures we implemented two variants of MMH:

1. The basic MMH construction with tree-hashing. This variant has a 32-bit output and collision probability of $\frac{1.5}{2^{30}}$ times the height of the tree.
2. A "high-security" version, where each hashing operation is repeated twice using "the Toeplitz matrix construction". This variant has a 64-bit output and collision probability of $\frac{2.25}{2^{60}}$ times the height of the tree.

For each version we performed two different tests: First we tested what happens when the message is long and resides in a memory buffer. We evaluated the hash function on a 4 Mbyte buffer and repeated it 64 times. This yields total length of 256 Mbyte = 2 Gbit. Below we refer to this test as the "message in memory" test. Then we performed another test to find how much of the running time is spent on cache misses. For that we modified the code so that whenever it access data, it always takes it from the same memory buffer (of size a few Kbytes). We refer to this test as the "message in cache" test.

To get a good assessment of the performance potential of our construction in different architectures, we tested our implementation on the following platforms

- A 150 MHz PowerPC 604 RISC machine running AIX.
- A 150 MHz Pentium-Pro machine running Windows NT.
- A 200 MHz Pentium-Pro machine running Linux.

The results which we got for these variants are summarized in the following table.

150 MHz PowerPC 604	message in memory	message in cache
64-bit output	390 Mbit/second	417 Mbit/second
32-bit output	597 Mbit/second	820 Mbit/second

150 MHz Pentium-Pro	message in memory	message in cache
64-bit output	296 Mbit/second	356 Mbit/second
32-bit output	556 Mbit/second	813 Mbit/second

200 MHz Pentium-Pro	message in memory	message in cache
64-bit output	380 Mbit/second	500 Mbit/second
32-bit output	645 Mbit/second	1080 Mbit/second

Table 1: Timing results for various implementations of MMH.

We also tested MMH on a Pentium machine. However, the integer multiplication in the Pentium is slow and therefore we obtained our best results by using the floating-point unit for the multiply-and-add operations.[6] This implementation achieves a rate of about 160 Mbit/second on a 120 MHz Pentium for the 64-bit variant (message in memory). This is somewhat faster than the performance reported in [Sh96] for the polynomial division function and in [BGV96] for MD5, but not as impressive as the other speeds reported above.

It is important to note that the above results are for bulk data processing. For particular applications, the actual effect of these faster functions depends on the details of the application, the length of authenticated (or hashed) data, etc. In particular, when used for message authentication the operation of encrypting the hash value (e.g., the generation of a pseudorandom one-time pad) can be a significant overhead for very short messages. However, we remark that MMH does not need of particularly long messages in order to achieve its superior performance relative to other universal hash functions (this is to be contrasted, for example, with bucket hashing [Ro95]).

3.4   Further issues and variants

Variants. Some further optimizations to our implementation can be achieved by introducing some changes to the definition of MMH. In particular, by exploiting some architecture-specific optimizations we have achieved performance improvements of about 10% over the above reported figures. In the Pentium-Pro 200 MHz, for example, this brings the 32-bit function speed to 1.2 Gigabit/second (at the cost of a slight increase in the collision probability of the function).

Mixing hash functions. Most of our description above concentrated on the basic MMH function as applied to fixed length messages (e.g., k-word long).

---

[6] This requires some modification in the definition of MMH.

The techniques described in Section 2.4 were used to implement the function for arbitrary length messages (the experimental results of section 3.3 correspond to this general construction). A drawback of this implementation is that the length of the keys not only depends on the parameter k but also on the height of the hash-tree (a different key is needed for each level in the tree even if that key is seldom used by the function). One approach to overcome this drawback is to apply the proposed function MMH only to the top levels of the tree (one or two levels) and then to use a different hash family to hash the result from these levels. The idea is to use for this second hashing a hash family that requires shorter keys even if it is slower than MMH. Since the data hashed by this function is much shorter than the original message (e.g., by a factor of $1/k^2$) the inferior performance of the second function would not be noticeable. A reasonable choice for the second hashing scheme can be the polynomial-evaluation-hashing. The reader is referred to [Sh96, AGS97] for a description and implementation details of this scheme.

Hashing short data items. Our hash functions are particularly flexible as for the way they deal with information of different sizes. While long streams of data can be processed as explained above, short strings of data can also be hashed very efficiently by just choosing a key which is not shorter than any such string. For example, a compiler that hashes a symbol table where no such symbol exceeds 1 Kbit in length can choose a 1 Kbit long key and hash all the symbols using that single key. In this case, there is no need for the hash tree technique. (Notice that shorter than 1 Kbit strings will just use the part of the key corresponding to their length.)

Padding to block boundaries. Another issue related to dealing with variable length messages is the need to pad data to some block boundary. This can be easily handled by appending some prescribed pad to the end of the data. In the case of message authentication it is particularly important that the padding will be unambiguous, namely, two different messages are mapped into different padded strings. (For example, a pad formed by concatenating a '1' followed by a suitable number of '0's could be appended to every message.)

Generation and sharing of long keys. Depending on the variant and application of MMH one may need long keys (e.g., a few Kbits). These keys can be generated using a strong pseudorandom generator. In particular, in the case of such a key being shared by two parties, only the seed to the pseudorandom generator needs to be exchanged (such a seed will be considerably shorter than the key, e.g. 100-200 bit long).

Byte ordering. For the purpose of hashing we identify streams of data with a sequence of integer numbers (e.g., 32-bit integers). However, different computer architectures load data bytes into words in different orders. Thus, when interested in inter-operability between different machines one needs to specify a particular loading order (little-endian or big-endian). Any such specification will favor one architecture or the other. We do not provide such a specification at this point. One thing to notice is that while all architectures provide instructions for switching their default order this conversion can cause a degradation in the performance of the function in these architectures. (See [BGV96, To95] for a related discussion.)

4   Analysis of the collision probability

In this section we analyze the collision probability of the Multilinear Modular Hash function. For simplicity, we concentrate on the parameters of MMH$_{32}$, however, the same analysis works for similar constructions using word-length other than 32 bits.

We start by analyzing the collision probability of a hybrid construction which is half-way between the ideal family MMH$_{32}^{*}$ and the actual construction MMH$_{32}$.

Definition 6. Set $p = 2^{32} + 15$ and $k = 32$. Define a family H_{32} of functions from $(\{0, 1\}^{32})^k$ to Z_p as follows

$$H_{32} \stackrel{\text{def}}{=} \left\{ \tilde{h}_{\mathbf{x}} : \left(\{0, 1\}^{32}\right)^k \to Z_p \ \middle| \ \mathbf{x} \in \left(\{0, 1\}^{32}\right)^k \right\}$$

where the function $\tilde{h}_{\mathbf{x}}$ is defined for any $\mathbf{x} = \langle x_1, \cdots x_k \rangle, \mathbf{m} = \langle m_1, \cdots, m_k \rangle,$

$$\tilde{h}_{\mathbf{x}}(\mathbf{m}) \stackrel{\text{def}}{=} \left[\left[\sum_{i=1}^{k} m_i x_i \right] \bmod 2^{64} \right] \bmod (2^{32} + 15)$$

Note that H_{32} is defined like MMH$_{32}$, but without the reduction mod 2^{32} at the end.

Lemma 7. H_{32} is an ϵ-AΔU family of hash function with $\epsilon \leq 2 \cdot 2^{-32}$.

Proof. Fix any $a \in Z_p$, and any two different message-vectors $\mathbf{m} \neq \mathbf{m}'$, and assume w.l.o.g. that $m_1 \neq m_1'$. We prove that for any choice of $x_2, \cdots, x_k$, $\Pr_{x_1}[\tilde{h}_{\mathbf{x}}(\mathbf{m}) - \tilde{h}_{\mathbf{x}}(\mathbf{m}') \equiv a \pmod{p}] \leq 2/2^{32}$, which implies the lemma.

Since $x_1 m_1 < 2^{64}$ for any value of x_1, then for any choice of $x_2 \cdots x_k$ the term $x_1 m_1$ adds at most one carry bit to the sum. Fix some choice of $x_2 \cdots x_k$ and denote $s \stackrel{\text{def}}{=} [\sum_{i=2}^{k} x_i m_i] \bmod 2^{64}$. We conclude that

$$\left[\sum_{i=1}^{k} x_i m_i \right] \bmod 2^{64} = x_1 m_1 + s - 2^{64} b \qquad \text{for some } b \in \{0, 1\}$$

Similarly we denote $s' \stackrel{\text{def}}{=} \sum_{i=2}^{k} x_i m_i' \bmod 2^{64}$, and we get

$$\left[\sum_{i=1}^{k} x_i m_i' \right] \bmod 2^{64} = x_1 m_1' + s' - 2^{64} b' \qquad \text{for some } b' \in \{0, 1\}$$

Notice that s, s' do not depend on the choice of x_1, but b, b' may depend on it. We stress this below by writing $b(x_1), b'(x_1)$. We can now write $\tilde{h}_{\mathbf{x}}(\mathbf{m}) - \tilde{h}_{\mathbf{x}}^*(\mathbf{m}')$ as

$$
\tilde{h}_{\mathbf{x}}(\mathbf{m}) - \tilde{h}_{\mathbf{x}}(\mathbf{m}') \pmod{p}
$$

$$
= \left[\sum_{i=1}^{k} x_i m_i \bmod 2^{64}\right] - \left[\sum_{i=1}^{k} x_i m_i' \bmod 2^{64}\right] \pmod{p}
$$

$$
= (m_1 - m_1')x_1 - 2^{64}[b(x_1) - b'(x_1)] + s - s' \pmod{p}
$$

Since $b(x_1) - b'(x_1) \in \{-1, 0, 1\}$ for all x_1, then

$$
\Pr_{x_1}\left[\tilde{h}_{\mathbf{x}}(\mathbf{m}) - \tilde{h}_{\mathbf{x}}(\mathbf{m}') \equiv a \pmod{p}\right]
$$
$$
= \Pr_{x_1}\left[(m_1 - m_1')x_1 - 2^{64}[b(x_1) - b'(x_1)] + s - s' \equiv a \pmod{p}\right]
$$
$$
\leq \Pr_{x_1}\left[(m_1 - m_1')x_1 \equiv a - s + s' - 2^{64} \pmod{p}\right]
$$
$$
+ \Pr_{x_1}\left[(m_1 - m_1')x_1 \equiv a - s + s' \pmod{p}\right]
$$
$$
+ \Pr_{x_1}\left[(m_1 - m_1')x_1 \equiv a - s + s' + 2^{64} \pmod{p}\right]
$$
$$
\leq 3 \cdot 2^{-32}
$$

This bound can be improved to $2 \cdot 2^{-32}$ by noticing that the difference $b(x_1) - b'(x_1)$ cannot assume simultaneously the values 1 and -1. Namely, if there exists a value of x_1 for which $b(x_1) = 1$ and $b'(x_1) = 0$ then there cannot be another value x_1' for which $b(x_1') = 0$ and $b'(x_1') = 1$, and vice-versa. Indeed, having $b(x_1) = 1, b'(x_1) = 0$ for a given x_1 means that $x_1 m_1 + s \geq 2^{64}$ while $x_1 m_1' + s' < 2^{64}$. Since the expressions $x_1 m_1 + s$ and $x_1 m_1' + s'$ are monotonic increasing in x_1, then there cannot be another value x_1' for which $x_1' m_1 + s < 2^{64}$ while $x_1' m_1' + s' \geq 2^{64}$. That is, there is no x_1' for which $b(x_1') = 0$ and $b'(x_1') = 1$. $\qquad\square$

We now show MMH$_{32}$ to be a good -AΔU universal family (relative to the mod2^{32} subtraction).

Theorem 8. MMH$_{32}$ is an ϵ-AΔU family of hash functions with $\epsilon \leq 6 \cdot 2^{-32}$.

Proof. Recall that MMH$_{32}$ is obtained from H_{32} by reducing the result modulo 2^{32}. That is, for any $\mathbf{x}, \mathbf{m}$, $h_{\mathbf{x}}(\mathbf{m}) = \tilde{h}_{\mathbf{x}}(\mathbf{m}) \bmod 2^{32}$.

Fix any value $v, 0 \leq v < 2^{32}$ and two different message-vectors $\mathbf{m}, \mathbf{m}'$, and let $\mathbf{x}$ be a key-vector so that $h_{\mathbf{x}}(\mathbf{m}) - h_{\mathbf{x}}(\mathbf{m}') \equiv v \pmod{2^{32}}$. Equivalently, we have $\tilde{h}_{\mathbf{x}}(\mathbf{m}) - \tilde{h}_{\mathbf{x}}(\mathbf{m}') \equiv v \pmod{2^{32}}$. Since the values $\tilde{h}_{\mathbf{x}}(\mathbf{m})$ and $\tilde{h}_{\mathbf{x}}(\mathbf{m}')$ are both between 0 and $p - 1$ we get that their difference (over the integers) lies between $-p + 1$ and $p - 1$. As $p = 2^{32} + 15$, we get

$$
\tilde{h}_{\mathbf{x}}(\mathbf{m}) - \tilde{h}_{\mathbf{x}}(\mathbf{m}') \in \begin{cases} \{v - 2^{32}, v, v + 2^{32}\} & 0 \leq v < 15 \\ \{v - 2^{32}, v\} & 15 \leq v \leq 2^{32} - 15 \\ \{v - 2 \cdot 2^{32}, v - 2^{32}, v\} & 2^{32} - 15 < v < 2^{32} \end{cases}
$$

That is, if $h_{\mathbf{x}}(\mathbf{m}) - h_{\mathbf{x}}(\mathbf{m}') \equiv v \pmod{2^{32}}$ then over the integers the difference $\tilde{h}_{\mathbf{x}}(\mathbf{m}) - \tilde{h}_{\mathbf{x}}(\mathbf{m}')$ can assume at most 3 values. But then this is also true for this difference when taken mod p. Lemma 7 tells us that for any value v' the probability (over the choice of $\mathbf{x}$) that $\tilde{h}_\mathbf{x}(\mathbf{m}) - \tilde{h}_\mathbf{x}(\mathbf{m}') \equiv v' \pmod{p}$ is at most $2/2^{32}$. And then the probability for any value of v that $\mathbf{x}$ solves the equation $h_\mathbf{x}(\mathbf{m}) - h_\mathbf{x}(\mathbf{m}') \equiv v \pmod{2^{32}}$ is at most 3 times larger. In other words,

$$\Pr_\mathbf{x}\left[h_\mathbf{x}(\mathbf{m}) - h_\mathbf{x}(\mathbf{m}') \equiv v \pmod{2^{32}}\right] \leq 3 \cdot 2 \cdot 2^{-32}.$$

□

5   Further Work

Recently, Mark Wegman has suggested to us the use of an unpublished universal hash function invented by Larry Carter and himself many years ago. This function is related to the construction, by the same authors, that we presented in section 2.4 and which we called MMH*. This "new" function is not linear and then we denote it by NMH* (for Non-linear).

Definition 9. Let p be a prime and let k be an even positive integer. Define a family NMH* of functions from Z_p^k to Z_p as follows

$$\text{NMH}^* \overset{\text{def}}{=} \{g_\mathbf{x} : Z_p^k \to Z_p \mid \mathbf{x} \in Z_p^k\}$$

where the functions $g_\mathbf{x}$ are defined for any $\mathbf{x} = \langle x_1, \cdots x_k \rangle, \mathbf{m} = \langle m_1, \cdots, m_k \rangle$, $x_i, m_i \in Z_p$,

$$g_\mathbf{x}(\mathbf{m}) \overset{\text{def}}{=} \sum_{i=1}^{k/2} (m_{2i-1} + x_{2i-1})(m_{2i} + x_{2i}) \bmod p$$

It is not hard to see that NMH* is Δ-universal. This function uses the same number of arithmetic operations as MMH* but requires half of the number of multiplications (at the expense of more additions). At least in machines where multiplication is significantly slower than addition its performance should be expected to be better than that of MMH*. We modify NMH* as we did with MMH* for improved performance and define NMH$_{32}$ as follows.

Definition 10. Set $p = 2^{32} + 15$ and $k = 32$. Define a family NMH$_{32}$ of functions from $(\{0,1\}^{32})^k$ to $\{0,1\}^{32}$ as follows

$$\text{NMH}_{32} \overset{\text{def}}{=} \left\{ h_\mathbf{x} : \left(\{0,1\}^{32}\right)^k \to \{0,1\}^{32} \,\middle|\, \mathbf{x} \in \left(\{0,1\}^{32}\right)^k \right\}$$

where the functions $h_\mathbf{x}$ are defined for any $\mathbf{x} = \langle x_1, \cdots x_k \rangle, \mathbf{m} = \langle m_1, \cdots, m_k \rangle$, as

$$h_\mathbf{x}(\mathbf{m}) \overset{\text{def}}{=} \left[\left[\left[\sum_{i=1}^{k/2} (m_{2i-1} \pm x_{2i-1})(m_{2i} \pm x_{2i}) \right] \bmod 2^{64} \right] \bmod (2^{32} + 15) \right] \bmod 2^{32}$$

(the symbol $\pm$ denotes addition modulo 2^{32}):

We can show in a similar way as we did for MMH_{32} that NMH_{32} is ϵ-AΔU for ϵ close to 2^{-32}.

We have not yet implemented this function. Report on such an implementation will be presented in the future.

Acknowledgments. We thank Mark Wegman for invaluable comments and suggestions on this work, and Robert Geva for his help with the Pentium-Pro implementation. We also wish to thank Matteo Frigo for suggesting the 'C' implementation in Appendix A and Dave Wagner for helping us to get the Pentium-Pro 200 performance measurements.

References

[AGS97] V. Afanassiev, C. Gehrmann and B. Smeets. Fast Message Authentication using Efficient Polynomial Evaluation Appeares in these proceedings.

[AS96] M. Atici and D. Stinson. Universal Hashing and Multiple Authentication *Advances in Cryptology – CRYPTO '96 Proceedings*, Lecture Notes in Computer Science Vol. 1109, N. Koblitz, ed., Springer-Verlag, 1996. pp. 16-30.

[BCK96] M. Bellare, R. Canetti and H. Krawczyk. Keying hash functions for message authentication. *Advances in Cryptology – CRYPTO '96 Proceedings*, Lecture Notes in Computer Science Vol. 1109, N. Koblitz, ed., Springer-Verlag, 1996. pp. 1-15.

[BKR94] M. Bellare, J. Kilian and P. Rogaway. The security of cipher block chaining. *Advances in Cryptology – CRYPTO '94 Proceedings*, Lecture Notes in Computer Science Vol. 839, Y. Desmedt, ed., Springer-Verlag, 1994. pp. 341-358.

[BGV96] A. Bosselaers, R. Govaerts, J. Vandewalle. Fast Hashing on the Pentium, *Advances in Cryptology – CRYPTO '96 Proceedings* Lecture Notes in Computer Science Vol. 1109, N. Koblitz, ed., Springer-Verlag, 1996. pp. 298- 312.

[Br82] G. Brassard. On computationally secure authentication tags requiring short secret shared keys, *Advances in Cryptology – CRYPTO '82 Proceedings*, Springer-Verlag, 1983, pp. 79–86.

[CW79] L. Carter and M. Wegman. Universal Hash Functions. J. of Computer and System Science 18, 1979, pp. 143-154.

[CW] L. Carter and M. Wegman. Private Communication.

[GMR88] S. Goldwasser, S. Micali and R. Rivest. A digital signature scheme secure against adaptive chosen-message attacks. *SIAM Journal of Computing*, vol. 17, no. 2 (April 1988), pp. 281-308.

[HJ96] T. Helleseth and T. Johansson. Universal Hash Functions from Exponential Sums over Finite Fields *Advances in Cryptology – CRYPTO '96 Proceedings*, Lecture Notes in Computer Science Vol. 1109, N. Koblitz, ed., Springer-Verlag, 1996. pp. 31-44.

[Kr94] H. Krawczyk. LFSR-based Hashing and Authentication. Proceedings of CRYPTO '94, Lecture Notes in Computer Science, vol. 839, Springer-Verlag, 1994, pp. 129-139.

[Kr95] H. Krawczyk. New Hash Functions for Message Authentication. Proceedings of EUROCRYPT '95, Lecture Notes in Computer Science, vol. 921, Springer-Verlag, 1995, pp. 301-310.

[Ra79]   Rabin, M.O., "Fingerprinting by Random Polynomials", Tech. Rep. TR-15-81, Center for Research in Computing Technology, Harvard Univ., Cambridge, Mass., 1981.

[Ro95]   P. Rogaway. Bucket Hashing and its application to Fast Message Authentication. Proceedings of CRYPTO '95, Lecture Notes in Computer Science, vol. 963, Springer-Verlag, 1995, pp. 15-25.

[Sh96]   V. Shoup. On Fast and Provably Secure Message Authentication Based on Universal Hashing *Advances in Cryptology – CRYPTO '96 Proceedings*, Lecture Notes in Computer Science Vol. 1109, N. Koblitz, ed., Springer-Verlag, 1996. pp. 313-328.

[St94]   D. Stinson. Universal Hashing and Authentication Codes. Designs, Codes and Cryptography, vol. 4, 1994, pp. 369-380.

[To95]   J. Touch. Performance Analysis of MD5. Proc. Sigcomm '95, Boston, pp. 77-86.

[St95]   D. Stinson. On the Connection Between Universal Hashing, Combinatorial Designs and Error-Correcting Codes. TR95-052, Electronic Colloquium on Computational Complexity, 1995.

[WC81]   M. Wegman. and L. Carter. New hash functions and their use in authentication and set equality. *J. of Computer and System Sciences*, vol. 22, 1981, pp. 265-279.

A   A 'C' Implementation of MMH

Below we describe a sample 'C' implementation of MMH, using the **long long** data type of gcc to handle 64-bit integers. Note that the code below does not include an implementation of the hashing tree, nor does it implement the reduced collision probability from Section 2.4 (Page 177). Rather, it is just a straightforward implementation of the basis MMH_{32} function, as defined in Definition 5.

In the following code, the 32-word message is stored in the **msg[]** buffer and the 32-word key is stored in the **key[]** buffer. The following is a straight line code rather than in a loop, to take maximum advantage of the optimizing capabilities of the compiler.

```
#define DO(i)  sum += key[i] * (unsigned long long) msg[i]

unsigned long basic_mmh(unsigned long *key, unsigned long *msg)
{
        signed long long stmp;              /* temporary variables */
        unsigned long long utmp;

        unsigned long long sum = 0LL;    /* running sum */

        unsigned long ret;                  /* return value */

        DO(0);      DO(1);      DO(2);      DO(3);
        DO(4);      DO(5);      DO(6);      DO(7);
        DO(8);      DO(9);      DO(10);     DO(11);
        DO(12);     DO(13);     DO(14);     DO(15);
        DO(16);     DO(17);     DO(18);     DO(19);
        DO(20);     DO(21);     DO(22);     DO(23);
        DO(24);     DO(25);     DO(26);     DO(27);
        DO(28);     DO(29);     DO(30);     DO(31);

        /********** return (sum % 0x10000000fLL); **********/

        stmp = (sum  & 0xffffffffLL) - ((sum  >> 32) * 15);  /* lo - hi * 15 */
        utmp = (stmp & 0xffffffffLL) - ((stmp >> 32) * 15);  /* lo - hi * 15 */

        ret = utmp & 0xffffffff;
        if (utmp > 0x10000000fLL) /* if larger than p - subtract 15 again */
            ret -= 15;

        return ret;
}
```

Fast Message Authentication Using Efficient Polynomial Evaluation

Valentine Afanassiev[1], Christian Gehrmann[2], Ben Smeets[2]

[1] Institute for Problems of Information Transmission
of the Russian Academy of Science, Bolshoj Karetnyj 19,
GSP-4, Moscow, Russia,
Email: afanv@ippi.ac.msk.su
[2] Department of Information Thechnology, Lund University,
Box 118, S-221 00, Lund, Sweden,
Email: {chris,ben}@it.lth.se

Abstract. Message authentication codes (MACs) using polynomial evaluation have the advantage of requiring a very short key even for very large messages. We describe a low complexity software polynomial evaluation procedure, that for large message sizes gives a MAC that has about the same low software complexity as for bucket hashing but requires only small keys and has better security characteristics.

Key words: Message authentication, universal hash functions, polynomial evaluation, software MAC generation.

1 Introduction

The verification of the authenticity of a text document or a datafile is one of the main applications of cryptographic techniques. A common used technique for this purpose is the application of a *message authentication code* (MAC). Basically we have two users called the sender S (or signer) and the verifier V. S and V share a secret random key string and a publicly known MAC. The MAC maps a message string to a shorter, so called, tag string. The sender calculates the tag corresponding to the message string and the shared secret key string and sends the message to V together with the tag. V accepts a received message if the received tag is the same as the tag for the received message and the secret key. A good MAC is designed to make it hard for an adversary to send own messages or substitute observed messages by new ones, without being detected by the receiver.

Usually one distinguishes between so called *unconditionally secure, computationally secure,* and *provable secure* authentication codes, [1, page 392]. Codes belonging to the first category are codes for which the security of the MAC is independent of the computational power of the adversary. The security of these codes is expressed in the probability of success of an deception attack. A MAC is called computationally secure if the adversary is faced with the difficulty that all *known* computational methods to perform an attack require a infeasible amount of computation. MACs for which it can be shown that an successful attack implies that some other, usually well-known and presumed hard, problem can be solved are called provably secure. Traditionally computational and provably secure codes have been considered to be more of practical interest. However, unconditionally secure codes can easily be turned into practical provable secure codes by using a finite pseudo random function as was shown in [2], [3]. Usually the MAC computation has to be done in software. This is no problem when the message is of small size. Designing good efficient MACs for large message sizes is a challenging problem.

Carter and Wegman [4] introduced the concept of universal families of hash functions. They can be used to construct unconditionally secure MACs as shown in, for example, [4] and [5]. They also have numerous applications outside cryptography. In this paper we study how to construct a good family of universal hash functions that have a low complexity and which are suitable for software implementation. This problem was also the subject of the paper by Rogaway [2] where he introduced bucket hashing. Bucket hashing is a clever way to construct a family of universal hash function. Using bucket hashing the authentication of a long message requires only about 9-13 machine instructions per message word. Thus bucket hashing leads to very fast MACs. However it requires a key string of about the same size as the message string. Even if this string will be computed from a main key by some strong random number generator it is desirable to keep the string short. In [6] constructions were given of universal families of hash functions based on a relation between authentication codes and error correcting codes. These constructions require very small key size even for long messages and small probabilities of deception (there exist multi-round authentication constructions with smaller key size [7], [8], [9]). The constructions in [6] require polynomial evaluation in a finite field.

OUR CONTRIBUTION: This paper describes an efficient procedure for evaluating polynomials over a finite field to be used in the construction of a fast MAC. We obtain a MAC that can authenticate messages in about 7-13 instructions per word. Furthermore, the random (key) string that is required is much smaller than for the bucket hashing construction. Moreover, the probability of deception in our construction is uniformly (for all keys) bounded, where as this probability is given as an average for bucket hashing.

We begin with describing the construction of a universal family of hash functions that we are going to investigate. In Section 3 we propose fast procedures for polynomial evaluation in large fields. We calculate the complexity of the proposed procedures. Finally we give examples of MAC computations and investigate how fast they can be done in software. We compare the evaluation procedures with bucket hashing.

2 Universal hash functions based on polynomial evaluation

2.1 MAC construction

A *family of hash functions* is a finite multi-set H of functions, each $h \in H$ having the same nonempty domain set A and range set B. In what follows we will assume A and B to be sets over binary alphabet $\{0, 1\}$. We recall the following definitions.

Definition 1 [5]. Let $\epsilon > 0$. A multi-set H of n functions from a set A to a q-set B is ϵ-almost universal$_2$ (ϵ - AU$_2$) if for every pair $a_1, a_2 \in A, a_1 \neq a_2$ the number $d_H(a_1, a_2) = |\{h \in H; h(a_1) = h(a_2)\}| \leq \epsilon \cdot n$.

Definition 2 [5]. Let $\epsilon > 0$. A multi-set H of n functions from a set A to a q-set B is ϵ-almost strongly universal$_2$ (ϵ - ASU$_2$) if:
1. for every $a \in A$ and $y \in B$, the number of elements of H mapping $a \mapsto y$ is n/q,
2. for every pair $a_1, a_2 \in A, a_1 \neq a_2$, and every pair $y_1, y_2 \in B$ the number of elements of H that map $a_1 \mapsto y_1$ and $a_2 \mapsto y_2$ is $\leq \epsilon \cdot n/q$.

Given a family of hash functions H we can directly construct an unconditionally secure MAC [4], [5] or a complexity-theoretic variant when used together with a pseudo random function [2].

We investigate a construction of ϵ - ASU$_2$ family based on a concatenation of an ϵ_1 - AU$_2$ family and an ϵ_2 - ASU$_2$ family. For such a concatenation the following can be shown.

Theorem 3 [5]. *Let H_1 be and ϵ_1 - AU$_2$ from A_1 to B_1 and let h_2 be an ϵ_2 - ASU$_2$ from B_1 to B_2. Then $H = H_1 \times H_2$ is an ϵ - ASU$_2$ from A_1 to B_2 with $\epsilon \leq \epsilon_1 + \epsilon_2 - \epsilon_1\epsilon_2$.*

We also need the following lemma.

Lemma 4 [6]. *Let π be some $\mathbb{F}_{q_0}$-linear map from $\mathbb{F}_Q$ onto $\mathbb{F}_q$, where $Q = q_0^m, q = q_0^r$ and q_0 a prime power. Then the following family of hash functions $H = \{h_{a,b}; h_{a,b}(x) = \pi(ax) + b\}$, where $a, x \in \mathbb{F}_Q, b \in \mathbb{F}_q$ is ϵ - ASU$_2$, with $\epsilon = 1/q$.*

For large message sizes the following construction realizes an ϵ - ASU$_2$ family which maps elements from a large set A to a relative small set B for given ϵ requiring a small value of $|H|$.

Construction [6]: Let $q = 2^r, Q = 2^m = 2^{r+s}, n = 1 + 2^s$ and π be the same as in Lemma 4. Let $f_a(x) = a_0 + a_1 x + a_2 x^2 + \cdots + a_{n-1} x^{n-1}$, where $x, y, a_0, a_1, \ldots, a_{n-1} \in \mathbb{F}_Q, z \in \mathbb{F}_q$ and

$$H = \{h_{x,y,z} : h_{x,y,z}(\mathbf{a}) = h_{x,y,z}(a_0, \ldots, a_{n-1}) = \pi(y f_{\mathbf{a}}(x)) + z\}.$$

Theorem 5 [6]. *H in the construction above is ϵ - ASU$_2$ with $\epsilon \leq 2/2^r, |A| = Q^n = 2^{(r+s)(1+2^s)}$ and $|H| = Q^2 q$.*

The construction is a concatenation of a $1/2^r$ - AU_2 family of hash functions obtained from a Reed Solomon code and a $1/2^r$ - ASU_2 family of Lemma 4.

The construction above also gives us directly an unconditionally secure MAC. Considering the vector $\mathbf{a} = a_0, \ldots, a_{n-1}$ as a message where $a_0, a_1, \ldots, a_{n-1} \in \mathbb{F}_Q$ and $x, y \in \mathbb{F}_Q$, $z \in \mathbb{F}_q$ as the key parts gives us the corresponding MAC. This means that for the unconditionally secure MAC we have a key size of $\log_2 |H| = \log_2(Q^2 q) = 2m + r$ and a tag size of $\log_2 |B| = \log_2 q = r$ bits. The probability of deception is less than $\epsilon \leq 2/2^r$.

The following construction is also useful if for given message size parameters the realized bound on ϵ is too large.

Theorem 6. *Let H be a ϵ - ASU_2 with n functions, domain set A, and range set B. Then the family H^2 with domain set A and range set $B \times B$ defined by*

$$H^2 = \{h = (h_1, h_2) \in H \times H; h : A \ni a \mapsto (h_1(a), h_2(a)) \in B \times B\} \quad (1)$$

is an ϵ^2 - ASU_2. Furthermore, $|H^2| = n^2$.

Proof. The domain and range sets of H^2 as well as $|H^2|$ follow from the definition of H^2. Furthermore, it also follows from the definition that condition 1) of Definition 2 is satisfied for H^2. Since the component functions h_1 and h_2 of $h \in H^2$ can be chosen independently there will be at most $(\epsilon \cdot n/|B|)^2$. But $n^2 = |H^2|$ and the range set of H^2 has cardinality $|B|^2$. Thus H^2 is an ϵ^2 - ASU_2.

2.2 A multiple message MAC

Assume we want to authenticate l messages. Let $\mathbf{a}_1, \ldots, \mathbf{a}_l$ be a message sequence of l messages. For $1 \leq i \leq l$ let

$$H_i = \{h_{x,y,z_i} : h_{x,y,z}(\mathbf{a}_i) = h_{x,y,z},(a_{i_0}, \ldots, a_{i_{n-1}}) = \pi(yf_{\mathbf{a}_i}(x)) + z_i\},$$

where $a_{i_0}, \ldots, a_{i_{n-1}} \in \mathbb{F}_Q$ and $x, y \in \mathbb{F}_Q, z_i \in \mathbb{F}_q$, be the hash function of the i-th message. It was shown in [4], [10] that this gives us an unconditionally secure multiple message MAC. The key parts x and y may in this construction be considered as "hidden" key parts and they can remain unchanged for all messages in the sequence. Only the part z_i has to be refreshed for each message. A complexity-theoretical secure MAC for multiple use can easily be obtained from the unconditionally secure by changing the key part z_i by the output of a pseudo random function, see for example [2].

Recently, in [11] a new method for multiple authentication was proposed. This method does not use a refreshed key part and, hence, it does not require the z_i (called counter in [11]) for multiple authentication. Consequently there is no problem if one or several messages are lost during transmission. However this method exhibits a growth of the key size which is quadratic in the number of messages to be authenticated! Our method can easily be extended to handle lost messages or synchronization loss by adding the index i to the message a_i to be authenticated.

We consider multiple message MAC as a way of balancing between the complexity of pre-calculations and the complexity of a MAC calculation. In the next section we propose an efficient method for the evaluation of a polynomial of a large degree over a large finite field.

3   Polynomial evaluation

3.1   The basic procedures and their complexity

The main step in the MAC construction is the evaluation of a polynomial

$$f_{\mathbf{a}}(x) = \sum_{i=0}^{n-1} a_i x^i \text{ where } a_i \in \{0,1\}^w$$

in some point of a finite field $\mathbb{F}_Q$. Let $Q = 2^m$, $w \leq m$ and $n < Q$. As the measure of computational complexity we will use the *binary complexity* and the *multiplicative complexity*.

Definition 7. The **binary complexity** $C_b(\Phi)$ of the calculation [12] of some function Φ is defined as the total (minimal) number of elementary bit (Boolean or bit-level) operations from the set $\{\vee, \wedge, \oplus\}$ used in the calculation of Φ.
The **multiplicative complexity** $C_m(\Phi)$ is defined as the total (minimal) number of multiplications over the finite field used in the calculation Φ.

We also can use the binary complexity reduced to a subset of $\{\vee, \wedge, \oplus\}$, for example, the number of modulo 2 additions $C_\oplus(\Phi)$.

Lemma 8. *For any pair of elements of a finite field $\mathbb{F}_Q$, $Q = 2^m$, the binary complexity of addition is $C_b(+) = C_\oplus(+) = m$ and an upper estimate of the binary complexity of multiplication is $C_b(*) = O(m^2)$.*

Remark: This lemma has been proven in several variants. The bound $C_b(+) = C_\oplus(+) = m$ is trivial. The bound $C_b(*) = O(m^2)$ has been obtained for bit-serial structures of multipliers over finite fields as the product of m clock-cycles and m functional gates ($\oplus$, for example) for a dual basis [13], for a standard basis [14] and for an optimal normal basis [15]. It is known as an estimate of the number of functional elements for bit-parallel structures over the standard basis [16], [17]. It is also known for systolic arrays and other structures. The lowest asymptotic upper bound $O(m \log^{1+\epsilon} m)$, $\epsilon > 0$ can be obtained through fast convolution that uses a FFT over a finite or a surrogate field. However, it becomes less complex only for very large m (more than 1000). A modification of Karatsuba's method (or recursive double length multiplication formula [21]) for fast polynomial multiplication [16] gives another upper bound, $O(m^{\log 3})$, that is applicable for the range approximately from 50 to 1000. So, for the most practical range of finite fields (< 100) we have to use a quadratic bound. It can be shown that $C_b(*) \leq 4m^2$ and $C_\oplus(*) \leq 2m^2$ for a multiplier with a parallel structure [16].

A well known procedure for the calculation of $f_{\mathbf{a}}(\alpha)$ for any element α of $\mathbb{F}_Q$ is

Horner's procedure:

For given $\alpha \in \mathbb{F}_{2^m}$ and message $f(x)$:
$$u_1 = a_{n-1}; u_{i+1} = u_i * \alpha + a_{n-i}; \quad i = 1, \ldots, n; \quad u_n = f_{\mathbf{a}}(\alpha).$$

This procedure takes at most n additions and exactly n multiplications in $\mathbb{F}_Q$. So, the upper estimate for binary complexity of Horner's procedure over $\mathbb{F}_Q$ is $C_b(Horner) \le nC_b(*) + nm = nO(m^2)$.

We will consider the **MinPol procedure** that has a minimal multiplicative complexity. But, first we have to recall the definition of minimal polynomials [15] in a finite field. The minimal polynomial $\mu_\alpha(x)$ of any element $\alpha \in \mathbb{F}_Q/\{0\}$, $Q = 2^m$, is

$$\mu_\alpha(x) = \Pi_{j=0}^{t-1}\left(x + \alpha^{2^j}\right)$$

where the integer $t = m$ or $t \mid m$ is the minimal solution of $\alpha = \alpha^{2^t}$. If $t \mid m$, then α is an element of a subfield of $\mathbb{F}_{2^m}$. A minimal polynomial is clearly irreducible over the ground field $\mathbb{F}_2$ [15].

The following procedure for the computation of $f_{\mathbf{a}}(\alpha)$ has the minimal multiplicative complexity over $\mathbb{F}_{2^m}$:

MinPol procedure:

For given $\alpha \in \mathbb{F}_{2^m}$ and message $f(x)$:
1. Calculate the minimal polynomial $\mu_\alpha(x)$ of α.
2. Calculate $r_\mu(x) = f_{\mathbf{a}}(x) \bmod \mu_\alpha(x)$
3. Calculate $f_{\mathbf{a}}(\alpha) = r_\mu(\alpha)$

The first step of the MinPol procedure takes $t - 1$ squarings in $\mathbb{F}_{2^m}$ for the calculation of the conjugated elements of α and $\le t(t-1)/2$ multiplications and additions in $\mathbb{F}_{2^m}$ for the polynomial product calculation. The second step (step 2) takes $\le (n-t)t$ additions in $\mathbb{F}_{2^m}$ and the last step (Horner procedure) takes t additions and t multiplications in $\mathbb{F}_{2^m}$. Total multiplicative complexity in $\mathbb{F}_{2^m}$ (steps 1 through 3) is $\le 2t + t^2/2$. Thus the multiplicative complexity is independent of the degree of the polynomial $f_{\mathbf{a}}(x)$.

The following upper estimate is valid for the binary complexity over $\mathbb{F}_{2^m}$ of the MinPol procedure

$$C_b(MinPol) < nm^2 - \frac{m^3}{2} + m^2 O(m^2) .$$

We see that it has the same main term, i.e., nm^2, but in the MinPol procedure this is a strong upper estimate while for the Horner procedure this term depends on the multiplication complexity for the field $\mathbb{F}_{2^m}$. Now we can formulate a new problem: *find an evaluation procedure with the main term of order nm instead of nm^2*. We will see that for some nontrivial subsets of elements in $\mathbb{F}_{2^m}$ this estimate is achievable.

3.2   A new low complexity procedure for polynomial evaluation

Definition 9. The Minimal W-nomial $\tau_{\alpha,w}(x)$, $\alpha \in \mathbb{F}_{2^m}$, is the solution with respect to n_i and t of the equation

$$\min_t \left(x^t + \sum_{i=1}^{w-2} x^{n_i} + 1 = 0 \bmod \mu_\alpha(x) \right), t > n_i > 0, w \leq W.$$

By definition $\tau_{\alpha,w}(\alpha) = 0$ and $\tau_{\alpha,w}(x) = \mu_\alpha(x)$ if and only if the weight of $\mu_\alpha(x)$ is $\leq W$. If there is a multiple solution for different $w \leq W$ then we have an optimization problem for the complexity estimate. For the moment our choice is for minimal t.

Now we give a construction of a new low complexity procedure based on Minimal W-nomials.

MinWal procedure:

For given $\alpha \in \mathbb{F}_{2^m}$ and message $f(x)$:
1. Calculate the MinPol $\mu_a(x)$
2. Search for MinWal $\tau_{\alpha,w}(x) = 0 \bmod \mu_\alpha(x)$
3. Calculate $r_\tau(x) = f(x) \bmod \tau_{\alpha,w}(x)$
4. Calculate $r_\mu(x) = r_\tau(x) \bmod \mu_\alpha(x)$
5. Calculate $f_a(\alpha) = r_\mu(\alpha)$

To estimate the expected complexity we assume that a MinWal of degree $t < n$ exists (Step 2 is successful). Then Step 3 takes $\leq (w-1)(n-t)$ additions and Step 4 takes $< (t-m)m$ additions in $\mathbf{F}_Q$. So if $t \lesssim n/(m-w+1)$ then we can expect that the main term of the binary complexity to be of order $\lesssim nwm$. The complexity of Step 2 and the search procedure depend both on w.

The existence of MinWals is a consequence of the existence of Hamming and other cyclic codes of minimal distance $\leq w$ when $\mu_\alpha(x)$ is the factor of a generating polynomial of the code. *We call an element α bad if it has no corresponding W-nomial.* If α is an element of a subfield $\mathbb{F}_{Q'}$ of $\mathbb{F}_Q$ then there exists a binomial of degree $Q' \leq \sqrt{Q}$ which can be used instead of 3-nomial. As we will show later, the maximal degree of a W-nomial is of order $Q^{1/w-1}$. If α is an element from the multiplicative subgroup of order L, then α is a root of a binomial of degree L. This binomial can be used instead of the W-nomial if $L < Q^{1/w-1}$. So the only open question is *the existence of irreducible polynomials of weight $> w$ and degree m (and $m-1$ for 4 or 5-nomials) such that they generate a cyclic code of minimal distance $> w$ and length $L > Q^{1/w-1}$.*

The class of irreducible shortened cyclic codes, i.e., generated by an irreducible polynomial, has been considered in [18], and other papers. It has been proven that the codes in this subclass satisfy the Gilbert bound. This result was generalized to the class of shortened cyclic codes in [19]. Taking into account these results we can expect that the irreducible shortened cyclic code related with a bad element α should be very short in comparison with the threshold $Q^{1/w-1}$. We are not able to prove a more exact statement about bad elements in $\mathbb{F}_Q$. However because the codes related to the bad elements are short with high probability, we expect, provided $L > Q^{1/w-1}$, that the probability to choose a bad element is extremely low for large Q.

As a demonstration of the fact that the search of W−nomials is not a too complicated problem we give the description of

MinTrinSearch procedure:

For a given minimal polynomial $\mu_a(x)$ and $r_i(x) = x^i, i = 0 \ldots m - 1$,
Calculate for $i = m, m + 1, \ldots$
1. $r_i(x) = x * r_{i-1}(x) \mod \mu_a(x)$ and
2. Find first $j < i$ with $r_j(x) + 1 = r_i(x)$.

It is clear that $i \leq T$, where T is the degree (minimal) of a MinTrin we expect to find in the search. If the storage of $r_i(x)$ is organized as a dichotomous (binary) tree then the complexity of the **MinTrinSearch** procedure is linear in the degree $t_3 \leq T$ of the Minimal Trinomial (if it exists). In fact its binary complexity is $O(t_3 m)$. A similar search procedure could be used with other polynomials of a limited weight. The main trick is to organize the residues in a tree structure. However, the search procedure for 4 or 5− nomials has complexity $O(t_4^2 m)$ or $O(t_5^2 m)$.

3.3   New estimates for the complexity of polynomial evaluation

In this section we will give arguments to obtain a simple lower and upper estimates for the maximal degree of MinWal.

Definition 10. The maximal degree T_w of the Minimal W−nomial (MinWal) $\tau_{\alpha,w}(x)$ over $\mathbb{F}_Q$ (cf. Definition 9), is

$$T_w = \max_{\mu_\alpha(x)} (\deg \tau_{\alpha,w}(x) : \tau_{\alpha,w}(x) = 0 \mod \mu_\alpha(x)),$$

where $\alpha \in \mathbb{F}_Q$ and $\mu_\alpha(x)$ is the MinPol of α.

Proposition 11. *Lower and upper estimates for the maximal degree T_w of Min-Wal $\tau_{\alpha,w}(x)$ over $\mathbb{F}_Q$, $Q = 2^m$, are*

$$L_{w,m} = ((w-1)! \, 2Q)^{1/(w-1)} <\approx T_w <\approx L_{w,m} \eta^{\frac{1}{w-1}} = U_{w,m},$$

where $\eta = 2m/\log m$.

Proof. Let $T_w = T$. A W−nomial of degree t is *normal* if it is written as $1 + \sum_{i=1}^{w-2} x^{n_i} + x^t$ with $1 \leq \ldots < n_{i-1} < n_i < \ldots < t$. The number $D_{w,T}$ of normal W−nomials of degree $t \leq T$ is

$$D_{w,T} = \sum_{t=w-1}^{T} \binom{t-1}{w-2} = \binom{T}{w-1} < \frac{T^{w-1}}{(w-1)!}$$

We are going to estimate $L_{w,m}$ as the necessary T such that this set of normal W−nomials includes all MinWals over $\mathbb{F}_Q$ and only partially their multiples by squaring and other MinWals over $\mathbb{F}_{2^m}$ for $m < n \leq T$. The estimate $U_{w,m}$ will be given as the sufficient T for the same condition.

Let M_m be the number of MinWals for the field $\mathbb{F}_Q$. Clearly each MinWal can be squared less than m times if $T < 2^m$. Assuming that one MinWal is related to one of the irreducible polynomials of degree $\leq m$ we have an estimate $M_m \approx \sum_{k \leq m} I_k$ where I_k is the number of irreducible polynomials of degree k. Taking into account that $\sum_{k \leq m} I_k \ll \sum_{k \leq T} I_k$ for $T \geq 2m$ we can estimate $L_{w,m}$ as the solution of $D_{w,T} > m M_m$ with respect to T. A well known asymptotic estimate for I_k is $2^k/k$ [15] and for $\sum_{k \leq m} I_k$ is $2^{m+1}/m$. So we can write

$$T > L_{w,m} = ((w-1)!\, 2Q)^{1/(w-1)} .$$

Now we assume that each reducible MinWal of degree t over $\mathbb{F}_Q$ has one factor of degree $\leq m$ and the others of degree $> m$. We can expect that any binary polynomial of degree $t \leq T$ has $\leq \eta$ different irreducible factors of degree $\leq T$. This holds for W−nomials and MinWals as well. Now we can declare that $\frac{1}{m} D_{w,T}$ is less than the number of normal W−nomials of degree $t < T$ and is less than total number of their different irreducible factors. Taking into account that $M_m \approx \sum_{k \leq m} I_k \ll \sum_{k \leq T} I_k$ for $T > 2m$ we can estimate the total number of different irreducible factors as ηM_m . Thus we have $\frac{1}{m} D_{w,T} <\approx \frac{2Q}{m}\eta$ and

$$T < U_{w,m} = (2\,(w-1)!\, Q\eta)^{1/(w-1)} .$$

The exact upper bound for the number of different monic divisors of a binary polynomial of degree $T < 2^m$ is proved in [19] in the form $\eta \leq \frac{m}{2+\log m}(1 + o(1))$. For our purposes we can simplify a little this bound to the form $\eta \leq \frac{2m}{\log m}$.

The last proposition is more a hypothetical than a strong result because we have to rely on some reasonable but unproven assumptions. Let us consider another upper estimates for T_w. The total number of W−nomials can be estimated as the number of codewords A_w of weight w of a Hamming code (excluding their cyclic shifts) times the number of different codes. So, the estimate $\binom{T_3}{w} \approx A_w \frac{2Q}{m} <\approx \frac{1}{T}\binom{T_3}{w}\frac{2Q}{m}$ leads to an upper bound $T_3 < \frac{2Q}{m}$. Because each cyclic code of the distance $\leq w$ gives only one MinWal (of weight w) we can estimate the number of MinWals through the number of cyclic codes of that distance. If we estimate the number of codes through all pairs of irreducible polynomials of degree m or less we have for 5-nomials $T_5 \leq \left(\frac{48Q^2}{m}\right)^{1/4}$.

The estimation of the complexity of a polynomial evaluation concludes this section. It is evident that the calculation of $f_a(\alpha)$ modulo the W−nomial needs $\leq (w-1)(n - T_w)$ additions in $\mathbb{F}_Q$, where T_w is the degree of the W−nomial. Returning to the **MinWal** and **MinPol** procedures we can estimate the binary complexity of the **MinWal** procedure for evaluation of a polynomial of degree $n \leq Q - 2$ in a point of the finite field $\mathbb{F}_Q$, $Q = 2^m$, as

$$C_b \leq m\,(n\,(w-1) + (m - w + 1)\, U_{w,m}) + O\,(m^3) .$$

field size	real values		bounds (max)	
Q	mean	max	$\sqrt{4Q}$	$\sqrt{4Q\eta}$
2^9	29	61	45	107
2^{10}	40	83	64	157
2^{11}	64	143	90	227
2^{12}	85	217	128	331
2^{13}	128	337	181	480
2^{14}	181	473	256	694
2^{15}	257	801	362	1003
2^{16}	409	1285	512	1448

Table 1. Complete statistics for the degree of trinomials compared with lower and upper bounds for maximal degree of trinomials for some small fields.

This estimate is of order $nm\beta$ when $U_{w,m} \approx \alpha n$, $\alpha = \frac{\beta-w+1}{m-w+1} < 1$, and β is a constant $w - 1 < \beta < m$.

A hash procedure that uses a new evaluation point for each message, would demand a message length of order of the square of the minimal 4−nomial or 5−nomial. In contrast, when using the evaluation for multiple MAC calculations the 4−nomial or 5−nomial could be considered as part of the "hidden" key. Thus they only have to be calculated once and can then be used to authenticate a large number of messages. We will discuss these aspects further in Section 4.

3.4   Experimental results

As we can see there is a gap between the lower and the upper estimate of the maximal degree of the MinWal. Furthermore, they are only estimates. We are therefore interested in investigating how tight our estimates are. We continue by discussing experimental results for the degree of minimal trinomial, quadronomial and pentanomial for different field sizes. We have calculated the complete statistics for the average and maximum degree of minimal trinomials for field size between 2^9 and 2^{16}. These values together with the corresponding values obtained from our lower, $L_{3,m} = \sqrt{4Q}$, and upper, $U_{3,m} = \sqrt{4Q\eta}$, $\eta = 2m/\log m$, estimates on the maximum degree of the minimal trinomial are listed in the Table 1. It can be seen from Table 1 that the mean degree of the trinomial is rather close to $L_{3,m}$.

It is not possible to calculate the complete statistics for fields of large sizes in reasonable time. To see how the degree of the minimal 3, 4 and 5− nomials are distributed for larger fields, we have chosen random elements from fields of size less or equal 2^{31}. We have investigated the distribution of minimal W−nomial degree for a logarithmic scale. We tested a *log-normal* approximation for the distribution of the minimal degrees and it turns out to give good agreement in χ^2 tests. The means of the degrees for the tests are close to our lower bounds

W	Field size	# elm.	Exper.degree mean	max	Lower $L_{3,m}$	Upper $U_{3,m}$	log degree mean	STD	$\Pr\{t > U_{3,m}\}$
3	2^{15}	compl.	257	801	362	1003	5.36	0.67	$1.0 \cdot 10^{-2}$
	2^{16}	compl.	409	1285	512	1448	5.81	0.68	$1.5 \cdot 10^{-2}$
	2^{19}	3200	1049	3517	1448	4331	6.79	0.64	$6.7 \cdot 10^{-3}$
	2^{25}	3200	8399	29505	8437	27684	8.88	0.64	$1.7 \cdot 10^{-2}$
	2^{31}	800	66937	210074	92682	327872	10.9	0.62	$2.2 \cdot 10^{-3}$

W	Field	# elm.	Exper.degree mean	max	$L_{4,m}$	$U_{4,m}$	log degree mean	STD	$\Pr\{t > U_{4,m}\}$
4	2^{15}	compl.	55	151	73	145	3.940	.39	$3.7 \cdot 10^{-3}$
	2^{16}	compl.	70	141	92	185	4.17	.37	$2.5 \cdot 10^{-3}$
	2^{19}	2400	138	298	185	384	4.85	.40	$3.0 \cdot 10^{-3}$
	2^{25}	2400	551	1164	738	1630	6.23	.42	$3.0 \cdot 10^{-3}$
	2^{31}	400	2179	4608	2954	6858	7.6	.42	$1.8 \cdot 10^{-3}$

W	Field	# elmn.	Exper.degree mean	max	$L_{5,m}$	$U_{5,m}$	log degree mean	STD	$\Pr\{t > U_{5,m}\}$
5	2^{15}	compl.	28	51	35	59	3.29	.266	$1.5 \cdot 10^{-3}$
	2^{16}	compl.	33	57	60	104	3.47	.277	$1.1 \cdot 10^{-5}$
	2^{19}	3200	56	106	168	304	3.98	.296	$2.4 \cdot 10^{-9}$
	2^{25}	2400	156	321	476	895	5.0	.316	$6.6 \cdot 10^{-9}$
	2^{31}	400	433	726	566	1066	6.0	.316	$1.1 \cdot 10^{-3}$

Table 2. Minimal degree of W−nomial ($W = 3, 4, 5$) for random elements or complete statistics ($\mathbb{F}_{2^{15}}$ and $\mathbb{F}_{2^{16}}$) in different fields, the corresponding bounds for W−nomial degree t and the probability for a random element to take a larger degree for the W−nomial than the $U_{w,m}$, given a normal distribution with the estimated mean and standard deviation for a logarithmic scale.

for the maximal degree of the polynomials. This can be seen for the examples in Table 2.

It can be seen in Table 2 that the variance varies a little with field size in the three different cases, but that it is about the same value independent of the field size. The experimental maximal value for the tested fields are not the true maximum value, but just the maximum value for the tested elements. As we can see the hypothesis on η has a good experimental support .

To explain the results presented in Table 2 we can use estimates for the average number η_{avr} of different irreducible factors of a polynomial of degree $t \leq 2^m$, $\eta_{avr} = \log m + c + O\left(\frac{\log m}{m}\right)$ [20] where c is some constant. Thus, by using η_{avr} instead of η we can estimate $T_{w,avr} \approx L_{w,m}\left(\eta_{avr}\right)^{1/w-1}$. Now we can estimate the span for the log of the W−nomial degree as

$$\log T_w - \log T_{w,avr} = \frac{\log m - 2\log\log m}{w - 1}.$$

This estimate can be related linearly to the standard deviation of the experimental distribution of the degree of the W−nomial. As we can see from numeric calculations the last estimate varies very little within a range $m \leq 256$ for a fixed w. This estimate is more sensitive to w for $w \leq 5$.

4 Software MAC computations

It is often necessary to compute the MACs in software on a workstation or personal computer. In this section we investigate how fast the previous polynomial evaluation procedure would be if it is implemented in software. We make a comparison with the Bucket hashing method.

The evaluation point or "hidden" key in the polynomial evaluation MAC generation could be unchanged for several messages. For the trinomial procedure the complexity of the calculation of trinomial is linear in the degree and would not give a main contribution to the overall complexity of the MAC calculation. The 4 or 5−nomial calculations in our evaluation procedures only depend on the evaluation point and can be precomputed and used as part of the "hidden" key. We will in the comparison with Bucket hashing not include the key generation in the calculations, but only the message authentication for a message given a secret key. As we will show below Bucket hashing demands a much longer key than the polynomial evaluation method and would in general for long messages be more complex to generate than the 4-nomial or 5-nomial for the evaluation procedure.

Assume we use a computer with $k-$bit architecture (typical $k = 32$ or $k = 64$). We also assume that the message is represented as a vector of words of k bits. Reducing modulo a $W-$nomial and modulo a minimal polynomial is implemented for this $k-$bit architecture with no relation to the given finite field. However at the last stage of the **MinWal** procedure we have to implement the arithmetic of the given field.

The latter has also implications for the cache hit rate when executing the algorithm on a usual computer. After the first reduction by the $W-$nomial which coincides with reading the message (e.g. from disk), most of the computations occur 'locally' which gives high cache hit rates.

As we have shown previously the modulo calculations demand $w-1$ additions for each message word using a $W-$nomial. For the typical structure of a CPU instruction based on a bank of internal register we have the following approximations for the number of instructions for the MAC calculations by **MinWal** procedure: $\# = 3w - 2$ instructions per word (7, 10 or 13 using the 3, 4 or 5-nomial, respectively). If the CPU XOR instruction includes internal register and an address in main memory and contains the read and write cycle then we have the lower approximation: $\# = 2w - 1$. This can be compared with the Bucket hashing method, which needs about $9 - 13$ instructions per word [2].

We have investigated the method for different parameters. According to the experimental results in the previous section the expected $W-$nomial degree are a little bit below the lower bounds for the maximal degree. Hence, we have used the lower bounds as an estimate for the mean of a log normal distribution for the different fields. Assume $\log t \in N(x, \sigma)$, i.e., a normal distribution with mean

x and variance σ. Then the mean x' of the log *normal* distribution is given by $x' = e^{x+\sigma^2/2}$. Hence, we would have the following estimate

$$x_w = \log L_{w,m} - \sigma_w^2/2$$

for the mean of the normal distribution. According to the experimental results the variance for the normal distributions for W−nomial, $W = 3, 4, 5$, are likely to be values around $0.4, 0.18$ and 0.1 respectively. We have used these values as estimates for the normal distribution of the degrees using a logarithmic scale. We have calculated parameters for using polynomial evaluation for different message lengths n and polynomial degrees T for the different field sizes in Table 3 for a probability of deception 2^{-30}, 2^{-40}, and 2^{-60} respectively. We have assumed that the computer word size is 32 bits. The degrees T are chosen to have a small estimated probability $\Pr\{t > T\}$, according to the approximations above. If we use the evaluation procedure for fewer number of words, the total complexity would of course be less, but the number of instructions per word would increase.

To lessen the effect that if we require a small ϵ we need large messages to get an efficient evaluation procedure we use the result of Theorem 6 to obtain a new MAC from a given MAC with ϵ_0 that has twice the key size, identical message size, and $\epsilon = \epsilon_0^2$. Although at first glance this will also double the time to compute the tag but it will in fact be less, say 50% only, if properly implemented. For some of the examples in the tables we have used this doubling of tag and key size to get an efficient MAC with the desired probability of deception.

In the Table 3 we give examples where we have chosen the minimal field size for the given probability of deception. We are able to freely choose a proper field for the MAC because the first steps of the evaluation procedure are independent of field size. When we calculated the table we assumed the 4− nominal and 5−nominal to be part of the key and hence the key size is given by key size$_w = 2m + r + (w - 1) \log_2 U_{w,m}$ where the last term could be omitted in the case of a 3-nomial. The tag size for the polynomial evaluation MAC calculation is for all examples much smaller than that for Bucket hashing. In [11] other examples with different parameters are given. There is also comparisons with other authentication codes. It can be seen in the table that for shorter messages the 5− nominal is better to use than 3−nominal or 4−nominal. It is not possible to have a low number of instructions per word for shorter messages and still have a low probability of deception ϵ. The short authentication tag and small key size for the evaluation procedure and its suitability for very fast implementation makes it much better to use than Bucket hashing for long messages.

5 Conclusion

We suggested an efficient polynomial evaluation procedure based on calculations modulo a 3, 4 or 5−nomial that leads to a fast MAC. The complexity of our procedure is closely related to the degree of these low weight polynomials. We derived estimates of lower and upper bounds on the degrees of these polynomials and compared these bounds with experimental results.

$\epsilon = 2^{-30}$	W-nomial method						Bucket hashing [2]	
n	field	W	tag size	key size	T	$\Pr\{t > T\}$	tag size	key size
2^{25}	41	3	2×16	2×98	2^{22}	0.1937	18784	$9.26 \cdot 10^8$
2^{30}	46	3	2×16	2×108	2^{25}	0.0790	595521	$3.50 \cdot 10^{10}$
2^{17}	33	4	2×16	2×133	2^{13}	0.0633	4256	$2.78 \cdot 10^6$
2^{23}	54	4	31	211	2^{20}	0.0633	11840	$2.15 \cdot 10^8$
2^{16}	47	5	31	197	2^{14}	0.0212	4128	$1.38 \cdot 10^6$

$\epsilon = 2^{-40}$	W-nomial method						Bucket hashing [2]	
n	field	W	tag size	key size	T	$\Pr\{t > T\}$	tag size	key size
2^{31}	52	3	2×21	2×125	2^{28}	0.0790	75040	$7.21 \cdot 10^{10}$
2^{17}	38	4	2×21	2×154	2^{15}	0.0191	12608	$3.40 \cdot 10^6$
2^{29}	70	4	41	271	2^{26}	0.0044	47296	$1.70 \cdot 10^{10}$
2^{13}	34	5	2×21	2×149	2^{11}	0.0050	12576	$2.24 \cdot 10^5$
2^{23}	64	5	41	261	2^{18}	0.06940	14400	$2.22 \cdot 10^8$

$\epsilon = 2^{-60}$	W-nomial method						Bucket hashing [2]	
n	field	W	tag size	key size	T	$\Pr\{t > T\}$	tag size	key size
2^{31}	52	3	3×21	3×125	2^{28}	0.0790	131200	$7.73 \cdot 10^{10}$
2^{17}	38	4	3×21	3×154	2^{15}	0.0191	126752	$4.82 \cdot 10^6$
2^{26}	57	4	2×31	2×223	2^{21}	0.0633	126880	$2.41 \cdot 10^9$
2^{13}	34	5	3×21	3×149	2^{11}	0.0050	126752	$4.20 \cdot 10^5$
2^{20}	51	5	2×31	2×209	2^{15}	0.0212	126752	$3.77 \cdot 10^7$

Table 3. Construction parameters using polynomial evaluation and Bucket hashing with a computer word size of 32 bits for different message lengths for a probability of deception less than 2^{-30}, 2^{-40}, and 2^{-60}. Using the MAC more than once is marked with $\times$ (for example 2×21) in the key size column.

We investigated how fast the evaluation can be made in software. In terms of speed our procedure can be compared with bucket hashing [2]. But, our method requires much shorter key than those based on bucket hashing. In Bucket hashing a key size of about the same size as the message is required. By using polynomial evaluation it is possible to reduce the key to a size of about as large as that of the tag. When 4 or 5-nomials are used for the evaluation, they have to be precomputed and should be a part of the key. The search for a 4 or 5-nomial can be of rather high complexity. However, the same 4 or 5-nomial can be used to calculate MAC's for several messages. Loosely speaking one can say that our polynomial evaluation procedure is efficient for large messages and for multiple authentication.

Our evaluation method can be used in other constructions of universal families of hash functions to get similar complexity reductions.

References

1. G.J. Simmons, "A survey of information authentication", in *Contemporary Cryptology, The Science of Information Integrity*, ed. G.J. Simmons, IEEE Press, New York, 1992.

2. P. Rogaway, "Bucket hashing and its application to fast message authentication", *Proceedings of CRYPTO '95*, Springer Verlag, pp. 29-42, August, 1995.

3. M. Bellare, J. Kilian, and P. Rogaway, "The security of cipher block chaining", *Proceedings of CRYPTO' 94*, Springer Verlag, pp. 341-358, August, 1994.

4. M. Wegman and L. Carter, "New hash functions and their use in authentication and set equality", *J. of Computer and System Sciences 22*, pp. 265-279, 1981.

5. D. Stinson, "Universal hashing and authentication codes", *Designs, Codes and Cryptography*, Vol. 4, pp. 369-380, 1994.

6. J. Bierbrauer, T. Johansson, G. Kabatanskii and B. Smeets, "On families of hash functions via geometric codes and concatenation", *Proceedings of CRYPTO '93*, Springer Verlag, pp. 331-342, 1994.

7. P. Gemmell and M. Naor, "Codes for interactive authentication", *Proceedings of CRYPTO '93*, Springer Verlag, pp. 355-367, 1994.

8. C. Gehrmann, "Cryptanalysis of the Gemmell and Naor multiround authentication protocol", *Proceedings of CRYPTO '94*, Springer Verlag, pp. 121-128, 1994.

9. C. Gehrmann, "Secure multiround authentication protocols", *Proceedings of Eurocrypt '95*, Springer Verlag, pp. 158-167, 1995.

10. T. Johansson, *Contribution to Unconditionally Secure Authentication*, Ph. D. thesis, Lund 1994.

11. M. Atici and D. R. Stinson, "Universal hashing and multiple authentication", *Proceedings of CRYPTO '96*, Springer Verlag, pp. 16-30, 1996.

12. J.E. Savage, "The complexity of decoders. Computational work and decoding time", *IEEE. Trans. Inform. Theory*, Vol. 17, pp. 77-85, January, 1971.

13. E.R. Berlekamp, "Bit-serial Reed-Solomon encoder", *IEEE. Trans. Inform. Theory*, Vol. 28, pp. 869-874, November, 1982.

14. M.A. Hasan and V.K. Bhargava, "Division and bit-serial multiplication over $GF(q^m)$", *IEE Proceedings-E*, Vol.139, No. 3, May, 1992.

15. D. Jungnickel, *Finite fields: structure and arithmetics*, Wissenschaftsverlag, Mannheim-Leipzig-Wien-Zurich, 1993.

16. V.B. Afanassiev, "On the complexity of finite field arithmetic", *Fifth Soviet-Swedish Int. Workshop on Inform. Theory, Moscow*, January, 1991.

17. E.D. Mastrovito, *VLSI Designs for Computations over Finite Fields $GF(2^m)$*, Internal Report LiTH-ISY-I, Linköping Univ., Sweden, 1988.

18. T. Kasami, "An upper bound on k/n for affine-invariant codes with fixed d/n", *IEEE Trans. Inform. Theory*, Vol. 15, pp. 174-176, January, 1969.

19. Ph. Piret, "On the number of divisors of a polynomial over $GF(2)$", Springer Verlag, Lecture Notes in Comp. Sci. 228, pp. 161-168, 1985.

20. I. E. Shparlinski, *Computational and algorithmic problems in finite fields*, Kluwer, Dordtrecht-Boston-London, 1992.

21. H. Riesel, *Prime Numbers and Computer Methods for Factorization*, Birkhuser, Boston-Basel-Stuttgart, 1985.

Reinventing the Travois: Encryption/MAC in 30 ROM Bytes

Gideon Yuval

Microsoft Research, Redmond, WA 98052, U.S.A.

Abstract. By using a large number of round, we hope to be able to scrounge an Sbox out of nowhere, in an environment for which even TEA and the SAFERs are gross overdesign.

1 Background

Some people in the software industry are looking into home-control systems, much preferably without stringing new wires. This raises a simple issue: I have no more access to my power-lines than a thief has; and ditto (even more so) for wireless.

Therefore, if we want the same security we now have by owning the wire, we need some kind of cryptologic authentication.

The CPUs considered for this are quite underpowered (by today's standards): 8051 or similar[1], 1KB flash EPROM, 64 bytes RAM, 128 bytes EEPROM, and a peak 1MHz instruction rate; that last figure is relatively very fast, since the wire is 10KBPS or less.

In the classic study of access-control weaknesses, Ali Baba could replay the "open" message, using the authenticator "simsim" (or "sesame"); since replay attacks were not blocked, it did not matter how strong the 40 thieves' crypto & authentication was on other fronts. We therefore have to use EEPROM to keep track of a serial number, and get rid of replayed ones.

To add to the problem, messages (including the authenticator) had better be kept down to 8 bytes or so, to give them some decent chance to get through all the line noise, with the wimpy power-supplies planned.

2 MACs & ciphers

To authenticate a message, between parties who share a secret, we need a keyed MAC. I notice that all the secure hashes in common use are keyed MACs for which the key is frozen at spec time, and used as a chaining variable for longer messages. I also notice that all these hashes/MACs are a block cipher use in Meyer/Davies feedforward mode. So it seems we need a block cipher.

The SAFER-SK family is great on the security front (unless the crooks get smarter than Lars Knudsen); but it needs 512 bytes ROM for its two S-boxes (even if we never decrypt), plus 1-2KB for the code. Getting that much space off a 1KB chip, which is also trying to get some useful work done, is obviously unrealistic.

TEA may indeed be a Tiny Encryption Algorithm on a 32-bit CPU; but on a chip for which 16-bit subtraction is already a design issue, it is liable to be rather less tiny.

In summary, we need a decent block-cipher that uses as little ROM and RAM as possible, except what is available anyway. The only resource we have in some abundance is CPU cycles - the 6.4msec it takes to ship an 8-byte message amount to 6,400 instructions at the peak rate.

3   Stealing the Sbox

Since the chip will also-have non-crypto code running on it, we can try to scrounge an Sbox in the memory used for that code. This Sbox is not designed by Coppersmith; it is not even designed by a semi-competent cryptologist; it is whatever bits are there when the assembler has done its thing.

But a 256 ∗ 8 Sbox (the obvious size for S/W crypto), is large enough to avoid some attacks; and on that kind of chip, the coders want the code to do useful work, and the chip designers want high code-density; so such an Sbox ought to have enough entropy to make life interesting.

Using an 8-byte key and and 8-byte plaintext, the resulting 8051 encryptor is

```
$title(small slow 51 encryptor)
$nomod51
$nopaging
$list
 name slow_51_encrypt
 sbox equ ????h
 text data 32
 key data 41
 ofRounds equ 32
 size equ 8

 cseg
 org ????
 mov dptr,#sbox

Allrounds:
 mov r3,#ofRounds

Oneround:
 mov r0,#text
 mov r1,#key ; 8-byte key, 8-byte text
 mov r2,#size; =8
 mov text+8,text ; to get wraparound logic

subround:
```

```
mov a,@r0
add a,@r1
movc a,@a+dptr ; dptr is frozen, pointing to table in ROM, which is
; just part of the code
inc r0
add a,@r0
rl a ; make sure bits in table get ''scrambled'' some
mov @r0,a
inc r1
djnz r2,subround

mov text,text+8 ; finish wraparound logic
djnz r3,Oneround
end
```

For the 51-challenged, a C decompilation follows:

```c
for(r=0; r<NumRounds; r++) {
  text[8]=text[0];
  for(i=0;i<8;i++) {
    text[i+1] =(text[i+1] + Sbox[(key[i]+text[i])%256])<<<1;
// rotate 1 left
 }
  text[0]=text[8];
}
```

The full .HEX file is

```
:1004B00090?????7B20782079297A08852028E627DD
:0D04C00093082623F609DAF6852820DBE8EC
:00000001FF
```

The .LST file-excerpt below may help correlating the two. Lines have been truncated to fit

```
14:    0200 90 C0 DE  mov dptr,#sbox
15:    0203    Allrounds:
16:    0203 7B 20    mov r3,#ofRounds
17:
18:    0205    Oneround:
19:    0205 78 20    mov r0,#text
20:    0207 79 29    mov r1,#key ; 8-byte key, 8-. . .
21:    0209 7A 08    mov r2,#size; =8
22:    020B 85 20 28  mov text+8,text ; to get . . .
23:
```

```
24:    020E    subround:
25:    020E E6    mov a,@r0
26:    020F 27    add a,@r1
27:    0210 93    movc a,@a+dptr ; dptr is . . .
28:    0211 08    inc r0
29:    0212 26    add a,@r0
30:    0213 23    rl a ; make sure bits . . .
31:    0214 F6    mov @r0,a
32:    0215 09    inc r1
33:    0216 DA F6    djnz r2,subround
34:
35:    0218 85 28 20    mov text,text+8 ; finish . . .
36:    021B DB E8    djnz r3,Oneround
```

The 32 *rounds* (each byte gets hit 32 times; we look up that Sbox 256 times) are there to cover up for the many weaknesses of the design. Also:

1. Not all bit-planes in the Sbox will be equally random, nonlinear, ? ; the rotate instruction spreads the XORing among all the bit-planes.
2. The key schedule is stupid, making parts of the encryptor commute with other parts (with or without rotating the key and text). This is a code-size issue. It also lets us ship the message while the authentication code is running.
3. The modulo-2 additions do not commute with the rotation; this should stop attacks like Biham and Ben-Aroya's on Lucifer, in which the XOR is moved up & down the flowchart.

On the other hand, this cipher lets carry-propagation do its thing at least as many times as TEA does, and uses no >8-bit operations (coding a 32-bit rotate, on a pure 8-bit chip like the 8051, is almost a major design-issue).

Using a public-domain BASIC/51 interpreter as a source for S-boxes, and checking out how many output bits change when one bit is changed in byte 7 of the input, we find

```
1 round  5+-2 bits change
2 rounds 24+-10 for the worst S-boxes, up to 28 for better ones
3 round  30+-5
4        32+-4
```

But the minimum over 1024 pairs usually goes something like

```
1 round  1
2        1
3        1
4        5
5 rounds 20, & stays there (+-)
```

So 5 rounds are needed before the outliers behave decently; and it seems unwise to go below 8 rounds if we want real security.

Since this cipher is intended to be used when SAFER would be way too expensive, I refer to it as TREYFER.

4 Protocol

To put this cipher into a protocol:

Messages can be repeated indefinitely; but they only get acted on once. The sender will repeat the message until it gets an ACK.

To get the useful content of a message, subtract the previous message from it (since we wait for an ACK, "previous" is well-defined between sender and receiver). Thus, 0 is never a useful content. Other contents are encoded so as to make common contents be small integers. This will make wraparound take a fairly long time; but not as long for (e.g.) "dim light to 75%" as for "open the garage door".

We now can have an 8-byte message that contains a 2-byte from/to field, a 4-byte counter/content field, and a 2-byte authenticator. All of these numbers can be juggled up & down. They will never give us Fort Knox security, but are likely to be more secure than other weaknesses around the house.

Since all messages are one encryption block, it seems the attacks by Preneel & van Oorschott[2] and by Bellare at al.[3] do not apply.

And, since we never code up a decryptor, we have a chance to export non-joke security without going to jail.

5 Standard S-box

If anyone wants to try his hands at breaking this kind of cipher, he should avoid his own code, and get something that someone else wrote. If nothing else is at hand, use the 1st 256 primes (all modulo 256), starting with 2 (the only even value in the table).

References

1. Hersch, Russ: 8051 Microcontroller FAQ, widely available on the Internet
2. Preneel, B. and van Oorschott, P., MDx-MAC and Building Fast Macs from Hash Functions, Crypto'95, p.1.
3. Bellare, M. Canetti, R. and Krawcyzk, H., Keying Hash Functions for Message Authentication, Crypto'96, p.1

All-or-Nothing Encryption and the Package Transform

Ronald L. Rivest

MIT Laboratory for Computer Science
545 Technology Square, Cambridge, Mass. 02139
rivest@theory.lcs.mit.edu

Abstract. We present a new mode of encryption for block ciphers, which we call *all-or-nothing encryption*. This mode has the interesting defining property that one must decrypt the *entire* ciphertext before one can determine *even one* message block. This means that brute-force searches against all-or-nothing encryption are slowed down by a factor equal to the number of blocks in the ciphertext. We give a specific way of implementing all-or-nothing encryption using a "package transform" as a pre-processing step to an ordinary encryption mode. A package transform followed by ordinary codebook encryption also has the interesting property that it is very efficiently implemented in parallel. All-or-nothing encryption can also provide protection against chosen-plaintext and related-message attacks.

1   Introduction

One way in which a cryptosystem may be attacked is by brute-force search: an adversary tries decrypting an intercepted ciphertext with all possible keys until the plaintext "makes sense" or until it matches a known target plaintext. Our primary motivation is to devise means to make brute-force search more difficult, by appropriately pre-processing a message before encrypting it.

In this paper, we assume that the cipher under discussion is a block cipher with fixed-length input/output blocks, although our remarks generalize to other kinds of ciphers. An "encryption mode" is used to extend the encryption function to arbitrary length messages (see, for example, Schneier [9] and Biham [3]).

In general, the work required to search for an unknown k-bit key to a known block cipher is 2^k in the worst-case, or 2^{k-1} on the average. Here (and throughout this paper) we measure the work by the number of elementary decryptions attempted, where an elementary decryption is a decryption of one block of ciphertext. For example, in the "electronic codebook" encryption mode the adversary needs to decrypt only the first block of ciphertext to obtain the first block of plaintext; this is usually sufficient to identify the correct key. (If not, the second block can be decrypted as well...)

Sometimes the size of the key space for one's encryption algorithm is fixed, "marginal," and can't be improved. For example, one can argue that a 56-bit DES key is marginal (see Blaze et al. [4]). Or, one may be encumbered by export regulations that restrict one to a 40-bit secret key. The question posed here is: *is there any way to significantly increase the difficulty for an adversary of performing a brute-force search, while keeping the key size the same and not overly burdening the legitimate communicants?*

We show that the answer to the question is *yes*.

2   Strongly non-separable encryption

The problem with most popular encryption modes is that the adversary can obtain one block of plaintext by decrypting *just one* block of ciphertext.

We illustrate this point with cipher-block chaining (CBC mode). Let the s blocks of the message be denoted m_1, m_2, ..., m_s. The CBC mode utilizes an initialization vector IV and a key K. The algorithm produces as output ciphertext c_i for $1 \leq i \leq s + 1$, where

$$c_1 = IV$$

and

$$c_{i+1} = E(K, c_i \oplus m_i) \text{ for } i = 1, 2, \ldots, s .$$

Thus

$$m_i = c_i \oplus D(K, c_{i+1}) \text{ for } i = 1, 2, \ldots, s ,$$

and so any one of the s message blocks can be obtained with the decryption of just one ciphertext block. This makes the adversary's key-search problem relatively easy, since decrypting a single ciphertext block is generally enough to test a candidate key.

Let us say that an encryption mode for a block cipher is *separable* if it has the property that an adversary can determine one block of plaintext by decrypting just one block of ciphertext. Thus, CBC mode is separable.

We wish to design non-separable encryption modes. More precisely, we wish to design *strongly* non-separable modes, defined as follows.

Definition. Suppose that a block cipher encryption mode transforms a sequence

$$m_1, m_2, \ldots, m_s$$

of s message blocks into a sequence

$$c_1, c_2, \ldots, c_t$$

of t ciphertext blocks, for some t, $t \geq s$. We say that the encryption mode is *strongly non-separable* if it is infeasible to determine even one message block m_i (or any property of a particular message block m_i) without decrypting *all* t ciphertext blocks.

3   All-Or-Nothing Transforms

We propose to achieve strongly non-separable modes as follows:

- Transform the message sequence $m_1, m_2, \ldots, m_s$ into a "pseudo-message" sequence $m'_1, m'_2, \ldots, m'_{s'}$ (for some $s' \geq s$) with an "all-or-nothing transform", and
- Encrypt the pseudo-message with an ordinary encryption mode (e.g. codebook mode) with the given cryptographic key K to obtain the ciphertext sequence $c_1, c_2, \ldots, c_t$.

We call encryption modes of this type "all-or-nothing encryption modes." A specific instance of this mode would be "all-or-nothing codebook mode," when the encryption mode used is codebook mode, (or "all-or-nothing CBC mode", etc.).

To make this work, the all-or-nothing transform has to have certain properties.

Definition. A transformation f mapping a message sequence $m_1, m_2, \ldots, m_s$ into a pseudo-message sequence $m'_1, m'_2, \ldots, m'_{s'}$ is said to be an *all-or-nothing transform* if

- The transformation f is reversible: given the pseudo-message sequence, one can obtain the original message sequence.
- Both the transformation f and its inverse are efficiently computable (that is, computable in polynomial time).
- It is computationally infeasible to compute any function of any message block if any one of the pseudo-message blocks is unknown.

We note that an all-or-nothing transformation must really be randomized, so that a chosen or known message attack does not yield a known pseudo-message, and so that a deterministic function which computes the first pseudo-message block is not available as a function to contradict the last requirement above.

We note that the all-or-nothing transformation is not itself "encryption," since it makes no use of any secret key information. It is merely an invertible "pre-processing" step that has certain interesting properties. The actual encryption in an all-or-nothing encryption mode is the operation that encrypts the pseudo-message resulting from the all-or-nothing transform. An all-or-nothing transform is a fixed public transform that anyone can perform on the message to obtain the pseudo-message, or invert given the pseudo-message to obtain the message.

Theorem 1. *An all-or-nothing encryption mode is strongly non-separable.*

"Proof": We assume that the underlying encryption mode is such that all cipher-text blocks must be decrypted in order to obtain all pseudo-message blocks. (If this were not the case, the encryption mode would not be efficient, and a more efficient reduced mode could be derived from it.) Thus, all ciphertext blocks must be decrypted in order to determine any (property of any) message block.   □

4   The Package Transform

The all-or-nothing scheme we propose here (the "package transform") is quite efficient, particularly when the message is long; the cost of an all-or-nothing transform is approximately twice the cost of the actual encryption. We shall also see that all-or-nothing encryption admits fast parallel implementations.

The legitimate communicants thus pay a penalty of approximately a factor of three in the time it takes them to encrypt or decrypt in all-or-nothing mode, compared to an ordinary separable encryption mode. However, an adversary attempting a brute-force attack pays a penalty of a factor of t, where t is the number of blocks in the ciphertext.

As an example, if I send you a eight-megabyte message encrypted in all-or-nothing CBC mode with a 40-bit DES key, the adversary must decrypt the entire eight-megabyte file in order to test a single candidate 40-bit key. This expands the work-factor by a factor of one-million, compared to breaking ordinary CBC mode. Since one million is approximately 2^{20}, to the adversary this feels like having to break a 60-bit key instead of a 40-bit key!

Using this scheme, it can clearly be advantageous for the communicants to "pad" the message with random data, as it makes the adversary's job harder.

We propose here a particular all-or-nothing transform, which we call the "package transform." We note that while it uses a block cipher itself as a primitive, no secret keys are used. (Instead, a randomly chosen key is used, and this key can be easily determined from the pseudo-message sequence.) The block cipher used in the package transform need not be the same as the block cipher used to encipher the pseudo-message (the package transform output), although it may be. (If it is the same encryption algorithm, note that we assume below that the key space for the package transform block cipher is sufficiently large that brute-force search is infeasible, while the motivation for the use of an all-or-nothing encryption mode was that the key space for the outer encryption algorithm was marginal. This situation can arise for variable-key-length block ciphers such as RC5. For concreteness, the reader may imagine that we are working with RC5 for both the package transform encryption algorithm and the outer encryption algorithm, with 128-bit input/output blocks, a 128-bit encryption key for the package transform, and a 40-bit key for the outer encryption transform.)

For this exposition, then, we assume that the key size of the package transform block cipher is the same as its block size; this assumption can easily be removed and is made here only for convenience in exposition. We also assume that the key space for the package transform block cipher is sufficiently large that brute-force searching for a key is infeasible. The scheme also uses a fixed publically-known key K_0 for the package transform block cipher.

Here is the package transform:

- Let the input message be $m_1, m_2, \ldots, m_s$.
- Choose at random a key K' for the package transform block cipher.
- Compute the output sequence $m'_1, m'_2, \ldots, m'_{s'}$, for $s' = s + 1$ as follows:
  - Let $m'_i = m_i \oplus E(K', i)$ for $i = 1, 2, 3, \ldots, s$.

- Let

$$m'_{s'} = K' \oplus h_1 \oplus h_2 \oplus \cdots \oplus h_s \ ,$$

where

$$h_i = E(K_0, m'_i \oplus i) \text{ for } i = 1, 2, \ldots, s \ ,$$

where K_0 is a fixed, publically-known encryption key.

The intent here is that the key K' be chosen from a large space (for example, chose K' as a 128-bit RC5 key). Since K' is not a secret shared key (it is disclosed in the pseudo-message), it is not restricted by the limitations of the following encryption mode.

The package transformation is similar to encrypting in counter mode, except that the key is randomly chosen rather than fixed, and the last pseudo-message block is the exclusive-or of the key and a hash of all previous pseudo-message blocks (computed as the exclusive-or of the encryptions of variants of these blocks under a fixed key, where the i-th variant is computed as the exclusive-or of i and the block). This technique ensures that simple modifications to the ciphertext, such as permuting the order of two blocks or duplicating a blocks, is highly likely to change the key K' computed by the receiver.

One could also define variant package transforms based on block-chaining techniques instead of counter mode.

It is easy to see that the package transform is invertible:

$$K' = m'_{s'} \oplus h_1 \oplus h_2 \oplus \cdots \oplus h_s \ ,$$

$$m_i = m'_i \oplus E(K', i) \text{ for } i = 1, 2, \ldots, s \ .$$

We also note that if any block of the pseudo-message sequence is unknown, then K' can not be computed, and so it is infeasible to compute any message block. (Formal proof omitted here, but we recall that the key K' is assumed to be drawn from an infeasibly large set, so that (for example) a meet-in-the-middle attack is not more efficient than decrypting all the ciphertext blocks.)

5   Discussion

A related well-known approach towards getting more security out of fixed number of key bits is to use encryption techniques that have a long "set-up" time (see Quisquater et al. [8], or Schneier's "Blowfish" algorithm [9]). This penalizes the legitimate user whenever he performs a key-change, whereas all-or-nothing encryption incurs a fixed penalty for each block encrypted. While this may seem to favor the increased set-up time approach, we note that

- An all-or-nothing transform is merely a pre-processing step, and so it can be used with already-existing encryption devices and software, without changing the encryption algorithm.

- Increasing the set-up time may still yield an algorithm that is efficiently implemented with a special-purpose brute-force chip, since there may be little need for inter-chip communications. On the other hand, the two-pass nature of all-or-nothing encryption may necessitate large amounts of input/output, something that usually slows down operations considerably.
- In any case, the approaches are complementary, and can easily be combined.

We note that all-or-nothing encryption modes are only defined here when the message to be encrypted is a finite sequence; an infinitely long message can not be encrypted in an all-or-nothing mode, whereas other modes such as CBC work perfectly well in this case. All-or-nothing encryption modes work very well in cases such as for encrypting packets in a network.

We observe, however, that one can begin encrypting in package CBC mode (or package codebook mode) before one knows the end of message sequence, since the inner package operation and the outer CBC (or codebook) encryption modes can both be implemented in a sequential manner. However, decrypting a package mode ciphertext more-or-less requires two passes and/or having the entire ciphertext available at once.

Package codebook mode is particularly interesting, since the outer codebook decryption and the inner package transformation can both be performed efficiently in parallel. (I don't mean that they are performed at the same time, but that each one separately admits an efficient parallel implementation.) With a sufficient number of encryption units, a message of length s can be encrypted or decrypted in time $O(\log s)$. This may be an advantage for the legitimate communicants in a high-speed communications scenario. Note that the same advantage is available to the adversary–although he has to decrypt the entire ciphertext, he can also do it in parallel. However, for the adversary this advantage is probably meaningless, since it is the total search time that is important to him, not the latency for performing a single decryption. Thus package codebook mode has much to recommend it from a performance perspective.

We note that all-or-nothing encryption modes can provide protection against differential attacks and other forms of attack that depend on chosen plaintext, since a randomized all-or-nothing transformation can effectively destroy any patterns in the actual input (the pseudo-message) to the underlying encryption operation.

In addition, an all-or-nothing transformation can be useful before RSA encryption, as it prevents various kinds of "related message" or other attacks (e.g. those of Coppersmith et al. [5]). Indeed, the package transform described here can be viewed as a special case of the "simple embedding scheme" proposed by Bellare and Rogaway [2] in their "optimal asymmetric encryption" preprocesing scheme (used before applying RSA encryption):

$$x \oplus G(r) \parallel r \oplus H(x \oplus G(r)) \ .$$

Here x is the message to be encrypted (like our message m), r is a randomly chosen quantity (like our key K'), $G(r)$ is a pseudo-random output (like our $E(K', 1)$, $E(K', 2)$, ...), and H is a hash function (like our $h_1 \oplus h_2 \oplus \dots h_s$).

The correspondence would be closer if we had proposed using $m'_{s'} = K' \oplus MD5(m'_1, \ldots, m'_s)$, which would also give some improved efficiency, but we wished to confine ourselves to just using the block cipher as a primitive operation. We are applying these ideas to symmetric block cipher modes of operation rather than asymmetric encryption, but the principles are essentially the same. However, it may also be the case that a rather different approach can be applied to achieve our goals with substantially greater efficiency than the approach suggested here or by Bellare and Rogaway's approach in general.

There are many approaches one might take towards devising all-or-nothing transforms. One might consider computing the pseudomessage as the concatenation of a description of a hash function h chosen randomly from a universal family of hash functions with a suitably large range, followed by the application of h to the message. Another approach that may work well is to use a scheme based on an FFT-like arrangement of randomized multipermutations (see Schnorr et al. [10]).

Or, one can base an approach on secret-sharing schemes. Actually, the package transform can be viewed as a s' out of s' secret-sharing threshold scheme; each of the s' pseudo-message blocks can be viewed as one "share" of the underlying message. Decrypting so as to obtain fewer than s' pseudo-message blocks yields no information at all about the underlying message. This is "computational secret sharing" (see [6]) since the shares are shorter than the message itself. Indeed, one can design all-or-nothing schemes based on Krawczyk's proposals.

An entirely different approach is given by Anderson and Biham [1], who design block ciphers (such as BEAR and LION) from scratch that seem to have an "all-or-nothing" property. Their approach is different because they design block ciphers with variable-length blocks to accomodate messages of varying lengths, whereas our focus is on designing an encryption mode for fixed-length block ciphers that provide an all-or-nothing property. Nonetheless, their schemes may be the method of choice in some situations.

We note that all-or-nothing encryption has *terrible* error-propagation properties: if *any* ciphertext block is damaged, then it is likely that *every* message block will be damaged. Thus, ciphertext should be transported with reliable transmission means. (One could interpose an error-correction phase between the all-or-nothing transformation and the encryption; this could help handle errors while only modestly decreasing non-separability.)

Using this error-propagation property to one's advantage, one can extend all-or-nothing mode by appending a suitable block of redundancy (such a block of all zeros, or the sum of all the previous message blocks) to the message before applying the all-or-nothing transformation. This redundancy can be verified and removed upon decryption. This helps to detect corrupted ciphertext.

As a variation on the idea of the previous paragraph, the redundancy block may be computed as the sum of previous message blocks and a secret value that is known only to the two parties communicating; this provides a form of message authentication. The redundancy block could of course also be computed with more conventional keyed hashing techniques.

The preceding paragraphs touch upon an important issue: that an encryption mode should provide *integrity* as well as *confidentiality*. Mao and Boyd [7] make this point well. Bellare and Rogaway prove that their simple embedding scheme provides non-malleability, for example.

6 Conclusion

We have presented an encryption mode—the all-or-nothing encryption mode—and a specific means of implementing it using the package transform. Other forms of all-or-nothing encryption are presumably yet to be devised.

We leave it as an open problem to devise an all-or-nothing encryption mode that is substantially more efficient than the scheme presented here. Is it possible, for example, to reduce the cost of implementing an all-or-nothing mode from a factor of three greater than CBC to just a factor of two greater?

Acknowledgments

I would like to thank Don Coppersmith, Oded Goldreich, Shafi Goldwasser, Mihir Bellare, Burt Kaliski, and the referees for helpful comments and conversations. Silvio Micali deserves special thanks for suggesting the term "all-or-nothing." David Wagner deserves thanks for pointing out significant bugs in earlier versions of this paper, and for pointing out the relationship between this work and the Bellare-Rogaway work on optimal asymmetric encryption. And thanks to Mihir Bellare for noting the relationship with secret-sharing schemes.

References

1. Ross Anderson and Eli Biham. Two practical and probably secure block ciphers: BEAR and LION. In Dieter Gollman, editor, *Fast Software Encryption*, pages 114–120. Springer, 1996. (Proceedings Third International Workshop, Feb. 1996, Cambridge, UK).
2. Mihir Bellare and Phillip Rogaway. Optimal asymmetric encryption—how to encrypt with RSA. In *EUROCRYPT94*, 1994.
3. Eli Biham. Cryptanalysis of multiple modes of operation. 1995. Pre-Proceedings of ASIACRYPT '94. Submitted to J. Cryptology.
4. Matt Blaze, Whitfield Diffie, Ronald L. Rivest, Bruce Schneier, Tsutomu Shimomura, Eric Thompson, and Michael Wiener. Minimal key lengths for symmetric ciphers to provide adequate commercial security: A report by an ad hoc group of cryptographers and computer scientists, January 1996. Available at http://www.bsa.org.
5. Don Coppersmith, Matthew Franklin, Jacques Patarin, and Michael Reiter. Low-exponent RSA with related messages. Technical Report IBM RC 20318, IBM T.J. Watson Research Lab, December 27, 1995. (To appear in Eurocrypt '96).
6. Hugo Krawczyk. Secret sharing made short. In Douglas R. Stinson, editor, *Proc. CRYPTO 93*, pages 136–146. Spring-Verlag, 1993.
7. Wenbo Mao and Colin Boyd. Classification of cryptographic techniques in authentication protocols. In *Proceedings 1994 Workshop on Selected Areas in Cryptography*, May 1994. (Kingston, Ontario, Canada).

8. J.-J. Quisquater, Yvo Desmedt, and Marc Davio. The importance of "good" key scheduling schemes (how to make a secure DES scheme with $\leq$ 48 bit keys). In H. C. Williams, editor, *Proc. CRYPTO 85*, pages 537–542. Springer, 1986. Lecture Notes in Computer Science No. 218.
9. Bruce Schneier. *Applied Cryptography (Second Edition)*. John Wiley & Sons, 1996.
10. C. P. Schnorr and S. Vaudenay. Black box cryptanalysis of hash networks based on multipermutations. In *EUROCRYPT94*, 1994.

On the Security of Remotely Keyed Encryption

Stefan Lucks

Institut für Numerische und Angewandte Mathematik
Georg–August–Universität Göttingen
Lotzestr. 16–18, D–37083 Göttingen, Germany
(email: lucks@math.uni-goettingen.de)

Abstract. The purpose of remotely keyed encryption is to efficiently realize a secret-key block cipher by sharing the computational burden between a fast untrusted device and a slow device trusted with the key. This paper deals with how to define the security of remotely keyed encryption schemes. Since the attacker can take over the slow device and actually take part in the encryption process, common definitions of the security of block ciphers have to be reconsidered.

Using random mappings, collision resistant hash functions, and stream ciphers as building blocks, the Random Mapping based Remotely Keyed (RaMaRK) encryption scheme is proposed. Also GRIFFIN is proposed, a fast new block cipher for flexible but large blocks. The RaMaRK scheme and GRIFFIN are provably secure if the underlying building blocks are secure.

1 Introduction

At the Fast Software Encryption conference 1996 in Cambridge, Blaze [5] proposed a new paradigm for secret-key block ciphers: *Remotely keyed encryption.* This means to share the workload for en- and decryption between a fast host and a slow card. The host is trusted with plaintexts and ciphertexts, but only the (hopefully tamper-resistant) card does know the key.

While Blaze's "remotely keyed encryption protocol" is a fresh and interesting approach to solve the paradoxical problem how to realize "high-bandwidth encryption with low-bandwidth smartcards", it also has some drawbacks. In Section 3.1 of his paper, Blaze himself mentions some security problems, but assumes "neither of these attacks is likely to pose a serious threat to most practical applications." In spite of this assumption, the current author believes that a block cipher realized by a remotely keyed encryption scheme should be as secure as usually demanded from other block ciphers.

One goal of this paper is to generalize the common notions of block cipher security. For block ciphers, one usually considers attacks where the attacker can choose ciphertexts and decrypt them, and/or choose plaintexts and encrypt them. For remotely keyed encryption, we have to consider the case when the attacker does take part in the encryption or decryption protocol as the host. A second goal is to come up with a new remotely keyed encryption scheme. This has to be provably secure with respect to the generalized notions of security if its building blocks are secure. As a side result, the block cipher GRIFFIN realized by our scheme is a fast and secure block cipher for large blocks—and of practical interest even outside the scope of remotely keyed encryption, see section 6.

Throughout this paper, "random" always means "according to the uniform probability distribution". If x is randomly chosen from the set S, we write $x \in_R S$. If x and y are independent random values from S, then $(x, y) \in_R S^2$. By $x \oplus y$ we denote the bit-wise XOR of x and y.

2   How to Attack the RKEP

In this section, we describe Blaze's *remotely keyed encryption protocol* (RKEP) [5] and point out some of its weaknesses.

For the RKEP, we need a collision resistant hash function $H : \{0,1\}^* \longrightarrow \{0,1\}^b$ and a secret-key block cipher with b-bit blocks. Given the key $K \in \{0,1\}^k$, the encryption function is $E_K : \{0,1\}^b \longrightarrow \{0,1\}^b$ and the decryption function is D_K. To cover the case when $k \neq b$, one needs a public mapping $M : \{0,1\}^b \longrightarrow \{0,1\}^k$. By $P = (P_1, \ldots, P_n) \in \{0,1\}^{nb}$ we denote a plaintext block, which can be divided into n subblocks $P_1, \ldots, P_n$ each of b bits. $C = (C_1, \ldots, C_n) \in \{0,1\}^{nb}$ is the corresponding ciphertext block. When encrypting P or decrypting C, an intermediate block $I = (I_1, \ldots, I_n) \in \{0,1\}^{nb}$ is considered.

Encrypting with the RKEP works as follows:

1. Given P, the host computes I by $I_i := P_i \oplus H(P_1)$ for $i \in \{2, \ldots, n\}$ and then $I_1 := P_1 \oplus H(I_2, \ldots, I_n)$. We abridge this to $I := \mathrm{Modify}(P)$.
2. I_1 is sent to the card, which knows the secret key K and computes $C_1 := E_K(I_1)$ and $K_P := M(E_K(C_1))$. The pair (C_1, K_P) is sent back to the host. For the sake of shortness, we write $(C_1, K_P) := \mathrm{Local}_K(I_1)$.
3. The host computes the remaining $n - 1$ subblocks $C_2 := E_{K_P}(I_2 \oplus C_1)$, $C_3 := E_{K_P}(I_3 \oplus C_2), \ldots, C_n := E_{K_P}(I_n \oplus C_{n-1})$ of the ciphertext C.

Decryption is similar:

1. Given the ciphertext C, the host sends C_1 to the card.
2. The card computes $I_1 := D_K(C_1)$, $K_P := M(E_K(C_1))$ and sends both values to the host. For short, we write $(I_1, K_P) := \mathrm{Local}_K^{-1}(C_1)$.
3. The host computes $I_2 := D_{K_P}(C_2) \oplus C_1$, $I_3 := D_{K_P}(C_3) \oplus C_2, \ldots, I_n := D_{K_P}(C_n) \oplus C_{n-1}$.
   Also, the host computes the plaintext P from I: $P_1 := I_1 \oplus H(I_2, \ldots, I_n)$ and $P_i := I_i \oplus H(P_1)$ for $i \in \{2, \ldots, n\}$—in short: $P := \mathrm{Modify}^{-1}(I)$.

Apart from enabling remotely keyed encryption, the RKEP can be seen as a block cipher with large (i.e. bn bit) blocks. Thus we can write $C := \mathrm{RKEP}_K(P)$ for encrypting with the secret key K, and $P := \mathrm{RKEP}_K^{-1}(C)$ for decrypting. For block ciphers, one usually requires: *If the attacker encrypts or decrypts at most q times, but has no further knowledge about the key K or plaintext-ciphertext pairs, there should be no more than q such pairs $(P^{(i)}, C^{(i)})$ with $C^{(i)} = RKEP_K(P^{(i)})$*

known to the attacker. As we will see below, the RKEP block cipher does not meet this requirement.

Now, we consider two attacks on the RKEP block cipher. The attacker, we call her Alice, has the unwitting help of Bob, who encrypts and decrypts for her. Attack I is a *chosen plaintext attack*, since given a plaintext P, Alice chooses two plaintexts $P' = (P'_1, \ldots P'_n)$ and $P'' = (P''_1, \ldots P''_n)$ different from P, receives the corresponding ciphertexts $C' = (C'_1, \ldots C'_n)$ and $C'' = (C''_1, \ldots C''_n)$ from Bob, and uses this knowledge to compute the ciphertext $C = \text{RKEP}_K(P)$ without more help by Bob. Attack II is a *two-sided attack*, i.e. the attacker must be able to encrypt *and* to decrypt.

Attack I: Given a "suspicious" plaintext P Bob is not willing to encrypt for Alice, Alice chooses two plaintexts P' and P'' which appear random. If Bob gives her $C' = \text{RKEP}_K(P')$ and $C'' = RKEP_K(P'')$, she can derive the pair (C_1, K_P) of values required to compute $C = \text{RKEP}_K(P)$ from this, without further asking Bob. The attack requires the following six steps:

1. Alice computes $I = \text{Modify}(P)$.
2. She chooses any $I' \neq I$, except for $I'_1 = I_1$.
3. She computes $P' = \text{Modify}^{-1}(I')$ and asks Bob for $C' = \text{RKEP}_K(P')$. (Note that $I'_1 = I_1$ and hence $C'_1 = C_1$.)
4. She chooses any $I'' \neq I$, except for $I''_1 = C'_1$.
5. She computes $P'' = \text{Modify}^{-1}(I'')$ and asks Bob for $C'' = \text{RKEP}_K(P'')$. (Now $C''_1 = E_K(I''_1) = E_K(C'_1) = E_K(C_1)$.)
6. She computes $K_P = M(C''_1)$.

Attack II: Given a ciphertext C Bob is not willing to decrypt for her, Alice chooses a ciphertext C' different from C and a plaintext P''. Similarly to the above attack, Alice can find the pair (C_1, K_P) she needs to compute $P = \text{RKEP}_K(C)$, once she is given $P' = \text{RKEP}_K^{-1}(C')$ and $C'' = RKEP_K(P'')$:

1. Alice chooses any $C' \neq C$, except for $C'_1 = C_1$.
2. She asks Bob for $P' = \text{RKEP}_K^{-1}(C')$.
3. She computes $I' = \text{Modify}(P')$. (Now she knows $I_1 = I'_1$.)
4. As in the last three steps of attack I, Alice computes $K_P = M(E_K(I_1))$.

For attack III, we will not consider attacks on the RKEP block cipher, but examine how to misuse a given smartcard. Here, Alice does not need Bob's help but can decrypt ciphertexts for herself, using a "decrypt only" smartcard. This is a reasonable demand e.g. for pay-TV applications, where (paying) customers are given a smartcard to decrypt broadcast data (i.e. the TV program). Customers should not be able to encrypt. Otherwise they could encrypt and broadcast video data for themselves, and thus forge the pay-TV program.

We consider a smartcard to compute the function Local_K^{-1}—but not Local_K. We assume that with significant probability it is possible to determine $E_K(C_1)$ from $M(E_K(C_1))$. This assumption appears to be reasonable, since Blaze only requires the public function M to map a b-bit string to a k-bit key string, but does not demand M to be either secret or a one-way function ([5], top of p. 35).

Attack III: Given a target plaintext P, Alice proceeds like this:

1. She computes $I := \text{Modify}(P)$.
2. She feeds I_1 into her smartcard to get
   $$\left(\text{dummy}, X\right) = \text{Local}_K^{-1}(I_1) = \left(\ldots, M(E_K(I_1))\right).$$
3. She determines C_1 with $X = M(C_1)$.
4. She feeds C_1 into the smartcard to get
   $$\left(Y, K_P\right) = \text{Local}_K^{-1}(C_1) = \left(D_K(C_1), M(E_K(C_1))\right).$$

Note that if $b > k$, exaustive search over the 2^{b-k} possible values C_1' with $M(C_1') = X$ is needed to find the unique C_1 with $I_1 = Y = D_k(C_1)$. Once Alice knows $(C_1, K_P) = \text{Local}_K(I_1)$, computing $C_2, \ldots, C_n$ is easy for her.

We will not discuss under which circumstances the above attacks become practical. Possibly, our attacks are of no relevance for the applications Blaze considered for the RKEP. However, the RKEP block cipher clearly has some properties, a *good* block cipher should not have.

3   When is Remotely Keyed Encryption "Secure"?

A *remotely keyed encryption scheme* is a protocol to distribute the computational burden for a B-bit block cipher between two parties, a *host* and a *card*. (The "card" could either be a smartcard connected to the host, or something quite different, e.g. an "encryption server" in a computer network.) The host knows plaintext and ciphertext, but only the card is trusted with the key.[1] The protocol is divided into two subprotocols, an *encryption protocol* and a *decryption protocol*.

Given a B-bit input, either to encrypt or to decrypt, such a subprotocol runs like this: The host sends a *challenge value* to the card, depending on the input, and the card replies a *response value*, depending on both the challenge value and the key. E.g. in the case of the RKEP encryption protocol, I_1 is the challenge value, and the pair (C_1, K_P) is the response value, while for RKEP decryption C_1 is the challenge value, and (I_1, K_P) is the response value. Exchanging challenge and response values can be iterated. During one run of a subprotocol every challenge value depends on the input and the previously given response values and the response values depend on the key and the previous challenge values. We may assume that neither the overall number of bits for the challenge values, nor the overall number of bits for the response values exceed β, where $\beta \ll B$.

For a key $K \in \{0,1\}^k$, the encryption protocol realizes the encryption function $\text{Encrypt}_K : \{0,1\}^B \longrightarrow \{0,1\}^B$ and the decryption protocol the decryption function $\text{Decrypt}_K : \{0,1\}^B \longrightarrow \{0,1\}^B$, such that for every plaintext $X \in \{0,1\}^B$ the equation $X = \text{Decrypt}_K(\text{Encrypt}_K(X))$ holds.[2]

---

[1] Such a scheme can also be seen as the card's "mode of operation".

[2] For $B^* \geq B$ this definition can be generalized to $\text{Encrypt}_K : \{0,1\}^B \longrightarrow \{0,1\}^{B^*}$ and $\text{Decrypt}_K : \{0,1\}^{B^*} \longrightarrow \{0,1\}^B$, as long as $X = \text{Decrypt}_K(\text{Encrypt}_K(X))$ is valid for all plaintexts $X \in \{0,1\}^B$. But for the purposes of this paper, we don't need that generalization.

In order to define the security of such a protocol, we first need to describe the security of block ciphers. In this paper, we only consider block ciphers with ciphertexts of the same size as the plaintexts. Encrypting with such a block cipher means to apply a key-dependent permutation g to the plaintext, decrypting to apply its inverse g^{-1} to the ciphertext.

In cryptography, one often considers a *chosen plaintext attack*, where attackers can encrypt plaintexts of their choice. Here, we consider an even stronger type of attack, the *two-sided attack*, often called "combined adaptive chosen plaintext/chosen ciphertext attack", where attackers are able to encrypt plaintexts of their choice and decrypt ciphertexts of their choice.

A block cipher is *forgery secure*, if after q encryptions resp. decryptions of chosen inputs, the attacker can know at most q plaintext-ciphertext pairs $(P^{(1)}, C^{(1)})$, ..., $(P^{(q)}, C^{(q)})$, but no $(P, C) \notin \{(P^{(1)}, C^{(1)}), \ldots, (P^{(q)}, C^{(q)})\}$, with $C = \text{Encrypt}_K(P)$ (without encrypting or decrypting again).

In the case of remotely keyed encryption, we need to consider attacks, where instead of encrypting plaintexts or decrypting ciphertexts the attackers act as the host when executing the encryption resp. decryption protocol.

Definition: A remotely keyed encryption scheme is *forgery secure*, if after q executions of the encryption resp. decryption protocol with arbitrarily chosen challenge values, the attacker can know no more than q plaintext-ciphertext pairs $(P^{(1)}, C^{(1)})$, ..., $(P^{(q)}, C^{(q)})$ which are *valid*, i.e. $C_i = \text{Encrypt}_K(P_i)$.

Another property of block ciphers is, when attackers can operate only in one direction, e.g. choose plaintexts and encrypt, they should be unable to solve the problem to compute in the reverse direction, i.e. to decrypt a given ciphertext.

Definition: A remotely keyed encryption scheme is *inversion secure*,

- if for attackers able to execute the encryption protocol it is infeasible to decrypt a randomly chosen ciphertext, and
- if for attackers able to execute the decryption protocol it is infeasible to encrypt a randomly chosen plaintext.

Often, the security of cryptographic primitives is defined as the non-existence of statistical tests to detect non-randomness properties. This point of view leads to an even stronger definition of the security of block ciphers than forgery-security: A B-bit block cipher is called *pseudorandom*, if for two-sided attackers it is infeasible to distinguish whether they are given a random permutation $p : \{0,1\}^B \longrightarrow \{0,1\}^B$ and its inverse, or the encryption function $\text{Encrypt}_K : \{0,1\}^B \longrightarrow \{0,1\}^B$ and the decryption Function Decrypt_K depending on the secret key K. Attackers who participate in an encryption or decryption protocol learn the card's response values and thus know that the encryption is not a random permutation. Hence when considering pseudorandomness, we can't consider attackers which execute the encryption or decryption protocol.

Definition: A remotely keyed encryption scheme is *pseudorandom*, if the block cipher it realizes is pseudorandom.

Note that the RKEP is neither forgery secure (cf. attacks I and II), nor inversion secure (cf. attack III), nor pseudorandom (cf. attacks I and II). Even if we only consider chosen plaintext attacks, the RKEP block cipher is neither forgery secure nor pseudorandom (cf. attack I).

4   The RaMaRK Encryption scheme

Now we describe the _Ra_ndom _Ma_pping based _R_emotely _K_eyed (RaMaRK) Encryption scheme. The name comes from one of our building blocks, a _fixed size random mapping_ $f : \{0,1\}^b \longrightarrow \{0,1\}^b$. For the proofs of security, we assume f to be a random function (a "random oracle" in the sense of Bellare and Rogaway [4]), i.e. for $s \neq t$: $(f(s), f(t)) \in_R \{0,1\}^{2b}$.

Except when b is tiny, one can't actually implement such random functions— one would have to store $b2^b$ bits. In practice, one assumes f to be a pseudorandom function depending on a small secret key and undistinguishable from a random function for everyone without knowledge of the key. This could be done e.g. by using a block cipher or a dedicated hash function. Note that realizing pseudorandom mappings from dedicated hash functions must be done with great care to be secure (cf. [9]). Also note that b must be large enough—performing close to $2^{b/2}$ encryptions has to be infeasible.[3] We use three building blocks:

1. Random mappings $f_i : \{0,1\}^b \longrightarrow \{0,1\}^b$, as described above.
2. A hash function $H : \{0,1\}^* \longrightarrow \{0,1\}^b$. H has to be _collision resistant_, i.e. it has to be infeasible to find any $t, u \in \{0,1\}^*$ with $u \neq t$ but $H(u) = H(t)$.
3. A pseudorandom bit generator (i.e. a "stream cipher") $S : \{0,1\}^b \longrightarrow \{0,1\}^*$. We restrict ourselves to $S : \{0,1\}^b \longrightarrow \{0,1\}^{B-2b}$. If the seed $s \in \{0,1\}^b$ is randomly chosen, the bits produced by $S(s)$ have to be undistinguishable from randomly generated bits.

   For theorem 3 we need an _additional property:_ If s is secret and attackers choose $t_1, t_2, \ldots \in \{0,1\}^b$ with $t_i \neq t_j$ for $i \neq j$ and receive outputs $S(s \oplus t_1)$, $S(s \oplus t_2)$, ..., it has to be infeasible for the attackers to distinguish these outputs from independently generated random bit strings of the same size. Hence, such a construction behaves like a random mapping $\{0,1\}^b \longrightarrow \{0,1\}^{B-2b}$, but actually is a pseudorandom one, depending on the secret s.

Based on these building blocks, we realize a remotely keyed encryption scheme to encrypt blocks of any size $B \geq 3b$, see figure 1. In contrast to Blaze's RKEP, B need not be a multiple of b.

We represent the plaintext by (P, Q, R) and the ciphertext by (A, B, C), where $(P, Q, R), (A, B, C) \in \{0,1\}^b \times \{0,1\}^b \times \{0,1\}^{B-2b}$. For the protocol

---

[3] If we only are given a random mapping $f' : \{0,1\}^{b'} \longrightarrow \{0,1\}^{b'}$ with b' just large enough that performing close to $2^{b'}$ encryptions is infeasible, we can apply Aiello's and Venkatesan's work [1] to construct a random mapping $f : \{0,1\}^b \longrightarrow \{0,1\}^b$ with $b = 2b'$. Provably $\Omega(2^{b/2})$ queries are needed to distinguish with $\Theta(1)$ probability between f and a truly random b-bit to b-bit function.

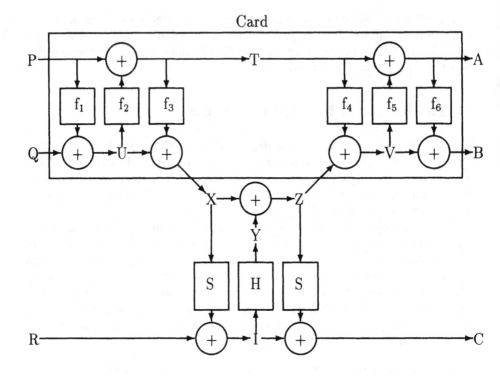

Fig. 1. The RaMaRK encryption protocol.

description we also consider intermediate values $T, U, V, X, Y, Z \in \{0, 1\}^b$, and $I \in \{0, 1\}^{B-2b}$. The encryption protocol works as follows:

1. Given the plaintext (P, Q, R), the host sends P and Q to the card.
2. The card computes $U = f_1(P) \oplus Q$, $T = f_2(U) \oplus P$, and sends $X = f_3(T) \oplus U$ to the host.
3. The host computes $I = S(X) \oplus R$, $Y = H(I)$, sends $Z = X \oplus Y$ to the card, and computes $C = S(Z) \oplus I$.
4. The card computes $V = f_4(T) \oplus Z$ and sends the two values $A = f_5(V) \oplus T$ and $B = f_6(A) \oplus V$ to the host.

Decrypting (A, B, C) is done like this:

1. The host sends A and B to the card.
2. The card computes $V = f_6(A) \oplus B$, $T = f_5(V) \oplus A$, and sends $Z = f_4(T) \oplus V$ to the host.
3. The host computes $I = S(Z) \oplus C$, $Y = H(I)$, sends $X = Z \oplus Y$ to the card, and computes $R = S(X) \oplus I$.
4. The card computes $U = f_3(T) \oplus X$ and sends the two values $P = f_2(U) \oplus T$ and $Q = f_1(P) \oplus U$ to the host.

One can easily verify that by first encrypting any plaintext using any key, then decrypting the result using the same key, one gets the same plaintext again.

5   The Security of the RaMaRK encryption scheme

Next, we prove the forgery security, the inversion security, and the pseudorandomness of the RaMaRK scheme.

Theorem 1. *A two-sided attack against the RaMaRK encryption scheme to find q valid plaintext-ciphertext pairs by $q-1$ protocol executions succeeds with at most the probability $q^2 2/2^b$.*

Proof. By P_i, Q_i, X_i, Z_i, A_i, and B_i we denote the challenge and response values of the ith protocol execution—independently of whether the encryption protocol is executed, i.e. (P_i, Q_i) and Z_i are challenge and X_i and (A_i, B_i) response values, or the decryption protocol is executed and the role of challenge and response values is reversed. Similarly are the subscripts for I_i, Y_i, T_i, U_i, and V_i used.

Since H is collision resistant, for every Y_i at most one I_i with $H(I_i) = Y_i$ can be known to the attacker and thus every triple (P_i, Q_i, Y_i) for the encryption protocol corresponds to at most one plaintext (P_i, Q_i, R_i) known to the attacker, and every triple (A_i, B_i, Y_i) corresponds to at most one ciphertext. It doesn't help attackers to repeat the challenge values of a previous protocol execution. For $i \neq j$ we can assume $(A_i, B_i, C_i) \neq (A_j, B_j, C_j)$ and $(P_i, Q_i, Z_i) \neq (P_j, Q_j, Z_j)$. After every protocol execution, the attacker can choose whether next to execute the encryption or the decryption protocol.

At first, we consider the **encryption protocol**. The attacker sends P_r and Q_r to the card. For every $i \in \{1, \ldots, r - 1\}$ we need to distinguish whether $P_i = P_r$ and $Q_i = Q_r$ or not.

Case I: $P_i = P_r$ and $Q_i = Q_r$. In this case, the corresponding internal values are equal, i.e. $T_i = T_r$ and $X_i = X_r$, but $Z_i \neq Z_r$ and thus $V_i \neq V_r$, $(A_i, A_r) \in_R \{0, 1\}^{2b}$, and $\text{prob}[A_i = A_r] \leq 1/2^b$. If $A_i \neq A_r$ then $(B_i, B_r) \in_R \{0, 1\}^{2b}$. Thus with a probability $\geq 1 - 1/2^b$, the ciphertext fractions (A_i, B_i) and (A_r, B_r) are independently chosen random values.

Case II: $P_i \neq P_r$ or $Q_i \neq Q_r$. If $P_i = P_r$ and $Q_i \neq Q_r$, then $U_i \neq U_r$. If $P_i \neq P_r$, then $f_1(P_i)$ and $f_1(P_r)$ are two independent random values, we write $(f_1(P_i), f_1(P_r)) \in_R \{0, 1\}^{2b}$, and so are U_i and U_r, thus $\text{prob}[U_i = U_r] \leq 1/2^b$.

If $U_i \neq U_r$, then $(T_i, T_r) \in_R \{0, 1\}^{2b}$, hence $\text{prob}[T_i = T_r] \leq 2/2^b$.

If $T_i \neq T_r$, then $(V_i, V_r) \in_R \{0, 1\}^{2b}$, $\text{prob}[V_i = V_r] \leq 3/2^b$, $\text{prob}[(A_i, A_r) \in_R \{0, 1\}^{2b}] \geq 1 - 3/2^b$, and $\text{prob}[A_i = A_r] \geq 1 - 4/2^b$. Thus with a probability not below $1 - 4/2^b$, the ciphertext fractions (A_i, B_i) and (A_r, B_r) are independently chosen random values.

Due to the symmetric construction of the RaMaRK scheme, similar arguments apply as well if the attacker runs the **decryption protocol**.

Finding q valid plaintext-ciphertext pairs with $q - 1$ protocol executions essentially means to find a plaintext (P_q, Q_q, R_q) where one can predict the corresponding ciphertext (A_q, B_q, C_q) without having used either $((P_q, Q_q), Z_q)$ or $((A_q, B_q), X_q)$ as challenge values for the encryption resp. decryption protocol.

We consider the second response value $w_i \in \{0, 1\}^{2b}$ the attacker receives during the ith protocol execution, i.e. (A_i, B_i) when encrypting and (P_i, Q_i) when decrypting. Note that after $q - 1$ protocol executions, the attacker needs to predict w_q. Random response values can't be predicted, the attacker can do no better than to deliberately guess one. There are $q(q - 1)/2$ sets $\{i, j\}$ with $i \neq j$ and $i, j \in \{1, \ldots, q\}$, hence the probability that after $q - 1$ protocol executions the attacker can either predict or guess w_q is at most $q^2 2/2^b$. □

Theorem 1, gives us an exact upper bound for the probability of success of forgery attacks against our scheme, depending on the number $q - 1$ of queries. For the sake of simplicity and shortness, in the following we omit such an exact treatment and assume that there are only $q = o(2^{b/2})$ queries. Note that the constants hidden by the asymptotics are quite small and can be found similarly to the proof of theorem 1.

Theorem 2. *The RaMaRK scheme is secure against inversion attacks.*

Sketch of proof: We restrict ourselves attackers who only can execute the encryption protocol. For reasons of symmetry, the proof for attackers only able to execute the decryption protocol is the same.

We consider the permutation $g : \{0, 1\}^{2b} \longrightarrow \{0, 1\}^{2b}$ with $g(P, Q) = (T, X)$. Given the ciphertext (T, X), our problem is to find a plaintext (M, N) with $g(M, N) = (T, X)$, where we are allowed to evaluate g, but not its inverse g^{-1}. We choose random functions f_4, f_5, and f_6, the value $R \in_R \{0, 1\}^{B - 2b}$, and compute the corresponding ciphertext (A, B, C). Now we can simulate the card's part of the encryption protocol. If one could find the plaintext (P, Q, R) corresponding to (A, B, C), one also could break g. But since $q = o(2^{b/2})$ and g is a Luby-Rackoff cipher, chosen plaintext attacks on it are infeasible [6]. □

Theorem 3. *The RaMaRK scheme is pseudorandom.*

Sketch of proof: Reconsider the proof of theorem 1. At the rth point of time, the attacker either chooses a plaintext (P_r, Q_r, R_r) for encryption, or a ciphertext (A_r, B_r, C_r) for decryption.

First, we consider the case when a plaintext is chosen and encrypted. For every $i \in \{0, \ldots, r - 1\}$, we need to distinguish whether $P_i = P_r$ and $Q_i = Q_r$, or not.

Case I: If $P_i = P_r$ and $Q_i = Q_r$ then $R_i \neq R_r$ and thus $I_i \neq I_r$, otherwise the attacker would not get any new information.

Case II: If $P_i \neq P_r$ or $Q_i \neq Q_r$ then $\text{prob}[T_i \neq T_r] \leq 2/2^b$, and if $T_i \neq T_r$ then $(X_i, X_r) \in_R \{0, 1\}^{2b}$ thus $S(X_i)$ and $S(X_r)$ are undistinguishable from independent $(B - 2b)$-bit strings chosen at random—and so are I_i and I_r.

For reasons of symmetry, the same arguments apply if a ciphertext is chosen and decrypted.

Now $q = o(2^{b/2})$, i.e. $q^2 \ll 2^b/2$, thus we can expect all I_1, I_2, ..., I_q to be pairwise distinct. Then all Y_1, Y_2, ..., Y_q are pairwise distinct, too—otherwise the collision resistant hash function H would be broken.

¿From the proof of theorem 1 we can deduce that a part of the output— (A, B) when encrypting and (P, Q) when decrypting—is undistinguishable from randomly chosen bit-strings with a probability not below $1 - q^2 2/2^b$. The pseudorandomness of the remaining part can be deduced from the *additional property* of the stream cipher S.   □

When designing cryptographic algorithms and protocols one often needs collision resistant hash functions with more properties [2]. For our proof of security, collision resistance of the hash function H is sufficient. But if H behaves like a random oracle [4], we can abandon the demand for the *additional property* of S.

6   The Block Cipher GRIFFIN

Even if remotely keyed encryption is not required, GRIFFIN, the block cipher realized by the RaMaRK scheme, is of some interest of its own.

In 1996, Anderson and Biham [3] proposed two block ciphers for flexible but large blocks, BEAR and LION. A similar proposal is the block cipher BEAST, see [8]. All three block ciphers depend on using a hash function H and a stream cipher S as building blocks, BEAST also needs a fixed size pseudorandom mapping.

With respect to chosen ciphertext attacks, BEAR, LION, and BEAST are pseudorandom (see [7] for a proof). But neither of the three ciphers is pseudorandom as defined in this paper, i.e. with respect to two-sided attacks. Luby's and Rackoff's attack on three-round Luby-Rackoff ciphers [6] works well for BEAR, LION, and BEAST. Therefore, Anderson and Biham [3] also proposed LIONESS, a block cipher pseudorandom even with respect to two-sided attacks. BEAR requires to evaluate H twice and S once, LION requires to evaluate S twice and H once, and LIONESS requires to evaluate H twice and S twice, where the inputs of H and the outputs of S are about as large as the cipher's blocks. As suggested by Anderson and Biham, we consider using the hash function SHA-1 for H and the stream cipher SEAL for S. SEAL appears to be faster than SHA-1, at least if one ignores SEAL's key set-up time. Thus, if the blocks are large, LION should be faster than BEAR, and LIONESS slower.

On a 133 MHz DEC Alpha machine (a "sandpiper") and for 1 MBit blocks, Anderson and Biham measured 13.62 MBit/sec for BEAR and 18.68 MBit/sec for LION. We expect LIONESS to operate at about 11.8 MBit/sec under the same conditions.

Due to theorem 3, GRIFFIN is pseudorandom with respect to two-sided attacks, too. For large blocks (e.g. $B = 1$ MBit), the effort to evaluate the fixed size pseudorandom mappings f_1, ..., f_6 is negligible. Then the speed of GRIFFIN is determined by evaluating S twice and H once. If we use SHA-1 and SEAL, and if the blocks are large, GRIFFIN can be expected to run at about the same speed as LION and thus significantly faster than its competitor LIONESS.[4]

References

1. W. Aiello, R. Venkatesan, "Foiling Birthday Attacks in Length-Doubling Transformations", in *Eurocrypt'96* (ed. U. Maurer), Springer LNCS 1070, 307–320, 1996.
2. R. Anderson, "The Classification of Hash Functions", in *Fourth IMA conference on cryptography and coding*, 83–93, 1993.
3. R. Anderson, E. Biham, "Two Practical and Provably Secure Block Ciphers: BEAR and LION", in *Fast Software Encryption* (ed. D. Gollmann), Springer LNCS 1039, 113–120, 1996.
4. M. Bellare, P. Rogaway, "Random Oracles are Practical: A Paradigm for Designing Efficient Protocols", in *First ACM Conference on Computer and Communications Security*, ACM, 1993.
5. M. Blaze, "High-Bandwidth Encryption with Low-Bandwidth Smartcards", in *Fast Software Encryption* (ed. D. Gollmann), Springer LNCS 1039, 33–40, 1996.
6. M. Luby, C. Rackoff, "How to Construct Pseudorandom Permutations from Pseudorandom Functions", *SIAM J. Computing*, Vol. 17, No. 2, 373–386, 1988.
7. S. Lucks, "Faster Luby-Rackoff Ciphers", in *Fast Software Encryption* (ed. D. Gollmann), Springer LNCS 1039, 189–203, 1996.
8. S. Lucks, "BEAST: A Fast Block Cipher for Arbitrary Blocksizes", in *IFIP Conference on Communications and Multimedia Security* (ed. P. Horster), Chapman & Hall, 144–153, 1996.
9. B. Preneel, P. van Oorschot, "On the Security of Two MAC Algorithms", in *Eurocrypt '96* (ed. U. Maurer), Springer LNCS 1070, 19–32, 1996.

---

[4] For similar reasons, BEAST outperforms LION. Under the same conditions as above, we expect BEAST to run at about 23.6 MBit/sec [8].

Sliding Encryption:
A Cryptographic Tool for Mobile Agents

Adam Young*,   Moti Yung**

Abstract. The technology of mobile agents, where software pieces of active control and storage (called mobile agents) travel the network and perform tasks distributively, is of growing interest as an Internet technology. Similarly, smartcard holders can be considered mobile users as they access the network at various points. Such mobile processing can be employed in large scale census applications in statistics gathering, in surveys and tallying, in reading and collecting local control information, etc.

This distributed computing paradigm where local pieces of data are getting accumulated in a mobile unit presents new information security challenges. Here, we point at some problems it poses and suggest solutions. The basic problem considered involves the design of a mobile agent that is capable of traversing an untrusted (curious) network while gathering and securing data from the nodes that it visits. We assume that some subset of the nodes may collaborate to track the agent, and we assume that snapshots of memory are taken at each node at times that are unpredictable to the agent. The data that is gathered must be securely stored within the agent and the adversarial nodes must remain oblivious to what is taken by the agent. In addition, the agent's movement throughout the network should be made difficult to trace. Furthermore, we assume that the agent is limited in storage capacity. To prevent the nodes from getting decryption capability, the agent must carry a public key for (asymmetric) encryption.

We present an economical solution that we call **"sliding encryption"**. This is a new mode of operation of public key cryptosystems that allows the encryption of small amounts of plaintext yielding small amounts of ciphertext. Furthermore, the encryption is performed so that it is intractable to recover the plaintext without the appropriate private key. We also describe how to modify sliding encryption so that the resulting ciphertexts are hard to correlate, thus making it possible to have mobile agents that are not easy to trace. Sliding encryption is applicable to mobile agent technology and may have independent applications to "storage-limited technology" such as smartcards and mobile units.

Key words: space-efficient encryption, sliding encryption, public key, RSA, mode of operation, mobile computing, network software agents, smartcards, WWW, network computing, applets, spiders, worms, viruses, cryptoviruses.

---

* Dept. of Computer Science, Columbia University, New York, NY, USA. Email: ayoung@cs.columbia.edu.

** CertCo New York, NY, USA. Email: moti@certco.com, moti@cs.columbia.edu

1 Introduction

Distributed agents is a relatively new technology which can be designed to perform many tasks. For example, an agent may travel from site to site in a network, performing local searches and gathering data on certain parameters (e.g., traffic load), so as to help control the network (e.g., redirecting the flow of packets to alleviate traffic). In an untrusted environment of non-malicious but curious nodes, such an agent needs to take measures to prevent traffic related data from falling into the wrong hands. If this data can be intercepted, it can be used for the purposes of traffic analysis, and if it can be tampered with, the performance of the system can be greatly hampered. Like the traffic data example, there are many distributed computations involving gathering of local pieces of data which require hiding the data from non-local nodes. The data eventually reaches the source of the agent. We will use the term *originator* to refer to the source entity that dispatches the agent to perform a certain task.

It is imperative that a data collecting and encrypting mobile agent conceal its decryption capability, thus our problem implies the use of public-key technology. This was pointed out in the context of the cryptovirus agent in [YY96]. The gathered data is made accessible exclusively to the originator when the gathered data is returned to it. The corresponding private key is not contained within the agent, and is kept secret by the originator.

In this paper we consider the problem of designing an agent that must gather small amounts of data from several nodes on a network, where each node is untrusted by the agent, and in particular the node may try to read information carried around by the agent. We further assume that the agent is restricted in the amount of data that it can store (e.g., on the order of kilobytes). As an example of this problem, consider the following. An agent contains a 128 byte public key, and must gather 1024 pieces of information from 1024 different nodes on a network. Suppose that it need only gather 4 bytes of information from each node. The greedy approach to this problem is to use a hybrid cryptosystem which selects a symmetric key per node, or to use the public key cryptosystem itself, both in ECB mode. In this case, ECB mode is most efficient. It follows that the agent must have the capacity to store 128k worth of data. However, the 128k of ciphertext would contain only 4k worth of plaintext. This is so, since each block of public key encrypted data has only a 4 byte value if data is encrypted directly. The rest of the block may be (should, in fact, be) random. The motivation for this work is therefore to study ways in which to public key encrypt small amounts of data, to yield small amounts of ciphertext, without compromising the overall security of the system. Note that due to the agent's mobility and the security constraints, it cannot accumulate a large block from many nodes and then encrypt it.

Specifically, we introduce what we call 'sliding' encryption that accomplishes exactly this. In a nutshell, sliding encryption is a way of enciphering a small amount of data within a larger block, and sliding away a small fraction of the result. This fraction constitutes the 'ciphertext' of the small piece of input plaintext. The sliding encryption scheme that we describe is based on RSA [RSA78], though the scheme is general enough to be used with any deterministic public key encryption algorithm.

2 Definitions and Background

Sliding encryption is a mode of operation of public key cryptosystems akin to using symmetric algorithms for stream ciphers. It is aimed at conserving space, rather than "fast" encryption (due to technological constraints the solution is public-key, and thus cannot be too fast). Recall that a self-synchronous stream cipher is a cipher in which each key block is derived from a fixed number n of preceding ciphertext blocks (e.g., [De83]). Sliding encryption has the same flavor, and it incorporates the idea of chaining from the mode of operation known as Cipher Feed-Back. However, in the scheme that we describe, each ciphertext block is derived from a state affected by all preceding ciphertext blocks, not just the preceding n blocks. It is possible to implement a sliding scheme that uses a fixed dependency of length n, but this is not applicable to our purposes. In this paper we do not concern ourselves with serial communications, but rather the problem of implementing efficient agents. Hence, the fact that all preceding encrypted data is lost if a ciphertext block is lost is not applicable to our problem (an interfering party can simply delete the entire agent if it is found).

Definition 1:

Sliding Encryption is a mode of operation of public key cryptosystems based on chaining, satisfying the following properties:

1. Encryption is granular, that is, a small amount u of plaintext bytes can be encrypted using a sufficiently large public key (We cannot delay the encryption by accumulating many small pieces together!)
2. It is computationally difficult to determine each u byte piece of plaintext without knowing the correct private key
3. The system is resistant against known plaintext attacks

The purpose of (1) is so that an agent that employs sliding encryption can gather small amounts of data from several nodes across the network, and public key encrypt each piece of data without wasting an excessive amount of storage space. Since we assume that an adversary may take a snapshot of the agent at any time, we cannot simply gather small amounts of data until we have enough to perform a space efficient encryption, because during this period a snapshot may be taken. The purpose of (2) goes without saying. Requirement (3) is needed in the case that an adversary has a good chance of guessing what the agent has gathered, and merely wishes to know what the agent has gathered. The sliding implementation that we present incorporates the notion of probabilistic encryption [GM84] so as to foil such an attack. Hence, the adversary must guess the random string used in the probabilistic encryption to verify that the guessed plaintext is correct.

At times throughout the rest of this paper we will refer to the agent as a probabilistic Turing machine (coin flipping computation). This model of computation is necessary, since a deterministic agent cannot hope to conceal, at the very least, the nature of the information which it gathers (its location, its structure, etc.), and cannot resist known plaintext attacks for small plaintexts. This randomization is achieved via inclusion of a function that generates random bits using events in the environment, such as AT&T's truerand function [MB95].

Related Work

Agents are loosely defined as personal software assistants with authority delegated from their users [Ch96]. Agent technology traces back to the early 1980's when the notion of a worm was invented at Xerox Palo Alto Research Center [Mc89]. John Shoch and Jon Hupp, two researchers at PARC, were interested in the concept of distributed processing, and thought that worms may be a novel way of accomplish distributed computing [Sl94]. Agent technology has evolved to include computer viruses, worms, web wanderers, etc. In the Crypto community they were suggested for factoring and other exhaustive search tasks [Wh89]; also, more recently, malicious agents called cryptoviruses have been proposed that make use of public key cryptography to mount attacks on their hosts [YY96]. Modern Internet technology gives rise to mobile applets and other mechanisms that exploit agents. An example of collective computation tasks and an algorithm allowing agents to "meet" after their deployment within a network has been discussed in [YuYu96].

Agents can be roughly categorized into those that are mobile and those that are immobile. Web wanderers, like Matthew Gray's WWW Wanderer (1993), are immobile, and gather information about the WWW for its user, while worms and distributed migratable tasks are examples of mobile agents. In this paper we will concern ourselves with the study of mobile agent technology in untrusted network environments. Also, smartcard technology which is readable, can nevertheless use encryption to perform secure and economical mobile data collection.

Remark: We note that there are numerous other security concerns with mobile agents. For example, validating the authenticity of the agent at the node and tracing (audit trail kept at or for the agent), assuming that the processing at the nodes is "as expected" and does not introduce "anomalies". Here we do not treat such security concerns, concerns which should be part of "network wide computing".

3   RSA Based Sliding Encryption

We will describe a sliding encryption implementation for m byte RSA keys (where m is a power of 2) [RSA78]. We will assume that the granular size of the plaintexts to be encrypted is (w.l.o.g.) fixed and that each is u bytes in length. Each piece of plaintext will be encrypted along with a v byte random string, where v is at least, say, 12 bytes in length. Let $t = u$ concatenated with v. We assume that t is

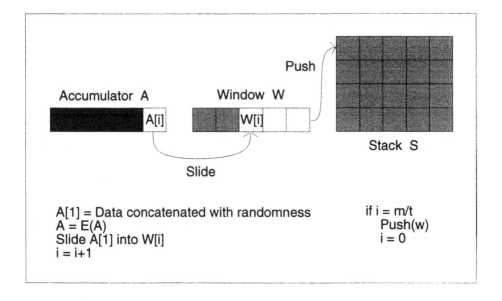

Fig. 1. Sliding Encryption Mode.

a power of 2 and $t \ll m$ and also that t divides m. As we shall see, the v random bytes that are added make the scheme a probabilistic encryption scheme. The v bytes constitute the random string used in the probabilistic encryption (analogous to an IV). The sliding encryption mode of operation (SE) makes use of a stack S, an m byte accumulator A, and an m byte window W. Stack elements $S[i]$ are m bytes in length.

The procedure Push(x) pushes an m byte quantity onto the stack. The function Pop(x) returns an m byte quantity from the stack. The function Empty(X) returns true iff stack X is empty. We assume the existence of an implicit state variable that points to the top of the stack. Accumulator elements $A[i]$ and window elements $W[i]$ are t bytes in length, where $1 \leq i \leq m/t$. $A[1]$ contains the least significant bytes of A, and $A[m/t]$ contains the most significant. The same applies for W. We assume the existence of a source of truly random bits (the bits must not be easily "guess-able"). Let the plaintexts be denoted by $a_1, a_2, ..., a_k$. Let the public key encryption function be denoted by $E()$, and let the corresponding private decryption function be denoted by $D()$.

Initially the stack is empty, and the accumulator is set to a random element in Z_n^*. To encrypt, new data is put in $A[1]$. We take a_1, The u lower order bytes of $A[1]$ are then set to be a_1. The v upper order bytes of $A[1]$ are set to be random. E is then applied to the accumulator, thus modifying all the elements of the accumulator. Note that an uneven "Feistel-like" preprocessing as in [BR94] which assures the accumulator is in Z_n^* and which pseudo-randomizes the accumulator based on the v random bits can take place here. In this case the definition of $E()$

and $D()$ will include this Feistel-like pre- and post-processing (the processing involves running v through a pseudorandom generator generating a pad that is EXORed with the rest of the bits then hashing the resulting value and EXORing the result with v.) Base on [BR94] the value encrypted is a "probabilistic encryption" if the preprocessing is via "random oracle like functions". We then set $W[m/t] = A[1]$. By setting $W[m/t] = A[1]$ we thereby 'slide' (shift) t-bytes of ciphertext from the accumulator into the window. To encrypt a_2, we set the u lower order bytes of $A[1]$ equal to a_2, and make the v upper order bytes of $A[1]$ random. We then apply E to A, and slide $A[1]$ into $W[(m/t) - 1]$. To encrypt a_3, we set the u lower order bytes of $A[1]$ equal to a_3, and make the v upper order bytes of $A[1]$ random. We then apply E to A, and slide $A[1]$ into $W[(m/t) - 2]$. This process is continued until $a_{m/t}$ has been slid into $W[1]$. At this point Push(W) is executed.

The next values $a_{(m/t)+1}$ through $a_{2m/t}$ are encrypted in the same way, and also pushed onto S. A count of the number of array elements in W that are currently in use is stored in the variable 'count'. This entire process is described by the following pseudo-code.

```
function InitializeSE
Input: Nil
Output: S, A, W, count

/* Initializes data structures to allow sliding encryption */
begin
count = 0        /* stores the number of elements of W in use */
S = empty
Initialize A to contain all random bytes
A = E(A)    /* make sure A is less than public key modulus */
end

function SlidingEncrypt
Input: RSA Public Key, S, A, W, count, a[1],a[2],...,a[k]
Output: S, A, W, count

/* Encrypts one or more plaintext blocks using RSA in SE mode */
begin
for i = count up to k-1 do
        set the v upper order bytes of A[1] to be random
        set the u lower order bytes of A[1] to be a[i]
        A = E(A)
        index = i mod m/t
        W[m/t - index] = A[1]   /* slide ciphertext into W */
        if index = m/t - 1 then
                Push(W)
                count = 0
        else
```

```
               count = count + 1
end
```

To perform sliding encryption, the function InitializeSE is first invoked to initialize the necessary data structures. SlidingEncrypt is then invoked for each plaintext value a_i. Decryption is exactly the opposite of encryption. We pop elements off the stack and slide elements from the window into the accumulator and decrypt them. The following function decrypts all values at once.

```
function SlidingDecrypt
Input: RSA Private Key, S, A, W, count
Output: a[k],a[k-1],...,a[1]

/* Decrypts all ciphertext blocks contained in A, W, and S. */
begin
if count > 0     /* first empty out the window */
        for i = (m/t - count + 1) up to m/t do
                set A[1] = W[i]
                A = D(A)
                output the u lower order bytes of A[1]
while Empty(S) = false do
        W = Pop(S)
        for i = 1 up to m/t do
                set A[1] = W[i]
                A = D(A)
                output the u lower order bytes of A[1]
end
```

Note that any deterministic public key functions E and D can be used. Probabilistic encryption where the ciphertext space is larger than the plaintext space will not work. For example, if ElGamal [El85] were used, the accumulator would double in size with each application of E, hence preventing efficient storage of ciphertext.

Note also that A's state can be modified to be a function of various portions of the history or be pseudo-randomized. In addition a number of data blocks can be processed together, e.g., there are 10 elements so nine are put together and one is put by itself with an indication that the previous block was size nine (so that decryption can be made).

To perform the cryptographic operations (namely, the RSA encryption above) the agent can utilize cryptographic libraries at the nodes or must include such cryptographic tools which have to be small in size themselves. See the discussion in the appendix.

4   Security of RSA Based Sliding Encryption

Consider the case of an agent that employs sliding encryption in an untrusted environment. The data a_i gathered by the agent is at risk of being included in a snapshot from the time the agent gathers it through the time at which E is applied to it. Consider the time before and after this event. In the case where the agent is a probabilistic automaton and its choice of which data to encrypt is probabilistic, and since we use probabilistic encryption the agent do not reveal its choice of data a_i to the observers. Namely, a_i cannot be inferred from the state of the agent prior to taking a_i. After a_i is encrypted, even assuming the adversary properly guesses that a_i was taken, the adversary must also guess v correctly, re-encrypt with v and a_i using a preimage of the agent, and compare with the current image. If v is on the order of 96 bits in length (12 bytes), then an exhaustive comparison will require $O(2^{95})$ encryptions on the average. An exhaustive search for a_i will require at least that. Clearly any previous data that is within the stack is at least as difficult to decrypt. (Recall that the Optimal Asymmetric Encryption like mechanism, turn each block encryption to "probabilistic encryption"). Thus, criterion 2 and 3 of the definition of sliding encryption is met. Clearly criterion 1 is met. The system therefore constitutes a sliding encryption scheme as defined in definition 1.

Now consider the efficiency of the scheme. For each piece of plaintext we add v bytes. For k pieces of plaintext, the scheme requires $O(kv)$ storage in addition to the $O(ku)$ bytes corresponding to the plaintext. The total storage is therefore $O(k(u+v))$. This contrasts with the greedy approach that requires $O(km)$, where $m \gg v$.

Note that overall we use more space than an ECB encryption would if we could delay the encryptions and treat the entire gathered information as a stream of data. But, recall that we cannot delay encryptions due to the risk of having the plaintext captured in a snapshot.

5   Applications

Sliding encryption can be used to efficiently store data in embedded cryptographic devices such as smartcards that have limited storage capacity. Depending on the "guess-ability" of the plaintext and the number of bytes in the plaintexts a_i, the number of additional v bytes of randomness may be tuned. In addition, the accumulator can be a special machine register that is specifically designed to have multi-precision arithmetic performed on it.

Another obvious application of sliding is for tracking the path traversed by an agent generating private audit trails. In this case, the a_i can be 32 bit IP addresses corresponding to the nodes on the Internet that the agent has traversed. We can take v to be 12 bytes. In this case t is a 16 byte quantity. By using a 512 bit RSA key (trying to minimize size overall), we save 48 bytes on each encryption of an IP address. For the sake of argument, suppose that b is 32. The agent therefore contains an accumulator of 64 bytes, a window of 64 bytes, and an array S of 2048 bytes, for a total of 2,176 bytes. The originator releases several such agents, and upon returning, the originator is able to see the last 32 Internet nodes that were traversed by the returning agent. In this example, the adversaries clearly know the plaintext.

5.1   Making Agents Less Traceable

Another security consideration is traceability, which is apart from the security of the plaintext itself and concerns concealing the agent's whereabouts from the environment (related to the last application). Suppose that a subset of the adversaries have realized that an agent passed through after the fact, and all of them took snapshots of the agent, but failed to take snapshots while the a_i's were present in the agent in plaintext form (this is just a working assumption). Suppose further that these adversaries are willing to collaborate and want to verify that the same agent passed through their nodes, and they want to learn in what order the nodes were traversed. So, here we are interested in what can be learned about the path of the agent based on the state information of the agent. The collaborating adversaries are equipped with a set of images of agents, and with each image is a corresponding time stamp. We will assume that the agent chooses the nodes that it traverses uniformly at random from the set of all nodes on the network.

We must assume that the agent can modify the version of itself that moves on just before moving to another node and that this cannot be caught by the current node itself (otherwise untraceability is impossible). Alternatively, we can assume that the subset of adversaries is disconnected in the network and the agent manages to modify itself before it reappears at a node controlled by this set. This problem is similar to problems of untraceability in anonymous remailing schemes [Cha81]. In this respect, a mobile agent needs to be able to, in some sense, anonymously "remail" *itself*. Note that we are trying to minimize the risk here rather than avoiding the problem all-together. Our major problem being the fact that an agent cannot "encrypt and decrypt" itself, since if it can, the nodes can also decrypt it; decryption is made available only to the originator of the agent.

To reduce traceability we need to pay careful attention to all the state information in the agent that is revealed to the adversaries after the fact. This state information consists of the accumulator values, the window values, the values associated with the stack, the count, and the stack pointer. As described, the stack will grow monotonically in size. This will help reveal the order in which the nodes are traversed. We must therefore conceal the size of the stack. We can minimize this problem if we are willing to fix the maximum amount of information that can be obtained, and if we adopt a 'most recently used' algorithm, for instance. In this case, the agent stores only the most recently gathered information. S is an array that initially contains random elements (that can be recognized after decryption), and a pointer to S is maintained. Let this pointer be denoted by 'ptr'. ptr is initially set at a random offset, and is incremented from the lowest index to the highest index. When the highest index element is filled with m bytes of ciphertext, the pointer wraps around back to the lowest index, thus overwriting any data previously stored there.

So far we have shown how to fix the size of S, thereby insuring that A, W, S, count, and ptr remain fixed in size over time. Now consider the contents of accumulator A. The accumulator cannot be used to link two images of agents, since each accumulator value is a probabilistic RSA encryption of the previous. So, it remains to consider W, S, count, and ptr. It turns that in the scheme described thus far, both W and S can be used to successfully track the agent. Let b denote the maximum number of m byte blocks that can be stored in S. If b adversaries are willing to collaborate, then they have a chance at determining a path of length b that the agent follows Similarly the window can be used to trace out paths of length $m/t - 1$. In this case the adversaries track the agent by comparing the contents of S and W, respectively.

The following is how the traceability problem with W can be minimized. Note that when each a_i is encrypted, v truly random bytes are incorporated into the encryption. These v bytes of entropy can be used to seed a cryptographically secure random number generator, to generate a stream of size (count$*t$) bytes. This stream is EXORed with the contents of W that is currently in use. A is then encrypted as usual, thereby concealing the seed v. $A[1]$ is then slid over to the window. We redo this process if the resulting value for W will end up being greater than or equal to the public modulus. The v random bytes are therefore used to conceal the previous contents of the window and to probabilistically encrypt a_i, thereby concealing the previous contents of the accumulator as well.

It remains to show how to conceal the contents of the array S. To conceal S, we take the same v random bytes and use them on S in much the same way as we did for W. Using a different cryptographically secure pseudo-random number generator, we proceed as follows. We generate a stream of length $b * m$ bytes and EXOR it with S. We can do so, since we will be overwriting the element currently pointed to in S anyway. We then proceed as usual by overwriting the value currently pointed to in S with the encryption of W, if necessary. We thus distribute the v random bytes over the array S as well as over W.

The random choices for the probabilistic encryptions are thus critical in decrypting the gathered data. Unlike before, they cannot simply be discarded. The pseudo-random key streams need to be reconstructed during decryption and EXORed with the ciphertexts in order to decrypt.

Note that if an agent moves from A to B, then both count and ptr will have increased by 1. Count reveals exactly how many elements of W are currently in use, and ptr reveals the next array element in S to overwrite. The values for count and ptr can be used to help identify the order in which nodes are traversed. It seems that there is no easy way to avoid revealing such information when we require maintaining all information in order, unless "generalized secure computation" is made possible. To reduce the problem we may "randomize" the task of the agents. Namely, we replicate agents in nodes and add dummy encryptions and 'detours' as described in [GT96], allowing also the nodes along the detour to encrypt "dummy" data elements randomly – thus making different copies look differently and forcing tracing adversaries to follow the detours.

6   Conclusion

We have shown a new mode of operation of deterministic public key cryptosystems. This sliding encryption technique can be used to encipher small amounts of data, yielding small amounts of ciphertext, while insuring that the data remains secure. This technique was shown to be applicable in smart card applications with limited storage capacity. We also showed how to make tracing of an information gathering agent somewhat difficult, by further randomizing its state and task.

References

[BR94]   M. Bellare, P. Rogaway, Optimal Asymmetric Encryption, Eurocrypt 94.

[Cha81] D. Chaum.   Untraceable Electronic Mail, Return Addresses, and Digital Pseudonyms. In *Communications of the ACM*, v. 24, n. 2, Feb 1981, pages 84–88.

[Ch96]   F. Cheong. Internet Agents: Spiders, Wanderers, Brokers, and Bots. New Riders Publishing, page 5, 1996.

[De83]   D. Denning. Cryptography and Data Security, Addison-Wesley, page 137, 1983.

[El85]   T. ElGamal. A Public-Key Cryptosystem and a Signature Scheme Based on Discrete Logarithms. In *Advances in Cryptology—CRYPTO '84*, Springer-Verlag, pages 10–18, 1985.

[GM84]   S. Goldwasser, A. Micali. Probabilistic Encryption. In *Journal of Computer and Systems Science*, v. 28, pages 270–299, 1984.

[GT96]   C. Gulcu, G. Tsudik. Mixing Email with BABEL. In *Proceedings of the 1996 Symp. on Network and Distributed System Security* ISOC, pages 2-16, 1996.

[MB95]   D. Mitchell, M. Blaze. truerand.c, AT&T Laboratories, 1995.

[Mc89]   J. McAfee. Computer Viruses, Worms, Data Diddlers, Killer Programs, and other Threats to Your System. St. Martin's Press, page 29, 1989.

[RSA78] R. Rivest, A. Shamir, L. Adleman. A method for obtaining Digital Signatures and Public-Key Cryptosystems. In *Communications of the ACM*, v. 21, n. 2, pages 120–126, 1978.

[Sl94]   R. Slade. Robert Slade's Guide to Computer Viruses. Springer-Verlag, page 49, 1994.

[Wh89]   S.R. White, Covert Distributed Processing with Computer Viruses. In *Proceedings of the Crypto 89*, pages 616-619.

[WN94]   D. Wheeler, R. Needham. Tiny Encryption Algorithm (TEA). In *Fast Software Encryption: second international workshop*, volume 1008 of Lecture Notes in computer science, Dec. 1994. Springer.

[YY96]   A. Young, M. Yung. Cryptovirology: Extortion-Based Security Threats and Countermeasures. In *Proceedings of the 1996 IEEE Symp. on Security and Privacy*, IEEE Computer Society Press, pages 129–140, 1996.

[YuYu96] X. Yu, M. Yung. Agent Rendezvous: A Dynamic Symmetry-Breaking Problem. In *Proceedings of the 1996 ICALP*. Lecture Notes in Computer Science, Springer, July 1996.

A   Cryptographic Mobile Agents

Currently, Cryptographic Application Programming Interfaces (CAPI) are available or in the process of being developed on many different computing platforms. Mobile agents for such platforms need not have their own multi-precision code and can utilize the node. However, there is evidence that such code can be implemented in compact form [YY96]. They compiled the portions of the GNU MP library that are needed to do 512 bit RSA encryptions with a public exponent of 3. The routines were compiled and run on a Macintosh SE/30. They made two notable optimizations.

1. First, they computed the reciprocal of the originator's public RSA modulus n, and included it within the agent. This allowed them to avoid including a MP division routine in the agent.
2. By choosing the public exponent of 3, a MP modular exponentiation routine was not needed since only two MP multiplications are required to encrypt.

The size of the compiled library, with C code and in-line assembly, is 4,372 bytes (cutting 70% of the general library). Note that using assembly language, it is of course possible to make this even smaller. Their agent also contained the TEA block cipher [WN94] in assembly, and it occupied a mere 88 bytes. TEA encrypted at a rate of 47k bytes/sec and can be employed for pseudorandom generation and hashing mentioned in this work, as well as for hybrid encryption. Thus, there is evidence to suggest that even when the agent has to implement the cryptographic procedures internally, the tools described in this paper can be implemented using relatively small space.

Fast Software Encryption: Designing Encryption Algorithms for Optimal Software Speed on the Intel Pentium Processor

Bruce Schneier
Counterpane Systems
101 E Minnehaha Parkway
Minneapolis, MN 55419
schneier@counterpane.com

Doug Whiting
Stac Electronics
12636 High Bluff Drive
San Diego, CA 92130
dwhiting@stac.com

Abstract. Most encryption algorithms are designed without regard to their performance on top-of-the-line microprocessors. This paper discusses general optimization principles algorithms designers should keep in mind when designing algorithms, and analyzes the performance of RC4, SEAL, RC5, Blowfish, and Khufu/Khafre on the Intel Pentium with respect to those principles. Finally, we suggest directions for algorithm design, and give example algorithms, that take performance into account.

1 Overview

The principal goal guiding the design of any encryption algorithm must be security. In the real world, however, performance and implementation cost are always of concern. The increasing need for secure digital communication and the incredible processing power of desktop computers make performing software bulk encryption both more desirable and more feasible than ever.

The purpose of this paper is to discuss low-level software optimization techniques and how they should be applied in the design of encryption algorithms. General design principles are presented that apply to almost all modern CPUs, but specific attention is also given to relevant characteristics of the ubiquitous Intel X86 CPU family (e.g., 486, Pentium, Pentium Pro). Several well-known algorithms are examined to show where these principles are violated, leading to sub-optimal performance. This paper concerns itself with number of clock cycles per byte encrypted—given a basic encryption algorithm "style." Factors of two, three, four, or more in speed can be easily obtained by careful design and implementation, and such speedups are very significant in the real world.

In the past, cryptographers have often designed encryption algorithms without a full appreciation of low-level software optimization principles associated with high-performance CPUs. Thus, many algorithms have unknowingly incurred significant performance losses which apparently could have been avoided at the design stage without impairing security. The goal of this paper to encourage greater awareness of these performance issues, and to encourage dialogue between cryptographers and programmers at the algorithmic design stage.

2 General Optimization Principles

By its nature, almost every known encryption algorithm includes a small inner loop where the vast majority of all processing time is spent. This inner loop is a natural candidate for assembly language implementation to achieve ultimate speed. With today's advanced CPU architectures, such optimization has become as much art as science. It is true that optimizing C compilers can produce very fast code in many cases. However, without understanding the architecture of the underlying CPU, it is easy to write C code which is apparently efficient but which cannot be compiled to utilize the CPU completely, particularly for such tight inner loops and for "register poor" CPUs (e.g., Intel Pentium). Lest cryptographers in particular feel slighted, note that these observations apply equally well to software optimization outside of cryptography; i.e., there are many high-level language programmers who do not understand how to take full advantage of a CPU.

The ensuing discussion needs to be prefaced with an important caveat: the fact that a particular operation is expensive in terms of CPU clock cycles does not mean a priori that the operation should not be used in an encryption algorithm. The associated slowdown may be deemed acceptable because of additional security afforded thereby, but the algorithm designer should be aware that such a tradeoff is being made.

Ideally, it is desirable to express an algorithm in a high-level language (e.g., C) in such a way that it can be compiled optimally to all target CPU platforms. Due to dramatic variances in CPU architectures and in the quality of compiler code generation, this ideal can rarely be achieved in practice; however, an understanding of how a CPU executes assembly language instructions helps the programmer comprehend how a given set of lines of code will run. There are purists who believe that optimizing compilers totally obviate the need for programming in assembly language. This is largely true, except in the exact case of interest here: designing an algorithm to wring every last ounce of performance out of a CPU. The compiler's job is in fact to translate high-level statements into assembly language, so the algorithm designer must understand (at least roughly) what the resulting compiled code will look like and how it will execute in order to achieve optimal speed. Alternately, the designer (or an associate) may code the algorithm directly in assembly language to see where the performance roadblocks are encountered.

Many modern CPUs have at their core a Reduced Instruction Set Computer (RISC) architecture. Basically, this means that most simple operations (memory load/store, register-to-register add/subtract/XOR/or/and, etc.) execute in one clock cycle. Even high-end CPUs with a Complex Instruction Set Computer (CISC) heritage, such as the Intel 486/Pentium and the Motorola 68K family, have migrated to a modified RISC approach, with only the more complex (and less frequently used) operations requiring multiple clocks. In other words, the RISC vs. CISC war is over, and RISC has won by assimilation. Even many embedded microcontrollers today use a RISC architecture, and this discussion applies equally well to them. To maximize performance, it is generally recommended that most instructions be selected from the RISC "subset." In order to execute assembly language instructions in a single clock, multiple levels of pipelining are used inside the CPU; that is, each instruction actually executes in several stages, requiring multiple clocks, but on each clock one instruction is completed. The more complex instructions (e.g., integer multiply or divide) can "stall" the pipeline, backing up the execution of subsequent instructions. Thus, it behooves the algorithm designer to know exactly which instructions are in the RISC subset. For example, on the Pentium processor, 32-bit integer multiplication (as in IDEA) requires 10 clocks[1], and from 13–42 clocks on the Intel 486.

Another important architectural improvement in high-end CPUs is superscalar execution; more than one instruction can be executed in a single clock cycle. For example, the Pentium processor can execute up to two RISC instructions per clock, in two separate pipelines, so careful design to allow complete instruction "pairing" on the Pentium can result in a doubling of speed. Such additional pipelines generally require even further lookahead in the instruction stream and can significantly increase the relative cost of any pipeline stalls. In addition, only certain types of instructions can be executed in parallel, which can introduce an extra (often non-intuitive) level of dependency between consecutive instructions or lines of C code in order to utilize fully the superscalar processing. Given two instructions, if the second one depends on the result of the first, the two cannot be executed in parallel. Thus, while superscalar architectures offer the possibility of substantial speed increases, they also dramatically increase the penalty for improper design.

As future CPUs achieve higher levels of superscalar execution, the rules for optimization will certainly become more complex. For example, the Pentium Pro can execute up to three instructions per clock. Almost all modern RISC CPU families (e.g., PowerPC, DEC Alpha, SUN Sparc) include processors of varying cost/performance levels, some of which are superscalar and some of which are not. Some of the principles discussed below do not apply to the non-superscalar CPUs, but it is assumed here that the algorithms of interest are intended to run well on a broad class of processor types. Fortunately, it has been empirically observed that, in most cases, an algorithm optimized for a high-end superscalar processor (e.g., Pentium) is also very close to optimal on a lower-end CPU (e.g., 486),[2] although unfortunately the converse is not usually true. The techniques discussed here should continue to apply in general, and high degrees of independence and parallelism should be included in a design to attempt to take full advantage of such future architectures

---

[1] It is actually one clock cycle slower to do a 16-bit multiply on the Pentium.

[2] Later we will see that RC4 is a counterexample.

Following are some general issues to be considered in designing an algorithm:

- Avoid conditional jumps in the inner loop. Any unpredictable change of the flow of control in the algorithm will normally flush the pipeline and cost extra clock cycles. In particular, any if/then/else or the C operator "?" will cause assembly language jump instructions to be executed. Sometimes table lookup can be used to remove the need for a jump without causing a pipeline flush. Jumps also increase vulnerability to timing attacks [Koc96].
- Unroll loops. Consider performing several iterations of the inner loop "inline" without looping, thus removing a jump (and a test for completion) at each iteration. This issue actually deals with implementation, not algorithmic design. However, at the design stage, it is useful to realize that the programmer implementing the algorithm will probably unroll the loop at least to a certain extent, so it is critical to understand any dependencies between the last instructions of one iteration and the first instructions of the next that may cause pipeline stalls.
- Avoid intrinsically expensive instructions. Among this category are multiplies, divides, and any other instructions which are slow on the processor(s) of interest. For example, on a Pentium, a variable rotate/shift instruction (i.e., where the rotation/shift amount is not known a priori) requires four clocks and cannot be paired with any other instruction; thus, the variable rotate alone costs as much as eight RISC subset instructions.
- Limit the number of variables. Many RISC processors have a fairly large general purpose register set (at least sixteen registers), but the Pentium has only seven general purpose registers. If too many variables are used extensively in the inner loop, they cannot fit in registers, so the performance will suffer greatly due to excessive memory accesses. Any memory access other than load/store (e.g., memory-to-register or register-to-memory add) will require multiple clocks.
- Limit table size. Although larger tables are better from a cryptographic standpoint, smaller is definitely better from a software speed standpoint. It is tempting to use large tables for S-boxes, but if those tables do not fit into the CPU's on-chip data cache, a substantial performance degradation may be incurred. In general, tables should be limited to no more than 4K bytes for today's CPUs.
- Allow parallelism. The general idea is to allow as many independent operations to be executed as possible. Unfortunately, this principle is often in direct opposition to the generally desirable goal of maximizing the "cascading" of bits, so there is usually a design tradeoff involved. From a high-level language, this guideline can be checked by breaking the code down into RISC subset operations and then seeing how the operations could be rearranged to perform as many as possible in parallel at each point. As an example, assuming that the variables a, b, and c all fit in registers, the C statement "a = (a + b) ^ c;" would be broken down into the two RISC statements "a += b; a ^= c;" In this case, notice that the second operation cannot be performed in parallel with the first. After breaking the statements down to this level, the

rearranged code often looks very unlike the original code, but this analysis is key to achieving high speed on superscalar architectures. On the Pentium and many of today's other superscalar CPUs, the goal is to have exactly two operations paired together at all times.

- Allow setup time for table indexing. This guideline is somewhat subtle and seems to be violated frequently in existing algorithms. When accessing tables (e.g., S-boxes) based on a function of the algorithm's variables, the table index must be computed as far ahead as possible. For example, on the 486, if an address is computed in a register on one cycle, and that register is used on the ensuing cycle to access memory, a one cycle penalty is incurred. On the Pentium, things get worse due to superscalar operation: the address should be computed at least two instructions ahead of when it is to be used; otherwise a one clock penalty is incurred. For example, the deceptively simple C statement "y = S[x & 0xFF];" on a Pentium would require at least two other instructions to be inserted between the "&" operation and the table access to utilize the CPU fully. Note that this address setup time is required regardless of whether the data is in the on-board CPU data cache.

These guidelines are fairly simple, but the performance gains resulting from following them can be quite impressive.

3   Analysis of Known Algorithms

In this section, several well-known encryption algorithms are analyzed in terms of the guidelines presented above. This scrutiny is not intended to be critical of the algorithms in question, but rather to illustrate the general principles with concrete examples. Only key portions of the inner loops of the algorithms are examined here.[3] The same principles could be applied to other portions of the algorithms (e.g., initialization), but the performance improvement is minimal. For each algorithm, some brief comments are also included about suitability of the algorithm for direct hardware implementation.

3.1   RC4

The inner loop of RSA's stream cipher RC4 [Sch96] is shown in C below:

```
unsigned char S[256];      /* initialized based on keys */

void RC4encrypt(unsigned char *p,int cnt)
    {
```

---

[3] It should be noted that most of the timings discussed below are approximate and are intended only to point out general performance issues; the only guaranteed way to obtain exact performance numbers is to implement and time the algorithm on the given platform.

```
int i,j,t;
unsigned char tmpI,tmpJ;

for (i=j=0;cnt;cnt--,p++)
    {
    i=(i+1) & 0xFF;                 /* update i */
    tmpI=S[i];
    j=(j+tmpI) & 0xFF;              /* update j */
    tmpJ=S[j];
    S[j]=tmpI;                  /* swap S[i], S[j] */
    S[i]=tmpJ;
    t=(tmpI+tmpJ) & 0xFF;
                    /* compute ``random'' index */
    *p ^= S[t];     /* XOR keystream into data */
    }
}
```

Entries in the S-box array $S[\ ]$ each consist of 8 bits and are initialized as a permutation of 0..255 based on the encryption key. In most assembly language implementations of RC4, the variables i and j are kept in 8-bit registers, so there is no need to mask with 0xFF as shown in the C code. This code compiles fairly efficiently, but there are some dependencies which limit performance on high-performance CPUs.

For example, j cannot be updated without first fetching $S[i]$, but only after i has been updated. The update of i, if performed literally as indicated, would cause an pipeline stall because of the setup time required for the table indexing. In a fully optimized (unrolled) version of this code, the increment of i would probably be deferred, only being computed once every 4 or 8 iterations; e.g., on every eighth iteration, i would be incremented by 8. In this case, the tmpI value would be loaded in a single instruction from $S[i + N]$, where N is a constant dependent on the "unroll" position. This method works around the address setup time, although a special "wrap" case has to be handled when i approaches 255. In any case, the fact that $S[j]$ is accessed immediately after updating j almost certainly results in a clock penalty in any implementation. A similar penalty probably applies for t and $S[t]$, although, with appropriate overlapping of the next iteration, this penalty may be avoided on some CPUs.

As an aside, a very surprising empirical result illustrates the difficulty in projecting optimization techniques onto future processors. A fully unrolled and optimized assembly-language version of RC4 has been measured as encrypting one byte roughly every seven clocks (averaged over a large block) on a Pentium, using byte registers for all variables to avoid the need for masking with 0xFF. For example, a 150 MHz Pentium can encrypt with RC4 at over 20 Mbytes/sec. However, the Pentium Pro processor contains an obscure internal pipeline stall which occurs when a register is accessed as a 32-bit quantity (e.g., EAX) soon after modifying an 8-bit subset (e.g., AL) of that register. The penalty in this case is a whopping six clocks (or more); as a result, the optimized Pentium code runs on a Pentium Pro at less than half the speed of a Pentium of the same frequency. One simple workaround in this case is to use 32-bit entries in $S[\,]$ and for all variables, necessitating the mask with `0xFF` and quadrupling the table size. Such a change is fairly simple, but the resulting code is no longer completely optimal on a Pentium.

The general performance problem with RC4 as designed is that almost every statement depends immediately on the statement before it, including the table index computation and the associated table accesses, limiting the amount of parallelism achievable. Fortunately, most (although not all) of these dependencies can be worked around without penalty in a fully unrolled optimized loop. It is extremely unlikely that an optimizing compiler would come close to such optimizations unless the C code was considerably unrolled, since the optimizations include carefully overlapping the start of the next iteration with the end of the previous one.

¿From a design standpoint, it would have been better to update i at the end of the loop to simplify unrolling. Even better would have been to use the previous value of j for the swap while updating the current value of j, thus allowing a full iteration for lookahead computation. It would probably also help if t could be computed at the beginning of the loop, or a previous value of t used, in order to allow maximum address setup time. Such changes typically cost nothing with appropriate register assignments, but they do allow much greater parallelism. It is possible that some of these proposed design changes might raise security issues, particularly with the initial few keystream bytes. If so, the inner loop should be run a few times as part of initialization, discarding the keystream value $S[t]$, to get rid of any such startup problems, still resulting in a significant speedup for the bulk of the encryption. Such changes probably would not have affected the overall security of the algorithm, although no cryptanalysis has been performed to verify this supposition.

Note that an RC4 implementation in hardware requires only 256 bytes of RAM, which is quite reasonable. However, each iteration also requires at least five RAM accesses (three reads, two writes), limiting the speed to no fewer than five clocks per byte, unless a dual-ported RAM is used, which is costly in terms of chip area. By contrast, a DES chip can easily run at two clocks per byte, even with each round implemented sequentially.

3.2   SEAL

SEAL [RC94] is a fast software stream cipher designed by Phil Rogaway and Don Coppersmith, and patented by IBM [CR95]. The inner loop uses four 32-bit registers, as well as 32-bit tables T with 512 entries and S with 256 entries. The table contents (more than 3K bytes) are initialized based on the encryption key in a fairly computationally intensive operation, so the algorithm does not seem to be appropriate if the key needs to change frequently. The relevant portion of the inner loop is given below:

```
long        T[512],S[256],N[4]; /* initialize based on keys */

void SEALencrypt(void)
    {
    int j;
    long a,b,c,d,p,q;

    /* some setup here */

    for (j=0;j<64;j++)
        {
        p =   a  & 0x7FC; b += T[p/4]; a=ROT9(a); b ^= a;
        q =   b  & 0x7FC; c ^= T[q/4]; b=ROT9(b); c += b;
        p =(p+c) & 0x7FC; d += T[p/4]; c=ROT9(c); d ^= c;
        q =(q+d) & 0x7FC; a ^= T[q/4]; d=ROT9(d); a += d;

        p =(p+a) & 0x7FC; b ^= T[p/4]; a=ROT9(a);
        q =(q+b) & 0x7FC; c += T[q/4]; b=ROT9(b);
        p =(p+c) & 0x7FC; d ^= T[p/4]; c=ROT9(c);
        q =(q+d) & 0x7FC; a += T[q/4]; d=ROT9(d);

        /* at this point, output:
        b+S[4*j], c^S[4*j+1], d+S[4*j+2], a^S[4*j+3]  */

        a+=N[2*(j&1)]; /* easily handled by unrolling */
        c+=N[2*(j&1) + 1];
        }
    }
```

The indexing with $p/4$ or $q/4$ is actually simple, since the index is multiplied by 4 (i.e., sizeof(long)). The major problem with this algorithm is the strong dependence between consecutive operations, allowing only minimal parallelism; in other words, on the Pentium, it is rare that both pipelines are utilized here. In addition, the table access (e.g., $T[p/4]$) follows immediately after the computation of the table index (p or q), thus incurring a clock penalty on a Pentium (and probably on most other RISC CPUs). For example, each of the first four lines of the inner loop involve a register-to-register move/add, a mask operation, a memory-to-register load from the table, a register-to-register add/XOR, a fixed rotate, and a register to register XOR/add. On a Pentium, with perfect pipelining and no address setup penalties, these four lines could ideally execute in sixteen clocks. It should be noted that the official Intel Pentium documentation claims that the opcode for rotation by a constant amount (e.g., 9 bits) can pair with another instruction, suggesting an ideal time of twelve clocks. However, it has been determined empirically [4] that only rotations by one bit can pair, thus costing

---

[4] See http://www.geocities.com/SiliconValley/9498/p5opt.html.

extra clocks. This anomaly underscores the importance of always checking the actual performance against the theoretical performance.

Instead, because of the problems mentioned above, these four lines actually require 24 clocks. By contrast, on a 486, the "ideal" speed of 28 clocks is actually achieved. The second set of four lines can be executed in 20 Pentium clocks, as opposed to the ideal of 10, while the 486 incurs no penalties and runs in 24 clocks. Obviously, this algorithm was designed with the 486 in mind, not the Pentium. In fact, at the same clock frequency, a Pentium is only slightly faster than a 486 at running SEAL. A fully optimized Pentium version should be able to encrypt at about 4 clocks per byte, which is quite a bit faster than RC4.

To maximize Pentium throughput, SEAL could be redesigned to stagger the usage of variables so that consecutive instructions can run in parallel. For example, q should generally be used as an index immediately after computing p (and vice versa), and the final statement of each of the first four lines (e.g., b ^= a) should modify a different variable than that being modified by the table entry (e.g., d ^= a, etc.). Such changes cannot be made without considerable thought and analysis, but it appears that minor modifications could produce an algorithm very similar in spirit (and hopefully in security) at nearly twice the speed on a Pentium, without slowing down the algorithm on a 486. It should also be noted that preserving the values of p and q across iterations of j should not incur any cost and would likely increase security somewhat.

SEAL does not appear very attractive for hardware implementation, principally because of the large size of the tables. A hardware implementation could fairly easily run at about one byte output every clock, which is faster than a simple DES implementation, but there is no obvious way to use pipelining to obtain much higher speed for SEAL, as is possible with DES.

3.3 RC5

RC5 [Riv95] is a Feistel-network block cipher (patent pending by RSA Data Security, Inc.). RC5 gets its strength from data-dependent rotations, which were first discussed in [Mad84]. The inner loop is shown in C below, where "<<<" indicates circular rotation:

```
long      S[2*ROUNDS+2];  /* initialized based on keys */
long      A,B;            /* values to be encrypted */

void RC5encrypt(void)
   {
   int j;

   A+=S[0];
   B+=S[1];
   for (j=1;j<ROUNDS;j++)
```

```
    {
    A = ((A^B) <<< B) + S[2*j];
    B = ((B^A) <<< A) + S[2*j+1];
    }
}
```

The algorithm looks simple and elegant. The recommended value for ROUNDS is 16. Observe that, since all table indices depend only on j, no table access penalties should occur in an unrolled version. At first glance, this algorithm should run very fast.

Unfortunately, on a Pentium processor, the cost of a variable rotation is very high (four clocks). Even worse, variable rotations cannot pair with any other instructions on the Pentium. Each round involves two memory loads, two adds, two XORs, and two rotations (which on the Pentium include an instruction to move the rotation amount in the CL). Thus, on a Pentium, assuming perfect pairing other than for the rotations, each round will require at least twelve clocks, whereas on an ideal superscalar RISC CPU the number would be only six clocks per round, given the order dependencies of the operations. At sixteen rounds, RC5 should encode at about 24 clocks per byte on a Pentium.[5] While this is fast compared to most block ciphers, it is disappointingly slow compared to what might be initially expected, again pointing out the danger in estimating the speed of an algorithm from a high-level language without understanding the underlying processor limitations. On a Pentium Pro, these pipeline problems appear to have been removed, so RC5 should run at or near its impressive theoretical software speed limit of six clocks per round, or 12 clocks per byte at sixteen rounds.

In hardware, RC5 should be able to execute one round in two (or perhaps four) clocks, thus giving a speed which is comparable to DES. However, it should also be noted that a full 32-bit variable rotation engine is expensive in hardware.

3.4   Blowfish

Blowfish [Sch94] is also a Feistel-network block cipher. It uses four tables, each consisting of 256 entries of 32-bits each, plus a separate table of eighteen 32-bit entries, for a total of more than 4K bytes. The table entries are initialized based on the keys, with a fairly large setup time, severely limiting the algorithm's usefulness if the key needs to be changed frequently. Here is the code for the inner loop of Blowfish:

```
long  P[18], S[4][256];   /* initialized based on keys */
long  A,B;                 /* values to be encrypted */

void BlowfishEncrypt(void)
    {
```

---

[5] RSADSI, Inc. has reference code that encrypts at 23 clock cycles per byte.

```
int j;

for (j=0;j<16;j++)                          /* 16 rounds */
    {
    A ^= P[j];
    B ^= ((S[0][ A & 0xFF] + S[1][(A >> 8) & 0xFF])
          ^ S[2][(A >> 16) & 0xFF])
          + S[3][(A >> 24) & 0xFF];
    swap(A,B);                    /* interchange values */
    }
/* some final operations here */
}
```

The algorithm looks simple, although the function of A computed using the S-boxes is slow because of the need to extract four bit fields and load four table entries. Each round can be executed on a Pentium in about 9 clocks, and there are no address setup penalties, but there is a pipeline stall of one clock due to the sequential nature of combining the S-box outputs, which could be alleviated by using xors to combine all the S-box entries. Using sixteen rounds, this equates to a throughput of about 18 clocks per byte.[6]

Blowfish is not attractive in dedicated hardware because of the large size of the tables involved. A simple implementation with sequential access to the four S-boxes would require about five clocks per round, while a version with parallel access could easily execute in two clocks per round, or four clocks per byte. The latter approach would almost certainly require on-chip RAM, making the cost quite high, and it is still considerably slower in hardware than a low-end DES hardware solution.

3.5 Khufu/Khafre

These algorithms, proposed by Ralph Merkle [Mer91a] and patented by Xerox [Mer91b], have some similarities to Blowfish, but each round does less, involving only a single S-box lookup, and thus is faster.

```
long  S[256];      /* Khufu: key dependent; Khafre: not */
long  A,B;                    /* values to be encrypted */

void Khufu(void)
    {
    int j;
    long tmp;
```

---

[6] CAST [AT93, Ada94] (designed by Carlisle Adams and patented by Northern Telecom [Ada96]) has a round function almost idential to Blowfish's. There are several variants of CAST; the current one seems to be a bit slower than Blowfish.

```
for (j=0;j<ROUNDS;j++)
    {
    tmp = B ^ S[A & 0xFF];
    B   = A <<< 8;
    A   = tmp;
    }
}
```

Because each round is so simple, there is no opportunity to pipeline the look-ups completely. Thus, each round consists of five instructions and requires five Pentium clocks and the speed is 20 clocks per byte for 32 rounds. However, this time could be halved to 2.5 clocks per round if two independent blocks are encrypted simultaneously, which makes sense for encrypting a stream of blocks in ECB or for interleaving two independent feedback streams: 10 clocks per byte. The Khafre algorithm is similar to Khufu, but the S-box is not key dependent, and the key is XORed into the encryption block after every eight rounds, thus slowing it down only slightly from Khufu. These algorithms can be implemented fairly simply in hardware at one clock per round.

3.6   IDEA and DES

As a point of comparison, it is useful to present a brief discussion of the other block algorithms, not designed for software.

IDEA [LMM91, ML91] uses a total of eight rounds, following by a final transformation. Each round includes six loads of 16-bit subkeys, four multiplies, four adds, and six XORs. On a Pentium, which has a dedicated hardware multiplier that can complete a multiply in 10 clocks (although the multiply instruction cannot pair with any other instruction), each round executes in about 50 clocks, so the entire algorithm requires more than 400 clocks per block, or over 50 clocks per byte. There are no address setup penalties, and any penalties due to lack of parallelism are swamped by the multiplication time. It should be noted that the Pentium Pro can perform integer multiplies at a throughput of one multiply per clock, but with a latency of four clock cycles for each result. Thus, IDEA should run considerably faster (perhaps by a factor of three or four) on a Pentium Pro than on a Pentium.

Because there are no sizable tables involved, IDEA in hardware might be appealing, but multipliers are not inexpensive. The speed of hardware would depend dramatically on how much cost could be absorbed. Given enough dedicated multipliers, each round could be implemented in only a few clocks, giving speeds comparable to DES.

Estimated performance numbers for DES [NBS77] will also be helpful as a point of reference. These performance numbers are derived from a rather conventional approach to DES software implementation, using eight S-box tables of 64 entries each, including an assumption that the key can be pre-processed to build tables that speed implementation. It is quite possible that a different approach, perhaps involving much larger tables, could speed implementation considerably.

The initial and final permutations do not appear to allow for much parallelism but should each execute in less than 35 clocks on a Pentium. Each round of the algorithm can be executed in about 18 clocks (this assumes you do two DES S-box lookups in a single 12-bit table), with very good pipeline utilization on a Pentium. Thus, the overall time for a DES block to be encrypted is about 360 Pentium clocks, leading to a throughput of about one byte every 45 clocks. Similar calcuations for triple-DES give a throughput of one 108 clocks per byte.

3.7   Comparison of Results

The following table summarizes the results of the previous sections, with all clock counts being given for a Pentium processor:

Algorithm	Type	Clocks/round	# rounds	Clocks/byte of output
RC4	Stream cipher	n.a.	n.a.	7
SEAL	Stream cipher	n.a.	n.a.	4
Blowfish	Block cipher	9	16	18
Khufu/Khafre	Block cipher	5	32	20
RC5	Block cipher	12	16	23
DES	Block cipher	18	16	45
IDEA	Block cipher	50	8	50
Triple-DES	Block Cipher	18	48	108

4   Proposed Directions for Further Investigation

Having dissected several well-known algorithms, it is hoped that the general design principles for software optimization are now fairly clear. Up to this point, however, only minor changes to the structure of existing algorithms have been considered. This section presents a few basic structures for consideration in future algorithm design. In a sense, the proposed idea is to use "RISC" encryption: lots of rounds, each designed to execute very efficiently on a high-end CPU. Each round consists of simple transformations that are extremely efficient on today's popular CPU architectures: add, subtract, XOR, rotate by a constant amount, and table lookup (i.e., S-boxes). None of the particular algorithms outlined here has been seriously cryptanalyzed, but it is hoped that these proposals can serve as a springboard for further investigation and research.

Below is sample source code for a 128-bit block encryption algorithm called "Test 1." In each round, one of the four 32-bit words is updated, including a round-dependent rotation constant (which may also be key-dependent) and an S-box lookup using bits from another of the words. The rotation constants are chosen within carefully designed constraints, as shown in the comments, to maximize diffusion of bits. For example, in the first round, note that $X3$ is merged into $X0$ both before and after the rotation. With properly selected rotation constants, each bit of $x3$ will affect each bit of $X0$ by the end of round 5, even ignoring the S-box lookups, and thus every bit of all three other words after seven rounds. In general, this "exponential" bit diffusion propagates a single bit change into 32 positions in five rounds, with the S-box lookups providing non-linearity. Observe that, for each set of five rounds, there are 1024 possible sets of rotation constants. After each ten rounds, a few rounds of a substantially different round function are inserted to inject some irregularity into the algorithm. The use of variables in the main round function is carefully chosen to allow full-speed encryption and decryption without pipeline stalls.

Each round can be performed in 3 Pentium clocks, with the intervening rounds each requiring 8 clocks each. With thirty rounds plus two intervening rounds, the entire block can be encrypted in about 9 clocks per byte. If the rotation constants are key-dependent, achieving this speed on a Pentium would require "compiling" a version of the code for each key, which can be easily accomplished as part of key initialization schedule.

```
                    /*   2a+1,4b+2,8c+4,16d+8,16              */
                    /*     ( --> exponential bit diffusion)   */
    #define int ROTCNT[30]= {      1,   2,   4,   8, 16,
                                   17,  26,  20,  24, 16,
                                   11,   6,  12,   8, 16,
                                   19,  14,  24,   8, 16,
                                    5,  22,  12,  24, 16,
                                   29,  18,  20,  24, 16
                            };
    /*   randomly generated from keys.  Every five rounds  */
    /*   has 10 bits of freedom in ROTCNT[] values.        */


    long S[1024];   /*   randomly generated from keys  */
    long P[8];      /*   randomly generated from keys  */
    long R[8];      /*   randomly generated from keys  */
                    /*              (five bits each)   */

    void Test1Encrypt(long *X0,long *X1,long *X2,long *X3)
    {
        /*   could start out with some input whitening here  */

        /*   ten initial rounds  */
        X0 = ((X0 + X3) <<< ROTCNT[ 0]) ^ (S[X2 & MASK] + X3);
        X1 = ((X1 + X0) <<< ROTCNT[ 1]) ^ (S[X3 & MASK] - X0);
        X2 = ((X2 - X1) <<< ROTCNT[ 2]) ^ (S[X0 & MASK] + X1);
        X3 = ((X3 - X2) <<< ROTCNT[ 3]) ^ (S[X1 & MASK] - X2);
```

```
X0 = ((X0 + X3) <<< ROTCNT[ 4]) ^ (S[X2 & MASK] + X3);
X1 = ((X1 + X0) <<< ROTCNT[ 5]) ^ (S[X3 & MASK] - X0);
X2 = ((X2 - X1) <<< ROTCNT[ 6]) ^ (S[X0 & MASK] + X1);
X3 = ((X3 - X2) <<< ROTCNT[ 7]) ^ (S[X1 & MASK] - X2);

X0 = ((X0 + X3) <<< ROTCNT[ 8]) ^ (S[X2 & MASK] + X3);
X1 = ((X1 + X0) <<< ROTCNT[ 9]) ^ (S[X3 & MASK] - X0);

/* pause after ten rounds to do something different  */
X0 =  (X0 << R[0]) ^ P[0];
X1 = ((X1 << R[1]) ^ P[1]) + X0;
X2 = ((X2 << R[2]) ^ P[2]) + X1;
X3 = ((X3 << R[3]) ^ P[3]) + X2;

/*  ten middle rounds  */
X2 = ((X2 - X1) <<< ROTCNT[10]) ^ (S[X0 & MASK] + X1);
X3 = ((X3 - X2) <<< ROTCNT[11]) ^ (S[X1 & MASK] - X2);

X0 = ((X0 + X3) <<< ROTCNT[12]) ^ (S[X2 & MASK] + X3);
X1 = ((X1 + X0) <<< ROTCNT[13]) ^ (S[X3 & MASK] - X0);
X2 = ((X2 - X1) <<< ROTCNT[14]) ^ (S[X0 & MASK] + X1);
X3 = ((X3 - X2) <<< ROTCNT[15]) ^ (S[X1 & MASK] - X2);

X0 = ((X0 + X3) <<< ROTCNT[16]) ^ (S[X2 & MASK] + X3);
X1 = ((X1 + X0) <<< ROTCNT[17]) ^ (S[X3 & MASK] - X0);
X2 = ((X2 - X1) <<< ROTCNT[18]) ^ (S[X0 & MASK] + X1);
X3 = ((X3 - X2) <<< ROTCNT[19]) ^ (S[X1 & MASK] - X2);

/* pause after ten rounds to do something different  */
X0 =  (X0 << R[4]) ^ P[4];
X1 = ((X1 << R[5]) ^ P[5]) + X0;
X2 = ((X2 << R[6]) ^ P[6]) + X1;
X3 = ((X3 << R[7]) ^ P[7]) + X2;

/*  ten final rounds  */
X0 = ((X0 + X3) <<< ROTCNT[20]) ^ (S[X2 & MASK] + X3);
X1 = ((X1 + X0) <<< ROTCNT[21]) ^ (S[X3 & MASK] - X0);
X2 = ((X2 - X1) <<< ROTCNT[22]) ^ (S[X0 & MASK] + X1);
X3 = ((X3 - X2) <<< ROTCNT[23]) ^ (S[X1 & MASK] - X2);

X0 = ((X0 + X3) <<< ROTCNT[24]) ^ (S[X2 & MASK] + X3);
X1 = ((X1 + X0) <<< ROTCNT[25]) ^ (S[X3 & MASK] - X0);
X2 = ((X2 - X1) <<< ROTCNT[26]) ^ (S[X0 & MASK] + X1);
X3 = ((X3 - X2) <<< ROTCNT[27]) ^ (S[X1 & MASK] - X2);
```

```
X0 = ((X0 + X3) <<< ROTCNT[28]) ^ (S[X2 & MASK] + X3);
X1 = ((X1 + X0) <<< ROTCNT[29]) ^ (S[X3 & MASK] - X0);
}
```

Below is a stream cipher called "Test2." Again, the cryptographically appropriate number of rounds and the size of the S-box are not known. Each round consists of a fixed rotate, a table lookup, and addition, and an XOR. The indices into the S-box table ($r[0]$ and $r[1]$) are continually updated, and use of the two values alternates between rounds to avoid address setup penalties. In each round, two of the 32-bit variables are used to update the third variable, and then a three-way "swap" occurs. By unrolling this loop appropriately, no instructions are ever executed to effect the swap. After each set of ROUNDS inner loops, a single 32-bit quantity is output. Since only partial information on the internal variables is output each time (i.e., 32 out of 96 bits), it is assumed that a much smaller value of ROUNDS is acceptable, perhaps in the range of 2-6. Again, the inner loop can execute in only four clocks per round on a Pentium; thus, with ROUNDS set to three, the algorithm can output a byte in less than every three clocks, more than double the speed of RC4.[7]

```
long S[SBOX_SIZE];          /* initialized based on key */
long X,Y,Z;                 /* initialized based on key */

void Test2Encrypt (long *p, long cnt)
    {
    long i,j,tmp;
    int r[2];
    r[0]=r[1]=0; /* could initialize based on keys also */

    for (i=0;i<cnt;i++,p++)
        {
        for (j=0;j<ROUNDS;j++)
            {
            r[j & 1] = (r[j & 1] + X) & (SBOX_SIZE-1);
            tmp = ((Z <<< ROTCNT) + Y) ^ S[r[j&1]];
            Z = X;
            X = Y;
            Y = tmp;
            }
        r[0] ^= r[1];
        *p ^= (X+S[r[0]]);                      /* output */
        S[r[0]] = Y;              /* update S box */
        }
    }
```

---

[7] If the Pentium behaved as documented, this would be faster by one clock per round.

Clearly there are many possible variations on these themes, including varying the rotation amount as a function of the round number, changing which arithmetic and logical operations are performed and in which order, etc.

5 Conclusion

The nature of encryption algorithms is that, once any significant amount of security analysis is done, it is very undesirable to change the algorithm for performance reasons, thereby invalidating the results of the analysis. Thus, it is imperative to consider both security and performance together during the design phase. While it is impossible to take all future computer architectures into consideration, an understanding of general optimization guidelines, combined with exploratory software implementation on existing architectures to calibrate performance, should help achieve higher speed in future encryption algorithms. The authors believe that it is possible to develop secure stream ciphers that encrypt data at two clocks per byte, and secure block ciphers that encrypt data at ten clocks per byte. Certainly security is most important when designing an encryption algorithm, but don't go out of your way to make it inefficient.

References

[Ada94]   C.M. Adams, "Simple and Effective Key Scheduling for Symmetric Ciphers," *Workshop on Selected Areas in Cryptography—Workshop Record*, Kingston, Ontario, 5–6 May 1994, pp. 129–133.

[Ada96]   C.M. Adams, "Symmetric cryptographic system for data encryption," U.S. patent 5,511,123, 23 Apr 1996.

[AT93]   C.M. Adams and S.E. Tavares, "Designing S-Boxes for Ciphers Resistant to Differential Cryptanalysis," *Proceedings of the 3rd Symposium on State and Progress of Research in Cryptography*, Rome, Italy, 15–16 Feb 1993, pp. 181–190.

[BGV96]   A. Bosselaers, R. Govaerts, and J. Vandewalle, "Fast Hashing on the Pentium," *Advances in Cryptology—CRYPTO '96*, Springer-Verlag, 1996, pp. 298–312.

[CR95]   D. Coppersmith and P Rogaway, "Software-efficient pseudorandom function and the use thereof for encryption," U.S. patent 5,454,039, 26 Sep 1995.

[Koc96]   P. Kocher, "Timing Attacks on Implementations of Diffie-Hellman, RSA, DSS, and Other Systems," *Advances in Cryptology—-CRYPTO '96*, Springer-Verlag, 1996, pp. 104–113.

[LMM91]   X. Lai, J. Massey, and S. Murphy, "Markov Ciphers and Differential Cryptanalysis," *Advances in Cryptology—CRYPTO '91*, Springer-Verlag, 1991, pp. 17–38.

[Mad84]   W.E. Madryga, "A High Performance Encryption Algorithm," *Computer Security: A Global Challenge*, Elsevier Science Publishers, 1984, pp. 557–570.

[ML91]   J.L. Massey and X. Lai, "Device for Converting a Digital Block and the Use Thereof," International Patent PCT/CH91/00117, 28 Nov 1991.

[Mer91a]   R. Merkle, "A Fast Software Encryption Function," *Advances in Cryptology—CRYPTO '90 Proceedings*, Springer-Verlag, 1991, pp. 476–501.

[Mer91b]   R. Merkle, "Method and apparatus for data encryption," U.S. patent 5,003,597, 26 Mar 1991.

[NBS77]   National Bureau of Standards, NBS FIPS PUB 46, "Data Encryption Standard," National Bureau of Standards, U.S. Department of Commerce, Jan 1977.

[Riv95]   R.L. Rivest, "The RC5 Encryption Algorithm," *Fast Software Encryption, Second International Workshop Proceedings*, Springer-Verlag, 1995, pp. 86–96.

[RC94]   P. Rogaway and D. Coppersmith, "A Software-Optimized Encryption Algorithm," *Fast Software Encryption, Cambridge Security Workshop Proceedings*, Springer-Verlag, 1994, pp. 56–63.

[Sch94]   B. Schneier, "Description of a New Variable-Length Key, 64-Bit Block Cipher (Blowfish)," *Fast Software Encryption, Cambridge Security Workshop Proceedings*, Springer-Verlag, 1994, pp. 191–204.

[Sch96]   B. Schneier, *Applied Crytography, 2nd Edition*, John Wiley & Sons, 1996.

A Fast New DES Implementation in Software

Eli Biham

Computer Science Department
Technion – Israel Institute of Technology
Haifa 32000, Israel
Email: biham@cs.technion.ac.il
WWW: http://www.cs.technion.ac.il/~biham/

Abstract. In this paper we describe a fast new DES implementation. This implementation is about five times faster than the fastest known DES implementation on a (64-bit) Alpha computer, and about three times faster than than our new optimized DES implementation on 64-bit computers. This implementation uses a non-standard representation, and view the processor as a SIMD computer, i.e., as 64 parallel one-bit processors computing the same instruction. We also discuss the application of this implementation to other ciphers. We describe a new optimized standard implementation of DES on 64-bit processors, which is about twice faster than the fastest known standard DES implementation on the same processor. Our implementations can also be used for fast exhaustive search in software, which can find a key in only a few days or a few weeks on existing parallel computers and computer networks.

1 Introduction

In this paper we describe a new implementation of DES[4], which can be very efficiently executed in software. This implementation is best used with a non-standard order of the bits of the DES blocks. This implementation does not suffer from high overhead of computing permutations of bits. Instead, we view a processor with (for example) 64-bit words, as a SIMD parallel computer which can compute 64 one-bit operations simultaneously, while the 64-bits of each block are set in 64 different words (of which the first bit is always of the first block, the second bit belongs to the second block, etc.).

The operations that DES uses are as follows: The XOR operation: in our view the XOR operation of the processor computes 64 one-bit XORs. The expansion and permutation operations: these operations do not cost any operation, since instead of changing the order of words (or duplicating words), we can address the required word directly. We remain with the S boxes. Usual implementations of S boxes use table lookups. However, in our representation, table lookups are very inefficient, since we have to collect six bits, each bit from a different word,

Cipher	Speed
DES (Eric Young's libdes)	28
Gost	8*
SAFER	22*
Blowfish	34*
Our DES Implementation	46
Our DES Implementation – triple DES	22
Our fastest DES	137
Our fastest DES – Triple DES	46

* Estimation, based on [9].

Table 1. The speeds of our implementations and of various ciphers on a 300MHz Alpha 8400 processor (in Mbps).

combine them into one index to the table, and after the table lookup take the four resultant bits and put each of them in a different word.

We observed that there is a much faster implementation of the S boxes in our representation: they can be represented by their logical gate circuit. In such an implementation each S box is typically represented by about 100 gates, and thus we can implement an S box by about 100 instructions.

We actually view the whole cipher by its gate circuit, and apply it in software. In this implementation we actually compute the circuit 64 times in parallel (as the size of the processor word), and thus can gain a high speedup even though we use very simple operations. In average, on 64-bit processors, each S box costs about 1.5 instructions for each encrypted block, while each instruction takes only one clock cycle.

The full circuit of DES contains about 16000 gates (including the key scheduling, which costs nothing), and thus we can compute DES 64 times in about 16000 instructions on 64-bit processors. In average we result with about 260 instructions for the encryption of each DES block. Conversion from and to the standard block representation takes (together) about 40 instructions per block, and thus encryption of standard representations with our implementation takes about 300 instructions. For comparison, our fast standard implementation of DES, described in this paper, requires about 634 instructions for each block.

Table 1 summarizes the speeds of our implementations, a standard fast DES implementation (Eric Young's libdes), and of various fast ciphers.

The same idea can be applied to other ciphers. Our implementation of these ciphers is efficient especially when the cipher does not use all the power of the machine instructions (i.e., when each instruction mixes only a few of the bits, such as S boxes or eight-bit additions on 32-bit processors), and when the word size of the processor is large (such as 64 bits, when the cipher use shorter registers). For example, our implementation of Feal[11] is expected to be about 2.5–5 times faster than direct implementations. Both variants of Lucifer[1,12] and GOST[10] can also be applied very efficiently using this implementation. Our implementation of ciphers which use more complex operations (such as multiplication, or large S boxes) requires more instructions to simulate the complex operations, and is thus less efficient.

In Section 3 we describe an optimized standard implementation on 64-bit computers. It uses the 64-bit registers of a 64-bit processor, and runs almost twice faster than the fastest implementation (designed for 32-bit architectures) on the same processor. It even runs faster than fast ciphers such as GOST[10], SAFER[2], and Blowfish[10]. The speed is gained by using the long 64-bit registers effectively — by all other means this is a standard implementation. We suggest a new DES-like cipher, to which we call *WDES*, based on the structure of this fast implementation, but is about 2.5 times faster.

In Section 4 we discuss using these fast implementations for exhaustive search, and conclude that it is applicable even today using existing general purpose parallel computers and computer networks.

2   The New Non-Standard DES Implementation

This implementation uses a non-standard representation of the data in software, and in particular it does not have any table lookup. Instead of encrypting many 64-bit words, one at a time, we encrypt simultaneously 64 words, and each operation encrypts one bit in each of the 64 words.

Actually, we view a 64-bit processor as a SIMD computer with 64 one-bit processors. This implementation simulates a fast DES hardware whose number of gates is minimal, and computes each gate by a single instruction. In particular, the S boxes are computed by their gate-circuit, using the XOR, AND, OR, and NOT operations, and the permutations and expansions do not require any instruction, since they can be viewed as only changing the naming of the registers. Although the S boxes are implemented in more instructions than in usual implementations, the parallelism of this implementation speeds up the implementation much more than the S box implementation reduces it. Moreover, some of the operations can be optimized out in some cases, such as if some parts of the S boxes are similar (same or complement).

We represent the S boxes by their gate circuit using the best-known XOR, AND, OR and NOT operations, optimized to reduce the total number of gates. Although the problem of finding the best such circuit is still open, we found the following optimization which requires at most 132 gates per DES S box, and only 100 gates in average. In the description we denote the six input bits by

	Instructions
Expansion	0
Key mixing	48
P	0
XOR with the left half	32
S boxes	$8 \cdot 100 = 800$ (in average)
load+store	$8 \cdot (6 + 6\text{load} + 4\text{load} + 4\text{store}) = 160$
Total per round:	1040

Table 2. The number of instructions in each round on Alpha.

	Total	Average per Block
IP,FP	0	0
16 rounds:	$16 \cdot 1040 = 16640$	260
		4 gates per bit
Conversion of representation	2500	40

Table 3. The number of instructions in DES on Alpha.

$abcdef$. We compute all the 16 functions of d and e into 14 registers (excluding the constant 0 or constant 1). It requires two NOTs ($\bar{d}, \bar{e}$) and 10 additional operations ($0, 1, d, e, \bar{d}, \bar{e}$ are already known). This computation is done only once for each S box. For each output bit of the S box we compute the result using these functions. We use six operations for each line of the S box and six operations to combine the results, together 30 operations for each output bit. In total we use at most $12 + 4 \cdot 30 = 132$ gates for each S box, but in average we need only about 100 gates per S box. Each combination of four values (the four values of b, c or the four values of a, f, e.g., combining the quarters of each of the four lines $((b = c = 0), (b = 0, c = 1), (b = 1, c = 0), (b = c = 1))$, or combining the four lines) are combined by (assuming the first case):

$$\left(\underline{f_{00}} \oplus c \cdot (f_{00} \oplus f_{01}) \right) \oplus b \cdot \left((\underline{f_{00}} \oplus f_{10}) \oplus c \cdot (f_{00} \oplus f_{01} \oplus f_{10} \oplus f_{11}) \right),$$

where the underlined values are known constants, and $f_{bc} = S(abcdef)$, where d, e are the actual values of the input (f_{bc} is one of the 16 values kept in registers above), and a, f are the values assumed for a, f, to be instantiated in the next step. More accurately, in the intermediate steps we compute the combinations of S box entries as suggested by the above equation (e.g., $f_{00}, f_{00} \oplus f_{01}, f_{00} \oplus f_{10}, f_{00} \oplus f_{01} \oplus f_{10} \oplus f_{11}$), rather than the various values of the entries themselves.

Tables 2 and 3 describe the maximum number of gates per round and for the full DES. Therefore, we expect the speed to be about $300 \cdot 2^{20}/4 = 75\text{Mbps}$ on

300MHz Alpha processors. In practice, we achieve speeds of about 137Mbps, since the processor (EV5) has two independent integer instruction units.[1]

Conversion between the standard and the non-standard representations can also be done in about 1250 instructions. Doing this twice, before and after encryption, takes about 2500 instructions, which are about 40 instructions for each encrypted block.

This implementation can actually be applied to any cipher, but the efficiency of the implementation depends on many factors, such as the efficiency of the original cipher, the word size of the processor, and the complexity of the operations that the cipher uses. The implementation is especially attractive to ciphers whose operations are simple (no multiplication for example), use only small S boxes (thus their gate complexity is small), or use small register sizes (thus cannot use the full power of modern processors). Examples of such ciphers are Lucifer[1,12], GOST[10] and Feal[11].

In the case of Feal, standard implementations require about 22 instructions for each application of the round function (4 loads, 2 load + 2 XORs for key mixing, 2 for XOR, 2 additions (S_0), 2 additions with carry (S_1; each might take two operations), 4 rotations and 4 XORS to mix with the left half of the data). The right-round cipher takes thus about $8 \cdot 22 = 176$ instructions (not counting the initial and final key mixing which can take a few additional instructions).

Our implementation requires 34 or 35 instructions for an eight-bit addition (one or two for the LSB, depends whether this is S_0 or S_1, 1 for the carry and 2 for the second bit. We need three additional instructions for computing each additional carry and two XORs for each additional bit: In total we need $1 + (1 + 2) + 6 \cdot (2 + .3) = 34$ instructions for S_0 and 35 for S_1). In total the F function requires $16 + 16 + 16 + 35 + 34 + 35 + 34 + 32 + 32 + 32 + 8 + 8 = 298$ instructions (16 XORs, 16 key loads+mixings, four S boxes, 32 loads, 32 stores, 32 mixings with the left half, and 8+8 extra loads+stores if necessary). The eight round Feal can then be implemented in $8 \cdot 298 + 64 + 64 = 2512$ instructions (64 for each of the initial and final key mixings). In average we get that only about $2512/64 = 39$ instructions per block, which is more than four times faster than standard implementations. Even if we do the conversion from/to the standard representation (which costs 40 instructions per block), our implementation takes only about $39 + 40 = 79$ instructions, which is more than twice faster than the standard implementations.

Both variants of Lucifer[1,12] and GOST[10] can also be applied very efficiently using this implementation.

---

[1] The next generation of this processor, EV6, will have four independent integer instruction units, thus we expect that with the same clock frequency the speed of our implementation will be about 250Mbps.

This implementation can be used for fast encryption and decryption, using the same key in all the 64 encryptions (i.e., the key words contain only 0 or -1), or for exhaustive search using the same plaintexts but different keys. We can also use different plaintexts with different keys, if it is of an advantage to the application.

This implementation can be used in three ways:

1. Encryption/decryption in standard representations, compatible to other DES implementations.
2. Encryption/decryption of large blocks, such as of disk clusters, or large communication packets. In this case, it is not important to use the standard representation, and thus our implementation is even faster, since conversion should not be done.
3. Application to exhaustive search.

It is easy to see that applications of this implementation in the ECB mode is very fast, but as usual in ECB modes, it suffers from many disadvantages. It would be preferable to use standard CBC, CFB and OFB modes with this implementation, but this is impossible due to their sequential order. However, it is possible to use this implementation for standard CBC decryption, since the whole data can be decrypted in parallel, and then each result can be mixed with the previous ciphertext. It is also possible to apply CFB decryption in a similar way. Therefore, this implementation can be used for fast decryption in standard modes, even when encryption is done by usual standard implementations.

64 parallel CBC encryption modes can be applied in this implementation by choosing 64 initial values for the 64 block encrypted simultaneously, and apply CBC on the full $64^2 = 4096$-bit blocks. In this case we can also encrypt under a different key in each of the 64 parallel CBC modes — it might be especially attractive when a server has to encrypt data to many clients in parallel.

This implementation is even faster when conversion from/to standard representation is not applied. In this case, DES is applied, but with a non-standard order of the plaintext/ciphertext bits. To protect against multiple occurrence of the same plaintexts (actually the 64 bits that enter one real DES in the non-standard representation) we should use new modes.

The ECB mode of this implementation takes the 4096 bits of the data, and encrypts them as is. A CBC-like mode can have an initial value of 4096 bits (which can be derived from a 64-bit value), and apply CBC on the 4096-bit cipher. This mode actually applies 64 standard CBC modes in parallel, one for each of the DES applications in the non-standard representation. An improvement of this mode can mix the bits of each register, for example by rotating register i (containing the i'th bits of the standard blocks) by i bits after adding i to the value of the register. A CFB-like and OFB-like modes can be designed in a similar way.

3   A Fast Standard DES Implementation on 64-bit Processors

DES can be applied very efficiently on 64-bit processors. Unlike on 32-bit processors, on 64-bit processors, the right half expanded to 48 bits can be stored in one word. Moreover, by substituting every group of six bits entering into the S boxes in a separate byte, we can directly access the S box table by referencing via a single byte.

We apply the initial and final permutations by lookup tables from each byte to 64-bits, and XORing the results of the various table lookups.

We apply each round by XORing the right half (represented as eight bytes, in each six bits are used) by a subkey (represented in the same way). Then, eight table lookups apply the eight S boxes, and the results are XORed. Each S box already includes the P permutation and the E expansion in its 64-bit result. Note that due to this representation, several (duplicated) bits of the two halves should be omitted by the final permutation.

Tables 4 and 5 describe the number of operations required by this implementation, with the number of instructions on an Alpha processor. We implemented this code in C on a 300MHz Alpha and got encryption speed of 46Mbps. Triple DES runs at 22Mbps (since some IP, FP's can be discarded). On the same processor, Eric Young's libdes (single-DES) runs at 28Mbps.

Some comments on this implementation:

1. The eight S boxes are applied in parallel, and thus pipelining can use it without pipeline stalls. In other ciphers and hash functions, like Feal[11], Khufu[3], Khafre[3], and MD4[7], MD5[8], SHA-1[5,6], each operation depends on the output of the previous operation, and thus might result with pipeline stalls, especially on newer or future processors which can compute several instructions simultaneously.
2. All the tables and the variables take together about 4Kbytes, and enter easily into the cache.
3. Still in DES the input of the next round depends on the output of the preceding one. Although in practice this does not slow the execution, we have another solution. In DES, the input of each S box depends only on the output of only six S boxes in the previous round. Thus, the code can be optimized to start computing the next round while still computing the preceding one.

	Operations	Number of Instructions
key XOR	1 load+XOR	2
EPS	8 table lookups	$8 \cdot 3 = 24$ (extbl, add, lookup)
XORing L with the S boxes	8 XORs	8
Total		34

Table 4. The number of instructions in each round on Alpha.

	Operations	Number of Instructions
IP	5 times (3 XORs, 2 shifts, 1 AND)	$5 \cdot 6 = 30$
E	Initial Expansion	26
16 rounds	each 34 instructions	$16 \cdot 34 = 544$
	Removal of expansion	4
FP	Final permutation	30
Total		634

Table 5. The number of instructions in DES on Alpha.

This can speed up implementations on pipelined processors, where we can compute several instances in parallel.

4. Unlike some (although not all) DES implementations, we implement each S box as one table lookup, rather than combining pairs of S boxes into one lookup. The latter is more than twice slower, since the tables become larger than the size of the on-chip cache.[2]

3.1 WDES

We can use this fast code to design a new, even faster, and more secure cipher, to which we call *WDES*. We convert the code by removing IP, FP, and changing the EPS operations (S boxes followed by P followed by E, as used in this implementation) into S boxes from 8 bits to 64 bits. These S boxes can be much better than the original, since each S box affects *all* the bits of *all* the S boxes in the next round (rather than one bit in only six S boxes).

WDES has 128-bit blocks, and it runs much faster than DES, with the same number of rounds (since the blocksize is larger, and the slow initial and final permutations are discarded): its speed is 106Mbps on the same processor as in Table 1.

---

[2] On Pentium, however, the latter is twice faster using the same C code.

4   Exhaustive Search on Powerful Computers and Networks

In this section we study the possibilities of exhaustive search on several kinds of machines and networks. We assume using the fast implementation described in the previous section.

Note that results similar to the ones described here hold also for breaking UNIX passwords, which are chosen from up to eight printable characters. In this case the password space has 96^8 passwords, while each password trial requires 25 encryptions (the salt should not be taken into account, since it is known to the attacker, and the encryption code can be justified to the specific value of the salt). Therefore, about $25 \cdot 96^8 \approx 2^{57}$ passwords should be tried, or about 2^{56} in average.

4.1   Special Purpose Computers

We can build a special purpose computer with very long registers, without the expensive operations (such as multiplication and floating point operations), and only with simple instructions, such as XOR, AND, OR, NOT. Assume that in a Pentium processor we remove the expensive operations, and use the extra chip space to increase the size of the registers to 1000 bits. Then, we need only 150 processors to search the keys exhaustively in one year in average (or six months in average using the attack based on the complementation property).

It is possible theoretically to build a machine with million-bit registers. Unexpectedly, we now know that such a machine was actually built with the support of the NSA: Cray Computers had announced in March 1995 about such a computer that can apply 2^{45} bit operations every second on a million one-bit processors (see Figure 1). This computer can compute 2^{45} bit-operations every second, and thus can compute about $2^{45}/16000 = 2^{45}/2^{14} = 2^{31}$ DES encryptions every second. Therefore, we can apply the searches on this machine with the following results:

Search of	Time	Notes
40 bits	512 sec=8.5 min, 4.25 min in av.	Exportable ciphers
43 bits	4096 sec=an hour, 1/2 an hour in av.	Linear Cryptanalysis
47 bits	2^{16} sec=a day, 12 hours in av.	Differential Cryptanalysis
56 bits	2^{25} sec=a year, 1/2 an year in av.	Full key search

Cray Computers has bankrupted, since nobody had bought this computer. Probably the NSA had a faster machine.

4.2   General Purpose Parallel Computers

It is known that Sandia National Labs has a parallel computer of 9000 Pentium-Pro 200MHz processors, each has two independent integer instruction units. This parallel computer can compute about $9000 \cdot 2 \cdot 200 \cdot 2^{20} \cdot 32 = 2^{47}$ bit operations every second. Thus, it can compute about $2^{47}/16000 = 2^{47}/2^{14} = 2^{33}$ DES encryptions in each second. Therefore, we can apply the searches on this machine with the following results:

Search of	Time	Notes
40 bits	128 sec=2 min, 1 min in av.	Exportable ciphers
43 bits	1024 sec=17 min, 8 min in av.	Linear Cryptanalysis
47 bits	2^{14} sec=5 hours, 2 hours in av.	Differential Cryptanalysis
56 bits	2^{23} sec=3 months, 7 weeks in av.	Full key search

When we apply the attack using the complementation property, exhaustive search of the full key space takes in average only about three weeks.

4.3   Internet and the DES Worm

We can use the Internet for our exhaustive search, just as RSA factorization teams are doing, and as recently attacks against the RSA-DES/RC5 challenge are applied. Assume that an average computer on the Internet is a single 32-bit 133MHz RISC processor (this averages slower and faster processors, as well as processors with several integer instruction units). Such a processor can encrypt about 2^{18} blocks every second. Therefore,

- Searching 40 bits takes about $2^{40}/2^{18}/2 = 2^{21}$ seconds in average, which are about three weeks on a single processor. 1000 computers can do it in about half an hours.
- Searching 43 bits takes about six months in average. 1000 computers can do it in six hours.
- Searching 47 bits takes about 8 years in average on a single processor. 1000 computers can do it in four days, and 4000 computers can do it in one day.
- Searching all the 56 bits takes about 4000 years in average on a single processor. 4000 computers can do it in a year (or in six months using the complementation property). It is practical to have this number of computers participating legally over the Internet: this is about the same number of computers as the RSA factorizations use. Million computers can do it in two days in average (or in one day using the complementation property).

At this point it is possible in practice to achieve participation of several thousands computers legally over the Internet. However, it is simpler, and faster to do it illegally[3]. A worm, for which we call the *DES worm*, can break into many computers over the Internet, and use their idle cycles for exhaustive search. The worm verifies that only one copy of it is executed on each computer (of course on computers with several processors it can execute several copies to increase performance). The DES worm makes sure it cannot be easily noticed: it does not need much memory anyway, and it is executed at the lowest possible priority, so it does not disturb other applications on the same computer.

If the DES worm can get hold of about a million computers over the Internet, and assuming that it get at least half of their cycles (people are usually not working over nights), the DES worm can find a key in four days in average (or in two days using the complementation property). Moreover, since most computers over the Internet are not used in weekends (which last over 60 hours from Friday evening to Monday morning), the DES worm can use all the cycles and find a key in one weekend.

4.4   In the Future

In five years the average computer on the Internet is expected to be much faster: We estimate that it will have a 500MHz 64-bit processor with two independent integer instruction units (this averages slower and faster processors, processors with many integer instruction units, and computers with several processors). Such a computer is expected to be about 16 times faster (in terms of bit-operations per second) than todays average computer. Thus, using a million computers, the DES worm is expected to finish the whole key search in about two hours.

5   Acknowledgements

We are grateful to Adi Shamir, Ross Anderson, Kevin McCurley and the referees for their various remarks and suggestions that improved the results and exposition of this paper. Some of this work has been done while the author was visiting the computer laboratory at the university of Cambridge, and in particular using their Alpha computer. This research was supported by the fund for the promotion of research at the Technion.

References

1. H. Feistel, *Cryptography and Data Security*, Scientific American, Vol. 228, No. 5, pp. 15–23, May 1973.

---

[3] The Author does not recommend to do it, but we should always be aware that such a threat exists.

2. James L. Massey, *SAFER-K64: A Byte Oriented Block Ciphering Algorithm*, proceedings of Fast Software Encryption, Cambridge, Lecture Notes in Computer Science, pp. 1–17, 1993.

3. Ralph C. Merkle, *Fast Software Encryption Functions*, Lecture Notes in Computer Science, Advances in Cryptology, proceedings of CRYPTO'90, pp. 476–501, 1990.

4. National Bureau of Standards, *Data Encryption Standard*, U.S. Department of Commerce, FIPS pub. 46, January 1977.

5. National Institute of Standard Technology, *Secure Hash Standard*, U.S. Department of Commerce, FIPS pub. 180, May 1993.

6. National Institute of Standard Technology, *Secure Hash Standard*, U.S. Department of Commerce, FIPS pub. 180-1, April 1995.

7. Ronald L. Rivest, *The MD4 Message Digest Algorithm*, Lecture Notes in Computer Science, Advances in Cryptology, proceedings of CRYPTO'90, pp. 303-311, 1990.

8. Ronald L. Rivest, *The MD5 Message Digest Algorithm*, Internet Request for Comments, RFC 1321, April 1992.

9. Michael Roe, *Performence of Block Ciphers and Hash Functions – One Year Later*, proceedings of Fast Software Encryption, Leuven, Lecture Notes in Computer Science, pp. 359–362, 1994.

10. Bruce Schneier, *Applied Cryptography, Protocols, Algorithms, and Source Code in C*, second edition, John Willey & Sons, 1996.

11. Akihiro Shimizu, Shoji Miyaguchi, *Fast Data Encryption Algorithm FEAL*, Lecture Notes in Computer Science, Advances in Cryptology, proceedings of EUROCRYPT'87, pp. 267–278, 1987.

12. Arthur Sorkin, *Lucifer, a Cryptographic Algorithm*, Cryptologia, Vol. 8, No. 1, pp. 22–41, January 1984.

OTC 03/07 1942 CRAY COMPUTER CORP. COMPLETES INITIAL TESTING

COLORADO SPRINGS, Colo., March 7 /PRNewswire/ – Cray Computer Corp. (Nasdaq: CRAY) reported today the successful test and demonstration on March 2, 1995, of an array of 256,000 single bit processors packaged using the company's multi-chip-module technology. This array is a major technical component of the CRAY-3/Super Scalable System (CRAY-3/SSS) that is being **jointly developed by the company, the National Security Agency** and the Supercomputing Research Center (SRC) which was originally announced on August 17, 1994. This test and demonstration completes the first of a number of major tasks required under the Development Contract.

Researchers from the SRC verified correctness of operation of the 256, 000 single bit processor array (approximately 4,000 individual Integrated Circuits), which is the first half of a 512,000 singe bit processor array called for in the development contract. This array is coupled to a CRAY-3. The CRAY-3/SSS utilizes the Processor-In-Memory (PIM) chips, developed by the SRC. Both **NSA** and SRC are **providing significant technical assistance** in both the software and hardware aspects of the system.

Once completed, the high performance system will consist of a dual processor 2,048 million byte CRAY-3 and a 512,000 single bit processor Single Instruction Multiple Data (SIMD) array with a 128 million byte memory. This CRAY-3/Super Scalable System will provide high-performance vector parallel processing, scalable parallel processing and the combination of both in a hybrid mode featuring extremely high bandwidth between the PIM processor array and the CRAY-3. The current schedule for completion of the Development Contract is the end of July 1995 including a 90 day public Internet access demonstration.

For suitable applications, a SIMD processor array of 1 million processors would provide up to 32 Trillion Bit Operations per Second and price/performance unavailable today on any other high-performance platform. The CRAY-3 system with the SSS option will be offered as an application specific product. The joint development contract is part of the Federal Government's High Performance Computing and Communications program.

Charles Breckenridge, executive vice president of Marketing at Cray Computer Corp. said, "The CRAY-3/SSS will provide unparalleled performance for many promising applications. We are pleased to participate in this **transfer of government technology**, and we are eager to help potential customers explore and develop appropriate applications."

Cray Computer Corp. is engaged in the design, development, manufacture and marketing of the CRAY-3, CRAY-3/SSS and CRAY-4 high- performance computer systems.

CONTACT: Charles Breckenridge, executive VP of Marketing, or Terry Willkom, president, of Cray Computer, 719-579-6464; or David Gould of Chip Shots Inc., 408-541-8706

Fig. 1. Cray Computer Corp. press release of March 1995.

Optimizing a Fast Stream Cipher for VLIW, SIMD, and Superscalar Processors

Craig S.K. Clapp

PictureTel Corporation, 100 Minuteman Rd., Andover, MA 01810, USA
email: craigc@pictel.com

Abstract. The mismatch between traditional cipher designs and efficient operation on modern Very Long Instruction Word, Single Instruction Multiple Data, superscalar, and deeply pipelined processors is explored. Guidelines are developed for efficiently exploiting the instruction-level parallelism of these processor architectures.

Two stream ciphers, WAKE-ROFB and WiderWake, incorporating these ideas are proposed. WAKE-ROFB inherits the security characteristics of WAKE, from which it is derived, but runs almost three times as fast as WAKE on a commercially available VLIW CPU. Throughput in excess of 40 MByte/s on a 100 MHz processor is demonstrated. Another derivative, WiderWake, whose security characteristics are not directly transferrable from WAKE runs in excess of 50 MByte/s on the same processor.

1   Introduction

Much of existing cipher design stems from an era when processors exhibited little or no instruction parallelism or concurrency. Given such processors, the route to achieving the ultimate in performance for software based encryption algorithms was to reduce to a minimum the total number of operations required to encrypt each symbol. Examples of algorithms that epitomize this strategy are RC4[1], SEAL[10], and WAKE[1].

The latest generation of processors gain much of their performance improvements by having deeper pipelines and a greater degree of available parallelism than their predecessors. This dictates additional design criteria for algorithms that are to fully benefit from these changes. Until now little attention has been paid to optimizing ciphers to run on such architectures.

We examine the performance limitations of some existing ciphers on the new CPU architectures and make suggestions for design practices to maximize speed. Some of these recommendations are contrary to practices currently favored in the design of ciphers for software execution.

Two related cipher families incorporating these ideas are presented as working examples of the potential for substantially increased throughput available on recently introduced processors.

---

[1] RC4 is a trademark of RSA Data Security Inc. Discussion of RC4 herein refers to the cipher described under that name in [11].

2   Exploiting Parallelism in Modern Processors

Instruction-level parallelism in modern processors is expressed in several forms:

- **Pipelining** - The execution pipe is split into several stages with registers in between. The clock rate can be increased because only a small amount of work is done in each stage of the pipe. It takes several machine cycles from the time an instruction enters the pipe until the time the result comes out, however, during that time additional instructions can be fed into the beginning of the pipe so long as they don't depend on results that are not yet available. Thus there is concurrency between instructions that are in different stages of the pipeline.

As well as being a technique used in single threaded processors, pipelining is typically used in the execution units of each the following architectures.

- **Superscalar** - Multiple execution units are implemented in one CPU. Several of them can accept instructions on each machine cycle. Assignment of instructions to the available execution units, and resolution of data dependencies between instruction streams, are tasks performed by the CPU at run-time.
- **Very Long Instruction Word (VLIW)** - Like superscalar, multiple execution units are implemented in one CPU. Several execution units can accept instructions on each machine cycle. Unlike superscalar however, VLIW instructions are assigned to specific execution units and data dependencies between instruction streams are resolved, at *compile-time*, i.e. the compiler generates assembly code that accounts for the CPU's pipeline delays in each path. So, two members of a VLIW CPU family having different pipeline delays may need different assembly code even if their instruction sets are identical.
- **Single Instruction, Multiple Data (SIMD)** - A single instruction stream is applied simultaneously to several data elements. This technique is especially favored for accelerating video and graphics processing. The Intel MMX instruction set recently added to the Intel Pentium line is one example.

Some authors have optimized their algorithms for 64-bit CPUs in their quest to extract more performance from recent processors, an example being the secure hash Tiger[2]. However, direct support for 64-bit arithmetic continues to be substantially limited to the niche market of high-end RISC CPUs.

Meanwhile, Intel-Architecture CPUs and other 32-bit processors for desktop and embedded applications are eagerly embracing a combination of superscalar, VLIW, and SIMD techniques. SIMD instructions have also been added to some of the high-end 64-bit RISC CPUs to accelerate video and graphics operations.

In SIMD architectures it is common for carry generation to take extra instructions, causing multiple precision arithmetic to be somewhat inefficient. For this reason we suggest that for widest applicability algorithms should avoid 64-bit arithmetic. $n \times 32$-bit SIMD compatibility is the preferred way of taking advantage of data paths wider than 32-bits.

A common assumption in the design of software-oriented algorithms is that table-look-ups are inexpensive operations provided that the table fits inside the processor's cache[1, 10, 12]. On modern deeply pipelined processors, memory accesses, even to the CPU's local cache, typically have a longer latency than simple arithmetic or logical operations. Also, in highly parallel CPUs it is rare to have as many concurrent ports into the data cache as the number of parallel execution paths. Consequently, table-look-ups on such processors can be much more expensive relative to arithmetic or logical operations than they would be on a non-pipelined single-threaded CPU. To mitigate this, algorithms should be designed so that other useful work can proceed in parallel with table-look-ups.

Bit-rotations within a word have found favor in a number of recent algorithms[6, 10], with some including *data-dependent* rotations[9]. Common SIMD instruction sets, MMX included, do not directly support rotations within each of the multiple data operands, so a rotation must be synthesized by merging the results of two simple shifts, at a cost of three instructions. On a SIMD machine, *data-dependent* shifts and rotates can be substantially more costly than their fixed-length counterparts since applying different shifts to each operand is commonly not supported.

So, for efficient SIMD operation, shifts are preferred over rotations, and fixed-length shifts and rotations are preferred over data-dependent ones.

VLIW and superscalar architectures do not necessarily show a penalty for rotations versus shifts or for data-dependent versus fixed, except that rotations may only be efficient for the native word length.

3   Parallelism in Existing Ciphers

A characteristic of many existing ciphers is that a symbol of plaintext undergoes numerous inherently sequential operations on its way to becoming ciphertext, and the next plaintext symbol cannot be processed until the preceding ciphertext symbol is known. Encryption using a block algorithm such as DES[7] in cipher-feedback (CFB) mode[4] is a classic example of this characteristic[2]. The DES rounds are inherently performed sequentially, and the level of parallelism within a DES round is small except for eight parallel S-box look-ups. Unfortunately, table-look-ups are one of the *least* parallelizable operations on a modern CPU.

Coarse-grained parallelism can be forced on a system by for instance interleaving several independent cipher streams, but for a *single* CPU to get more throughput this way the associated replicated instruction streams need to map efficiently to the *fine-grained* instruction-level parallelism of modern processors. As a minimum, interleaving requires replication of the state variables that differ with each instance of the cipher, such as chaining variables or initialization vector dependent look-up-tables. The enlarged amount of state necessary to execute multiple instances of a cipher concurrently may *reduce* performance if it causes the cache capacity to be exceeded, or if it causes variables that otherwise could be held in CPU registers to instead be accessed from memory.

---

[2] While *encryption* in CFB mode is inherently a serial process, *decryption* using a block cipher in CFB mode offers unlimited block-level (coarse-grained) parallelism. This parallelism arises from the same absence of feedback loops in the decryption computational flowgraph that accounts for this mode's finite error extension characteristic.

Table 1. Performance characteristics of various encryption algorithms

	RC4	SEAL	WAKE-CFB
Number of bits ciphered per iteration	8	128	32
Number of 32-bit operations per iteration	15	68	24
Number of 32-bit operations per byte	15	4.25	6.0
Number of 32-bit operations in critical path (Add/XOR/mask, table-look-up, other)	6 (2, 2, 2 stores)	35 (27, 8, 0)	17 (13, 4, 0)
Number of cycles in critical path	10	51	25
Normalized critical path	10 cycles/byte	3.2 cycles/byte	6.25 cycles/byte
Apparent parallelism (ops-per-iteration / ops in critical path)	2.5x	1.94x	1.41x
Critical-path efficiency (ops in critical path / cycles in critical path)	0.60	0.69	0.68
P-factor (Apparent parallelism x critical-path efficiency)	1.5	1.33	0.96
Benchmarked performance on 32-bit VLIW CPU	10.6 cycles/byte	3.5 cycles/byte	6.38 cycles/byte

To determine the upper limit on performance of a cipher on a suitably parallel processor architecture we attempt to identify the software *critical path* through the algorithm. This is the path through the algorithm from one output symbol to the next, that has the largest *weighted* instruction count, the weighting being the number of cycles of latency associated with each type of instruction.

For instance, on most processors the result of a simple operation like an addition or XOR can be used in the subsequent cycle - these instructions are said to have a one cycle latency. However, reading from memory, even when the data is in the CPU's local cache, will typically take several cycles. Data read from cache commonly suffers a two or three cycle latency on modern deeply pipelined processors, while one recent introduction has it is as high as five cycles. Table-look-ups are inherently memory references, since even if the table is so small that it could reside in the generous register space of some modern processors, it cannot be placed there because the instruction sets do not support indirect referencing of registers.

Table 1 compares the theoretical and benchmarked encryption performances of RC4, SEAL, and WAKE on a 32-bit CPU with a RISC-like instruction set. The critical paths assume a memory-read latency of three cycles. All other operations are assigned a latency of one cycle. The total number of operations per iteration includes reading a plaintext input buffer, applying the cipher, and writing the ciphertext to an output buffer. For purposes of comparison loop overhead has been ignored when counting operations since it is essentially common to all the algorithms and can generally be reduced to insignificant levels by sufficient loop unrolling.

The 'apparent parallelism' listed in the table is the ratio of the total number of operations per iteration of the cipher to the number of operations in the critical path. It gives an indication of the most execution units that the algorithm could take *some* advantage of.

Another metric - the critical-path efficiency, tells what fraction of the cycles in the critical path are actually used to issue an instruction. In a pipelined processor the unused cycles here represent an opportunity for performing additional operations without resorting to more execution units or slowing the computation. Ideally an algorithm will have enough parallelism to exploit these 'bubbles' in the critical-path pipeline and *still* make use of multiple execution units.

A gauge of the most execution units that can be *fully* exploited by an algorithm can be obtained by taking the ratio of the total number of operations per iteration of the cipher to the number of cycles in the critical path. This is the same as the apparent parallelism times the critical-path efficiency, and is referred to in the table as the P-factor (for parallelism-factor). Note that since the critical path is a function of both the algorithm *and* the associated instruction latencies, the values for these metrics will change if the processor under consideration has different latencies from those assumed here. The instruction latencies used for the table are those of the processor actually used for benchmarking.

In order for an algorithm to fully exploit a superscalar, VLIW, or SIMD processor the P-factor needs to be substantially greater than unity. Ideally it should be no less than the number of parallel execution paths available in the target CPUs.

For some recently introduced processors the number of parallel execution paths is in the range of four to eight. Clearly, with P-factors of less than two, the encryption algorithms examined here cannot efficiently exploit the resources of such processors.

4   Design Strategy for a New Cipher

The strategy for developing a new cipher was to attempt to apply these principles to an already existing fast cipher, hopefully in a way that could leverage the security claims of the original. Two candidates were considered as the starting point for this exercise, RC4 and WAKE.

RC4 offers opportunity for speed-up both by use of wider data paths, and by an increase in concurrency. Schneier[11] suggests a modification to RC4 to exploit wider data paths, but only at the unacceptable expense of an exponential growth in the associated look-up-table that precludes extension to 32-bits. In some ways ISAAC[5] can be viewed as an extrapolation from RC4 to 32-bit and 64-bit data paths while not particularly addressing RC4's modest P-factor.

WAKE is already very fast, makes good use of a 32-bit datapath, avoids rotations, and its regularity lends itself to the possibility of efficient mapping to SIMD architectures. Its only weakness is a lack of concurrency principally brought about by cascaded table-look-ups. In addressing this limitation, another attraction over RC4 is WAKE's lack of self modification of its look-up-table.

WAKE was chosen as the candidate cipher for modification.

5   WAKE

In [1] Wheeler introduced the cipher WAKE. It uses a mixing function, $M(x, y, T)$, that combines two 32-bit inputs, x and y, into one 32-bit output with the aid of a key-dependent 256×32-bit look-up-table, T. By constraining the values in the upper byte lane of the otherwise 'random' entries of T to form a permutation of the numbers 0 to 255, the mixing function is made reversible in the sense that knowledge of the output word and one of the two input words is sufficient to uniquely specify the other input word. The mixing function, and its inverse, are shown in Fig. 1.

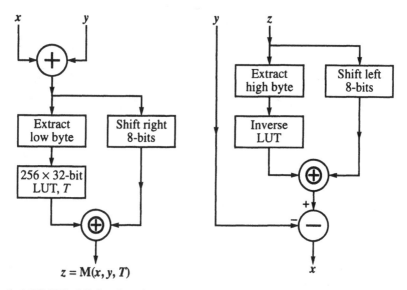

Fig. 1 a) WAKE's Mixing function   b) Inverse mixing function

WAKE consists of cascading four of these mixing functions with registered feedback around each one and overall feedback around the group. Four stages are chosen as the minimum number needed for complete diffusion. Fig. 2a shows WAKE in cipher-feedback mode as originally described.

In addition to its use in cipher-feedback mode, Wheeler suggests that WAKE is suitable for the production of a pseudo-random sequence for use as a stream cipher by XORing with the plaintext. This mode, shown in Fig. 2b and referred to here as WAKE-OFB, is used as the basis for the new ciphers since it conveniently circumvents the complaint that WAKE in cipher-feedback mode is susceptible to a chosen-plaintext attack.

Using the assumptions given for table 1, the mixing function has a critical path of 6 cycles, three of them for the table-look-up and one each for the other three operations in the path. WAKE-OFB cascades four mixing functions, for a total critical path of 24 cycles.

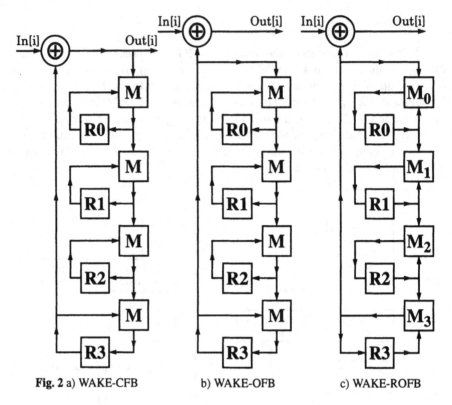

Fig. 2 a) WAKE-CFB   b) WAKE-OFB   c) WAKE-ROFB

6   Taking a Step Backwards

Cryptographic security of a pseudo-random sequence, as used for a stream cipher, demands that no one part of the sequence can be predicted from any other part of the sequence. This security property makes no distinction between the forward and time reversed versions of the pseudo-random sequence. Thus, if a *reversible* pseudo-random number generator (PRNG) produces a cryptographically secure sequence then that same generator running in reverse must also produce a cryptographically secure sequence.

WAKE's next-state function is designed to be reversible in order to maximize the cipher's expected cycle lengths, thus when used in OFB mode it forms a reversible PRNG.

Fig. 2c shows the flowgraph for WAKE-OFB run in reverse which we will refer to as WAKE-ROFB. It differs from the forward version in that the direction of data through the registers is reversed, resulting in its stepping backwards through its states. Strictly speaking, to achieve the *actual* time-reversed sequence the mixing functions should also be replaced by their inverses. However, for convenience we can just as easily leave them unchanged, reasoning that the forward or inverse mixing functions will provide equally good but different encipherment, just as Wheeler in [1] reasons that performing an arithmetic right shift on signed 32-bit operands inside the mixing function is as good, but different from, performing a logical shift as is done in his reference implementation. In any case, the forward and inverse mixing functions have identical critical-path lengths so choosing one over the other does not affect the performance analysis.

An interesting result of this reversal of the state machine is that the longest path from the output of one register to the input of another is now through only two mixing functions (M_3 and M_0), instead of through all four as in the original flowgraph. At first this might suggest that the critical path has been halved. However, closer inspection reveals that the critical path has in fact been reduced by a factor of *three* since at any given time three out of the four mixing functions can be evaluated concurrently, and three out of four registers updated, so over a period of four mixing function evaluations each of the registers can be updated three times, as illustrated by the following pseudo-code:

- Starting with known values for R0 through R3:
    - Evaluate M_1, M_2, and M_3 in parallel, update R1, R2, and R3
    - Evaluate M_0, M_2, and M_3 in parallel, update R0, R2, and R3
    - Evaluate M_0, M_1, and M_3 in parallel, update R0, R1, and R3
    - Evaluate M_0, M_1, and M_2 in parallel, update R0, R1, and R2
- All four registers have now been updated exactly three times. Each update of R3 allows another word to be ciphered. Repeat the sequence until done.

Thus, simply by running WAKE-OFB backwards we can achieve a threefold increase in parallelism while claiming identical security.

Wheeler offers that WAKE's security can be enhanced by increasing the number of stages from the four given, albeit at a reduction in speed. Indeed, for the original WAKE, both the total computation and, more importantly, the critical path, are essentially proportional to the number of stages, with the result that performance inherently declines as stages are added.

Now let's consider adding stages to WAKE-ROFB. Just as 4-stage WAKE-ROFB had three times as much parallelism as WAKE-OFB, a time reversed 5-stage version has *four* times as much parallelism as its non-reversed counterpart. Where 4-stage WAKE-ROFB had a critical path of $\frac{4}{3}$ mixing function evaluations, 5-stage WAKE-ROFB has the slightly shorter critical path of $\frac{5}{4}$ mixing function evaluations.

Counter to our intuition, we note that *adding stages can actually speed-up the cipher*, even though the total amount of work per word ciphered increases in proportion to the number of stages. The caveat here is of course the need for the processor to exhibit adequate parallelism. However, additional stages always harm the cipher's performance on a *single-threaded* CPU, and also on a CPU with parallelism once all of its parallelism has been exploited. For this reason, and also because further stages give progressively diminishing return in critical-path reduction, it is unattractive to extend beyond five stages.

The principal attraction of the 5-stage version is that its fourfold parallelism is a better fit for typical SIMD datapaths than the 4-stage version's threefold parallelism. Similarly, optimal loop-unrolling for the 5-stage version involves a factor of four, while for the 4-stage version loop-unrolling by a multiple of three is most efficient, which may be inconvenient for some applications.

7   Adding to the Flowgraph - WiderWake

To further increase WAKE's parallelism and shorten its critical path we investigate inserting additional pipeline stages in the algorithm's flowgraph. The trick is to do so without compromising its security characteristics. Our goal is to reduce the critical path to just *one* mixing function evaluation, while keeping the total computation down to that of 4-stage WAKE so that single-threaded CPU performance is not impaired.

For algorithms that include overall feedback - WAKE included, adding pipeline stages cannot be done without changing the nature of the algorithm. This means that the modified version does not necessarily inherit the security characteristics of the original scheme.

As an example, Fig. 3a shows a rearrangement of WAKE with its registers now acting as pipeline stages. This simple expedient increases the available parallelism by a factor of four over WAKE since now all four mixing functions can be evaluated concurrently.

However, it may be observed that unlike the original WAKE, the version of Fig. 3a has no direct way of determining its *previous* state. Indeed, some states may not have a *unique* previous state, while others may have no previous state at all. The lack of bijectivity in this version's state transition function would, due to the birthday paradox, result in its expected cycle lengths being dramatically shorter than those of the original (i.e. inferior security). This illustrates the caution necessary in making modifications to the flowgraph.

In order to achieve our critical-path goal, every mixing function output must be registered before becoming a mixing function input. Given this, we note that for the flowgraph to be reversible we additionally require that at least one of the mixing functions has an input node in common with the input of a register (this guarantees that when the flowgraph is reversed there becomes at least one mixing function with registers defining two of its three nodes). Since this input cannot be that of any of the mixing function output registers without violating our critical-path constraint, we conclude that for an n-stage flowgraph we need a minimum of $n + 1$ registers in order to meet our objectives.

One such arrangement, with the minimum five registers needed for four stages, is shown in Fig. 3b.

This general topology, in which an additional register is added in the feedback loop of one or more of the stages, we will refer to as *WiderWake*.

We identify specific instances of the topology by the total number of stages and the number of modified feedback loops, following the naming convention illustrated by Fig. 3b and Fig. 3c.

The available parallelism is determined by the overall number of stages, it does not change with the number of feedback loops that are modified.

Each modified feedback loop adds another 32-bits of state to the state machine, increasing the complexity of the generator, and potentially its security.

With four-stage WiderWake we can modify at most three of the four stages (WiderWake$_{4+3}$). This is because if all four stages are modified we get a common factor of two between the number of registers in the outer-loop (4 registers) and

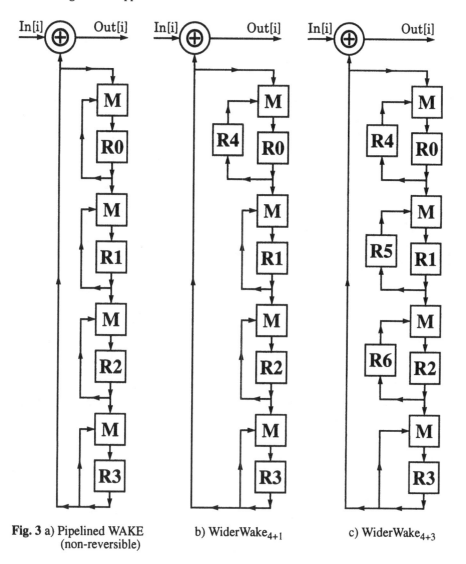

Fig. 3 a) Pipelined WAKE
(non-reversible)

b) WiderWake$_{4+1}$

c) WiderWake$_{4+3}$

each minor loop (2 registers each). This causes the state machine to degenerate into two interleaved 128-bit generators instead of being a single 256-bit generator. The same problem does not afflict generators having an *odd* number of stages, however it is questionable whether modifying the stage which is tapped for the generator's output does anything to enhance security since it does not increase the amount of hidden state.

The most useful variants are suggested to be WiderWake$_{4+1}$, and Wider-Wake$_{4+3}$. The former has the benefit of using the fewest registers, and so may be best suited for backward compatibility with older register-impoverished processors. The latter has the highest generator complexity given the minimum number of four stages.

Table 2. Performance characteristics of WAKE and WiderWake

	WAKE-OFB		WAKE-ROFB		WiderWake	
	4-stage	5-stage	4-stage	5-stage	4+1	4+3
Total generator state	128-bits	160-bits	128-bits	160-bits	160-bits	224-bits
Hidden state	96-bits	128-bits	96-bits	128-bits	128-bits	192-bits
Bits ciphered per iteration	32	32	32	32	32	32
32-bit ops. per iteration	24	29	24	29	24	24
32-bit ops. per byte	6.0	7.25	6.0	7.25	6.0	6.0
Number of 32-bit operations in critical path	16 (12, 4, 0)	20 (15, 5, 0)	5.33 ($^{12}/_3$, $^4/_3$, 0)	5.0 ($^{15}/_4$, $^5/_4$, 0)	4 (3, 1, 0)	4 (3, 1, 0)
Cycles in critical path	24	30	8	7.5	6	6
Normalized critical path	6.0 cycles/byte	7.5 cycles/byte	2.0 cycles/byte	1.9 cycles/byte	1.5 cycles/byte	1.5 cycles/byte
Critical-path speed-up	1x (ref.)	0.8x	3.0x	3.2x	4.0x	4.0x
Apparent parallelism	1.5x	1.45x	4.5x	5.8x	6.0x	6.0x
Critical-path efficiency	0.67	0.67	0.67	0.67	0.67	0.67
P-factor	1.0	0.97	3.0	3.87	4.0	4.0
Benchmarked performance on 32-bit VLIW CPU	6.23 cycles/byte	7.73 cycles/byte	2.32 cycles/byte	2.28 cycles/byte	1.90 cycles/byte	1.85 cycles/byte
Throughput at 100 MHz	16.1 MB/s	12.9 MB/s	43.1 MB/s	43.9 MB/s	52.6 MB/s	54.1 MB/s
Relative speed	1x (ref.)	0.81x	2.69x	2.73x	3.28x	3.37x

8 Performance

A Philips TriMedia processor was chosen for performance comparisons between the algorithm variants as a vehicle that amply demonstrates the potential for improvement given adequate parallelism.

The TriMedia processor is a VLIW CPU containing five 32-bit pipelined execution units sharing a common set of registers. All execution units can perform arithmetic and logical operations, but loads, stores, and shifts are only supported by a subset of them.

Table 2 compares the characteristics of WAKE in output-feedback mode with the algorithm variants proposed herein. Encryption and decryption have identical characteristics. Assumptions used are the same as those given for table 1.

The benchmark conditions were a 100 MHz TriMedia processor running performance-optimized C-code to encrypt a buffer several times larger than the on-chip data cache. Comparable source-code optimizations were applied to all cases. The advertised benchmark performance includes all loop-overhead and cycles lost to cache misses, memory accesses, etc. No off-chip cache was present.

The benchmarked performance of 5-stage WAKE-ROFB is seen to marginally exceed that of the 4-stage version, demonstrating that extra stages can improve performance for this topology in practice as well as theory.

WiderWake$_{4+3}$ shows very similar performance to WiderWake$_{4+1}$, as predicted. The slight improvement of WiderWake$_{4+3}$ over WiderWake$_{4+1}$ can be attributed to the former having fewer critical-length paths than the latter, thus easing the compiler's task of scheduling the code to simultaneously minimize all critical-length paths.

WAKE-ROFB and WiderWake achieve speed-ups over WAKE-OFB of better than 2.5x and 3x respectively. With performance of better than 2 cycles per byte, the WiderWake variants allow encryption or decryption at speeds in excess of 50 MByte/s on a 100 MHz processor. WAKE-ROFB exceeds 40 MByte/s in both 4-stage and 5-stage versions.

9   Cipher-Feedback Mode

Any of these generators could in principle be used in CFB mode while maintaining their speed, by taking the overall feedback from the ciphertext instead of direct from the final stage, just as in WAKE-CFB. However, the user is cautioned against doing so without further study. In CFB mode none of these generators can claim direct equivalence to WAKE's security. In particular, unlike WAKE-CFB where each ciphertext symbol affects the next ciphertext symbol, the fed-back ciphertext in all these new versions would take several cycles before it again influences the stream. So, as a minimum there might need to be some special processing to avoid weaknesses at the start and end of the stream.

In fairness to WAKE, it should be pointed out that WAKE-CFB itself offers fourfold parallelism in *decrypt* mode, an advantage not shared by WAKE-OFB. This comes about because the overall feedback of WAKE-CFB's encryption flow-graph becomes feed-forward during decryption, thereby breaking the outer feed-back loop. The critical path then becomes just that of the minor loops around each stage, i.e. just one mixing function evaluation. After the first four cipher-text symbols have 'filled the pipe' all four mixing functions can be evaluated in parallel as each new ciphertext symbol arrives. That WAKE-CFB's decryption parallelism is not *unlimited* is a reflection of the fact that WAKE-CFB suffers *infinite* error extension under decryption, unlike a block cipher operating in CFB mode.

10   Table Initialization and Stream Re-synchronization

For a complete stream cipher definition two further components - look-up-table initialization (key scheduling), and stream re-synchronization (Initialization Vector processing), need to be specified.

Wheeler supplies an ad-hoc table-generation routine based on a 128-bit key (referred to as the *table-key*). We retain this routine with minor modifications to remove ambiguity from the original definition. On a single-threaded CPU this routine runs in about the time it takes to encrypt 1000 bytes with any of the 4-stage variants. However, this routine is much less capable of exploiting instruction-level parallelism than our optimized cipher variants, so that on a CPU having substantial instruction-level parallelism table initialization can take more like 2000 to 3000 byte-encryption times. Improving on this is an area for future study.

Re-synchronization is achieved by setting the registers to a new state that is a hashed combination of the table-key and an initialization vector (IV). We propose a 64-bit IV for compatibility with common block ciphers operated in OFB mode. Our minimalist resync procedure consists of seeding the WiderWake state machine with a simple combination of the table-key and the IV, and stepping the generator until they are satisfactorily mixed, discarding the generator's output along the way. For an n-stage generator we choose to step the generator $2n$ times. This is enough so that all registers achieve avalanche at least once, and the output register achieves avalanche at least twice. For WiderWake $_{4+1}$ this resync process takes about as long as ciphering 32 bytes.

The resistance of this simple resync procedure to related-key cryptanalysis is unproven. This may represent an exposure if resync intervals are especially frequent, or if an attacker has the ability to force resyncs at will[3]. The first defence against such a weakness would be to step the generator more times during resync. Alternatively a secure hash such as SHA-1[8] could be used to robustly combine the table-key with the IV, however on a CPU that can exploit WiderWake's parallelism this is as costly as ciphering several hundred bytes.

11   Conclusion

We have illustrated the opportunities for performance gains through exploiting the instruction-level parallelism of current generation CPUs. We suggest that efficiency on these processors should be among the design criteria for new software-oriented ciphers.

The example ciphers are presented without a supporting security analysis. Cycle lengths have been determined experimentally for several models having shorter word-lengths and found to be as expected. It remains to be established what length of cipher stream can be safely exposed between IV changes and between table-key changes. Cryptanalysis is invited.

References

1. D. J. Wheeler, "A Bulk Data Encryption Algorithm", *Fast Software Encryption (Ed. R. Anderson), Lecture Notes in Computer Science No. 809*, Springer-Verlag, 1994, pp. 127-134
2. R. J. Anderson and E. Biham, "Tiger: A Fast New Hash Function", *Fast Software Encryption (Ed. D. Gollmann), Lecture Notes in Computer Science No. 1039*, Springer-Verlag, 1996, pp. 89-97
3. J. Daemen, R. Govaerts, and J. Vandewalle, "Resynchronization Weaknesses in Synchronous Stream Ciphers", *Advances in Cryptology - EUROCRYPT '93 Proceedings*, Springer-Verlag, 1994, pp. 159-167

4. ISO/IEC 10166, "Information Technology - Modes of operation for an n-bit block cipher algorithm", International Organization for Standardization / International Electrotechnical Commission, 1991

5. R. J. Jenkins, "ISAAC", *Fast Software Encryption (Ed. D. Gollmann), LNCS 1039*, Springer-Verlag, 1996, pp. 41-49

6. B. S. Kaliski and M. J. B. Robshaw, "Fast Block Cipher Proposal", *Fast Software Encryption (Ed. R. Anderson), LNCS 809*, Springer-Verlag, 1994, pp. 33-40

7. NBS FIPS PUB 46-1, "Data Encryption Standard", National Bureau of Standards, U.S. Department of Commerce, Jan 1988

8. NIST FIPS PUB 180-1, "Secure Hash Standard", National Institute of Standards and Technology, U.S. Department of Commerce, April 1995

9. R. L. Rivest, "The RC5 Encryption Algorithm", *Dr. Dobb's Journal*, v. 20, n. 1, January 1995, pp. 146-148

10. P. Rogaway and D. Coppersmith, "A Software-Optimized Encryption Algorithm", *Fast Software Encryption (Ed. R. Anderson), LNCS 809*, Springer-Verlag, 1994, pp. 56-63

11. B. Schneier, *Applied Cryptography, Second Edition*, John Wiley & Sons, 1996, pp. 397-398

12. B. Schneier, "Description of a New Variable-Length Key, 64-bit Block Cipher (Blowfish)", *Fast Software Encryption (Ed. R. Anderson), LNCS 809*, Springer-Verlag, 1994, pp. 191-204

Appendix A.   WiderWake$_{4+1}$ Reference Implementation

The following C-code is a functional reference only. It does not represent all the performance optimizations embodied in the benchmarked source-code.

```
/* WiderWake4+1,  Version 1.0, functional reference */
typedef  unsigned long UINT32;    /* 32-bit unsigned integer */
/* Array sizes - UINT32 T[256], t_key[4], IVec[2], r_key[5]; */
/* addition is modulo 2^32,    >> is right shift with zero fill */
#define M(x,y,T)    (((x)+(y)) >> 8) ^ T[((x)+(y)) & 0xff]

void ofb_crypt(UINT32 *In, UINT32 *Out, int length, UINT32 *T, UINT32 *r_key)
{
    UINT32 R0, R1, R2, R3, R4, R0a, R1a, R2a, R3a;
    int i;
    R0=r_key[0];  R1=r_key[1];  R2=r_key[2];  R3=r_key[3];  R4=r_key[4];
    for (i = 0; i < length; i++)
    {
        R3a = M(R3, R2, T);    /*                                      */
        R2a = M(R2, R1, T);    /*    All four mixing functions         */
        R1a = M(R1, R0, T);    /*    can be evaluated in parallel       */
        R0a = M(R4, R3, T);    /*                                      */
        Out[i] = In[i] ^ R3;   /* Execution can overlap with mixing functions */
        R4 = R0;  R3 = R3a;  R2 = R2a;  R1 = R1a;  R0 = R0a;
    }
    r_key[0]=R0; r_key[1]=R1; r_key[2]=R2; r_key[3]=R3; r_key[4]=R4;
}
```

```
void resync(UINT32 *T, UINT32 *t_key, UINT32 *IVec, UINT32 *r_key)
  /* mix t_key with initialization vector and discard first eight words */
{
    UINT32 temp[8];  /* bit-bucket */

    r_key[0] = t_key[0] ^ IVec[0];  r_key[1] = t_key[1];
    r_key[2] = t_key[2] ^ IVec[1];  r_key[3] = t_key[3];
    r_key[4] = IVec[0];

    ofb_crypt(temp, temp, 8, T, r_key);
}


void key_sched(UINT32 *t_key, UINT32 *T)  /* expand t_key to look-up-table T */
{
    UINT32 x, z, p, t0;
    static UINT32 tt[8] = { 0x726a8f3b, 0xe69a3b5c, 0xd3c71fe5, 0xab3c73d2,
                            0x4d3a8eb3, 0x0396d6e8, 0x3d4c2f7a, 0x9ee27cf3 };

    for (p = 0; p < 4; p++) { T[p] = t_key[p]; }

    for (p = 4; p < 256; p++)
        { x = T[p-4] + T[p-1];  T[p] = (x >> 3) ^ tt[x & 7]; } /* (UINT32)x */

    for (p = 0; p < 23; p++) { T[p] += T[p+89]; }

    x = T[33];  z = (T[59] | 0x01000001) & 0xff7fffff;

    for (p = 0; p < 256; p++)
        { x = ( x & 0xff7fffff) + z;  T[p] = (T[p] & 0x00ffffff) ^ x; }

    x = (T[x & 0xff] ^ x) & 0xff;  t0 = T[0];  T[0] = T[x];

    for (p = 1; p < 256; p++)
        { T[x] = T[p];  x = (T[p ^ x] ^ x) & 0xff;  T[p] = T[x]; }

    T[x] = t0;
}
```

Appendix B.   Test Case

```
void test(void)
{
    UINT32 t_key[4] = { 0x12345678, 0x98765432, 0xabcdef01, 0x10fedcba };
    UINT32 IVec[2] = { 0xbabeface, 0xf0e1d2c3 };  /* Initialization Vector */

    UINT32 text[4] = { 0x1234abcd, 0xa0b1c2d3, 0x1a2b3c4d, 0x55667788 };

    UINT32 T[256], r_key[5];
    int i;

    key_sched(t_key, T);           /* Schedule key */
    resync(T, t_key, IVec, r_key);  /* Initialize generator state using IV */

    for (i = 0; i < 256; i++)      /* Encrypt text buffer 256 times */
        { ofb_crypt(text, text, 4, T, r_key); }

    for (i = 0; i < 4; i++) { printf("0x%08lx ", text[i]); } printf("\n");
}
/* final text[] == { 0x94739922, 0xb251752f, 0x1de1f2fe, 0x405f83dd } */
```

Author Index

Springer
and the
environment

At Springer we firmly believe that an international science publisher has a special obligation to the environment, and our corporate policies consistently reflect this conviction.

We also expect our business partners – paper mills, printers, packaging manufacturers, etc. – to commit themselves to using materials and production processes that do not harm the environment. The paper in this book is made from low- or no-chlorine pulp and is acid free, in conformance with international standards for paper permanency.

Lecture Notes in Computer Science

For information about Vols. 1–1192

please contact your bookseller or Springer-Verlag